ELEMENTS OF CARTOGRAPHY

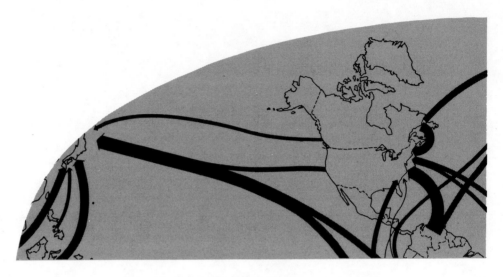

Arthur H. Robinson

Lawrence Martin Professor of Cartography
Director, University Cartographic Laboratory

Randall D. Sale

Associate Professor of Geography
Associate Director, University Cartographic Laboratory

Both of the Department of Geography,
University of Wisconsin, Madison

John Wiley & Sons, Inc.

ELEMENTS OF CARTOGRAPHY

THIRD EDITION

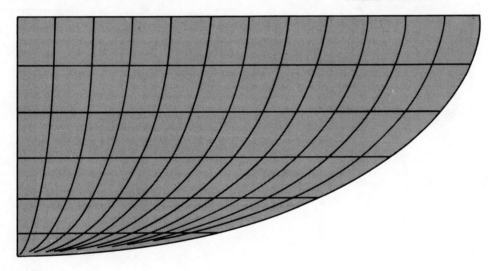

New York • **London** • **Sydney** • **Toronto**

Library of Congress Catalog Card Number: 69–19232
SBN 471 72805 5
Printed in the United States of America

PREFACE

Cartography, like all fields of learning, currently enjoys a remarkable rate of development that shows no sign of decreasing. Techniques, and even points of view, that were taken for granted a few years ago are now being superseded by new methods and concepts. This third edition attempts to incorporate these modernizations with the time-honored fundamentals of a "scientific art" that has been actively practiced for more than two millenia.

The study of spatial distributions is increasingly significant, both in the number of disciplines concerned and in the technical support provided by modern data gathering, machine processing, and new analytical methods. As a consequence, the demands upon cartography to map and display spatial variations and relationships also has increased. It is not just a matter of more maps being wanted by greater numbers of people: the complexity and quality of the maps needed have increased at the same time. In other words, the users of maps and the makers of maps are becoming steadily more sophisticated. Not many years ago, the map-maker was doing, in the main, what his counterparts of a generation earlier had been doing. The accumulation of data, their processing, their conversion to graphic form, the preparation of art work for the camera, and all the intricate aspects of cartographic design from projections to typography had not changed radically. Not so today; cartography is in the midst of a revolution, and few aspects of this complex field have escaped the forces of change. Scale concepts, cartographic generalization, compilation, the use of air photographs and other forms of remote sensing, scaling concepts, data manipulation, electronic processing and computer mapping, graphic design, typography, and construction and reproduction methods, each an integral part of cartography, have all been affected to an unusual degree.

There is a natural tendency for innovation to claim the spotlight, sometimes to the point where even professionals in the field tend to lose sight of fundamental objectives. The authors have attempted to resist the inclination to dwell on the new for the sake of its newness and, instead, have tried to integrate the up-to-date concepts and methods (as best they can in an essentially elementary book) with the fundamental elements of cartography. Basically, a map is both (a) a scientific tool with much utility for many kinds of research and for many kinds of technical application and (b) a form of graphic communication. The central operations in cartography can be simply stated: they are the accumulation of data and the designing of their display so as to make available or communicate the spatial information in the most accurate and efficient manner possible under the circumstances. This places cartography as a special class of what has been called recently "graphicacy" by Balchin and Coleman, and as such it joins the other skills of communication: "numeracy"—mathematical communication, "articulacy"—communication by speech, and literacy—communication by the written word. All such skills have a great variety of structural variations as well as many constraints. This book attempts to show the possibilities of cartographic "authorship" within the technical and conceptual limits of the field.

The importance of the recent developments in cartography have required that the book be entirely rewritten, while experience with its use in instruction has suggested a basic reorganization. As always, the concepts of scale and a coordinate system are fundamental to mapping and they remain the initial subject matter. Traditionally, such material has been followed by a treatment of the intricacies of map projection, a subject generally terrifying to the majority of neophytes. Based on instructional experience and past-instructional reactions, we have rearranged the sequence of material to progress through the methods of compilation and symbolization first. By the time the student has absorbed these, he is much more receptive to the fundamental importance of the nature and employment of map projections. The treatment of the actual construction of projections by hand, so to speak, has been placed in an appendix, a most appropriate place for it in these days of computer technology. As in the second edition, the elements of design, typography and lettering, and map reproduction and construction are treated in the latter portion of the book.

Several sections have been greatly enlarged and new material added: a chapter dealing with air photographs for compilation and control has been included; the subject of generalization in cartography has been greatly expanded; the up-to-date concepts of scaling observational data (nominal, ordinal, interval, and ratio) have been incorporated; the treatments of map construction and reproduction methods have been completely modernized; and the elements of design and typography include the latest experimental results. The illustrations have been thoroughly overhauled, with many new ones added and many redrawn.

We acknowledge a great debt to our students, colleagues, and the members of the cartographic profession. Ideas, points of view, innovations, and the like are commonly rooted to a greater or lesser degree in the work of others, and to sort out all our indebtedness is clearly impossible. Special appreciation is extended to Henry W. Castner, Roy Chung, Jerry B. Culver, Barbara S. Bartz, Mei-Ling Hsu, and George F. McCleary, Jr. Our colleague at the University of Wisconsin, Joel Morrison, has been most helpful. The illustrative work of James J. Flannery of the University of Wisconsin, Milwaukee, done for the previous editions, is reflected in this edition. Colleagues elsewhere, both on this continent and abroad, have contributed greatly. Suitable acknowledgment has been made in the appropriate places. Professor George F. Jenks of the University of Kansas read the entire manuscript and made many useful suggestions. Nevertheless, we alone bear the responsibility for the contents.

The number of people directly and indirectly involved in the production of a book is a constant source of wonder. They range from wives to colleagues, designers to typists, and editors to draftsmen. All are indispensible, and it always seems unfair to have so few names on the title page, since authorship depends so much on their help.

Arthur H. Robinson
Randall D. Sale

CONTENTS

ELEMENTS OF CARTOGRAPHY

THE ART AND
SCIENCE OF CARTOGRAPHY

1

MAN HAS ONLY limited abilities to observe directly those phenomena that interest him. Some things are very tiny and we must use complex optical and electronic means (a microscope, for example) to enlarge them so as to understand their configuration and structural relationships. At the other end of the continuum some things are so large that we must somehow reduce them for the same purposes. Cartography is a technique fundamentally concerned with reducing the spatial characteristics of large areas—a portion of the earth, the moon, or even the whole earth—to a form that makes them observable. The map allows man to extend his normal range of vision, so to speak, and makes it possible for him to see the broader spatial relations that exist over large areas.

A map is much more than a mere reduction, however. If well made, it is a carefully designed instrument for recording, calculating, displaying, analyzing and, in general, understanding the interrelation of things in their spatial relationship. Nevertheless, its most fundamental function is to bring things into view.

The Map an Essential Tool

A large-scale map of a small region, depicting its land forms, drainage, settlement patterns, roads, geology, or a host of other detailed distributions, makes available the knowledge of the relationships necessary to carry on many works intelligently. The building of a road, a house, a flood-control system, or almost any other constructive endeavor requires prior mapping. At a smaller scale, maps of soil erosion, land use, population character, climates, income, and so on, are indispensable to understanding the problems and potentialities of an area. At the smallest scale, maps of the whole earth indicate generalizations and relationships of broad earth patterns with which we may intelligently consider the course of events past, present, and future.

The increasing complexity of modern life with its attendant pressures and contentions for available resources has made necessary increasingly detailed studies of land utilization, soil characteristics, disease migrations, population and settlement distributions, and numberless other social and economic factors. The geographer, preeminently, as well as the planner, historian, economist, agriculturalist, geologist, and others working in the basic and applied physical science fields, long ago found the map a useful and often indispensable aid to research and communication. Almost everyone uses maps frequently. For example, the familiar and excellent road map, standard equipment of any driver in the United States, is prepared and given away in such quantities that every year each adult in this country could have his own personal copy.

To attempt to catalog with precision the infinite number of kinds and uses of maps is an impossible task. *Anything* that man can observe, tangible or otherwise, can be mapped in its two- or three-dimensional distribution. The uses of maps may vary widely: the historian may plot the routes travelled by Marco Polo or Alexander in order to evaluate their cultural influences; the engineer may measure carefully from a detailed map in order to analyze the runoff potential in a portion of a city for the purpose of planning adequate storm sewers; and the space scientist and astronaut may require detailed maps of the moon in order to plan and carry on explorations. The scales required may vary from a page-size map of the world presenting airline routes focusing on Chicago to a foot-square insurance map of a single city block showing individual buildings, including the one in which the airlines map was made.

The most meaningful listing of varieties of maps is one based on utility, and many such catalogs have been attempted. They include the standard divisions such as topographic maps, nautical charts, thematic maps, economic maps, historical maps, and other broad use categories. However, they all fail to impress on the reader that all maps are related, and that uses, scales, varieties, and sizes all shade, imperceptibly, one into another.

As far as maps of the earth are concerned, the "sky is the limit," but insofar as cartography is concerned, there truly is no limit. The earth, the moon, the sun, the planets—they are all the concern of the cartographer.

The Beginnings of Cartography

No one knows when the first cartographer prepared the first map, which was no doubt a crude representation of locations drawn in soft earth or scratched on a rock. Certainly the spatial concern was relatively narrow. Islanders of the South Seas are said to have constructed at an early date charts of reeds and sticks to record the relative positions of islands. Perhaps the oldest authentic map that survives is a clay tablet nearly five thousand years old having to do with some land holdings in Babylon. Without question, the valley of the Nile was carefully mapped in order to record property lines. Aside from the islanders' navigation maps and others like them, it is probable that the earliest "permanent" maps were records of land ownership. This same kind of map still survives as one important

use-category of cartography. They are made in a variety of scales, and today they are called cadastral maps; they record property boundaries much the way they did several thousand years ago (Fig. 1.1). One of the principal uses of cadastral maps is to provide a basis upon which to assess taxes—which may account for the fact that they have always been with us.

Concern with areas larger than the immediate locality developed long before the Christian era. The spherical shape of the earth was inferred early from the difference

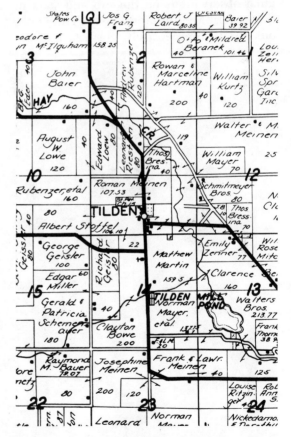

Figure 1.1 A section of a cadastral map showing ownership of land. The bold numbers are section numbers and the heavy lines are paved roads. (From the *Chippewa County, Wisconsin Plat Book, 1966*, courtesy Rockford Map Publishers.)

in the altitudes of stars at different places, from the fact that shorelines and ships seemed to "come over" the horizon as one moved across the sea, and even from the assumption that the sphere was the most perfect form. The development of civilization allowed more frequent travel, and a greater interest in faraway places was accompanied by increased thought about ways of presenting the relationships of areas on maps. Estimates of the size of the earth were made by ancient scholars such as Eratosthenes and Posidonius from angular observations on the sun and stars in the eastern Mediterranean area. The methods used were entirely correct, but the required assumptions and the precision of their observations was not. Nevertheless, although we cannot be absolutely sure of their results in terms of modern units of measure, the estimates seem not to have been very far off. From the descriptions in the manuscripts that exist there is evidence that many maps were made during the time of the classical Greeks, but none of the actual maps appears to have survived. Fortunately for our understanding of the cartography of the ancient period, there is, however, a record that apparently clearly reflects the stage to which cartography had developed by the end of the Greek period. These are the writings of Claudius Ptolemy.

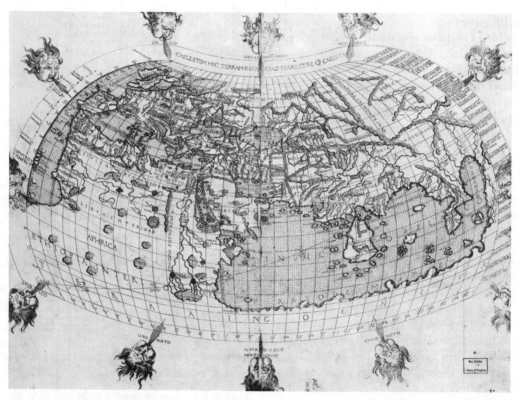

Figure 1.2 One of "Ptolemy's" maps as constructed from his written directions and descriptions. The area shown extends from the Atlantic on the west to beyond the closed Indian Ocean. This world map reflecting the knowledge of the second century A.D. was better than any other when Ptolemy's works were reintroduced to Europe in the fifteenth century. (From the Library of Congress collection.)

In Alexandria, Egypt, which had become the intellectual center of the western world, there had been established a great library and a cultural climate where a scholar could study and write. Claudius Ptolemy, who lived during the first and second centuries A.D., brought together all that was available concerning the known areas of the earth and wrote a long manuscript which included, among other things, a treatise on cartography. He described how maps should be made; he gave directions for dealing with the problem of presenting the spherical surface of the earth on a flat sheet; he recognized the inevitability of deformation in the process; and in many other ways he described what was known about cartography at that time. Ptolemy also made a list of numerous places in the world he had gleaned from the writings of others and from travelers, and gave his best estimate of their positions in latitude and longitude terms. To illustrate this he is thought to have provided a series of maps of local and larger areas (Fig. 1.2). This is not certain, however, for if he did they have since been lost. He did, however, give a detailed account of how they were made. Ptolemy's writings were lost to the western world for more than a thousand years, but fortunately they were preserved in the Arab world and later came to light again in Europe. From his descriptions the Ptolemaic maps were reconstructed, and they had a profound influence upon European geographical and cartographical thinking during the Renaissance.

It is worth observing here that Ptolemy was a compiler of maps, not a surveyor. Although there is no clear distinction even today, a compiler is one who obtains his data to be mapped from a variety of secondary sources, ranging from official records to the resources of libraries, while the topographic surveyor obtains his data from field measurements and air photographs. The compiler, like Ptolemy, usually is concerned with preparing small-scale maps of large areas, while the surveyor works toward producing large-scale maps of small areas; the skills, methods, and problems involved are quite different. Both compiling and surveying only contribute to the essential cartographic process, namely, arraying the data and designing the finished map.

The Dark Ages

After the enlightened period, which culminated with the writings of Ptolemy, cartography, along with other scientific and intellectual matters, entered the period called the Dark Ages and suffered a steady decline. For a time the practicality of the Romans showed in engineering and administrative types of maps; but concern with the general theory and practice of cartography all but died out. Furthermore, objective geographical thinking about faraway places was replaced by fancy and whimsy. Maps of the "known world" were produced; but, whereas during the Greek period these had been based upon observation and reason, they now came to be but media for preserving the results of fanciful speculation, such as the writings of Solinus, and literal interpretations of Biblical passages.

Many allegorical types of "world" maps were prepared containing symbolic representations of the earth, such as rectangular maps probably based on the scriptural reference to the "four corners of the earth." But even more common were circular maps (also probably because of a Biblical reference) with Jerusalem at the center. These maps reflected the threefold division of the earth among Shem, Ham, and Japheth, the sons of Noah. They are called *T in O* maps, for they were designed with the Mediterranean the upright part of the *T*, the Don and Nile rivers forming the crosspiece, and the whole inside a circular ocean (Fig. 1.3). The farthest area (known at all) was, of course, the orient. It became traditional to locate Paradise in the difficult of access, far eastern area and to put it at the top of the map. From this practice we have derived the term *to orient* a map, that is to turn it so that the directions indicated are understood by the reader—but today to orient a map means to arrange it so that map and earth directions are parallel or so that north is at the top.

On these maps often were placed mythical places, beasts, and dangers such as the kingdom of the legendary Gog and Magog, who were nonbelieving menaces to the Christian world. This kind of cartography hung on for a long time, and some of the later maps that follow this general pattern are very detailed and ornate. One of them is the map from the Hereford cathedral in England (Fig. 1.4).

There was, however, one bright cartographic light in the comparative darkness

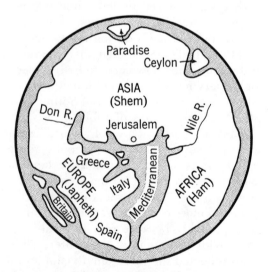

Figure 1.3 The layout of the *T* in *O* maps. Jerusalem was usually located at the center and Paradise in the top central section, from which comes the term "to orient" a map. Compare with Fig. 1.4.

of the period, namely, the sailing charts prepared to accompany the sailing directions, or *peripli*, which apparently existed in considerable numbers. These charts, called *portolani*, or harbor-finding charts, were the products of the experience of a large amount of navigation and coastwise sailing in the Mediterranean and adjacent areas (Fig. 1.5). Characteristically, they were covered by a systematic series of unlabelled, intersecting rhumb lines radiating from compass roses. The lines and shapes of the bounding coasts of the seas were remarkably accurate but unfortunately this accuracy did not seem to extend to maps of the land behind the coasts until toward the end of the Dark Ages. Although many of them must have been made, none earlier than approximately the thirteenth century has survived. Undoubtedly they were jealously guarded by their owners, which may account for their rarity.

The first few centuries of the second millenium A.D. saw a change taking place in intellectual standards which was, of course, reflected in the cartography. The transformation began slowly; but the Crusades, missions, the travels of merchants like the Polos, and the generally increased movement of peoples and goods began to awaken interest in the outside world and the fancy and superstition characteristic of the Dark Ages began to recede. With the discovery, translation, and printing of Ptolemy's writings and maps in the fifteenth century, items which had lain dormant for a thousand years, a new interest in geography and cartography developed in Europe.

The Renaissance in Cartography

The Age of Discovery, the monumental achievements of Columbus, Vasco de Gama, Cabot, Magellan, Elcano, and others at the end of the fifteenth and the beginning of the sixteenth centuries kindled such an interest in the rapidly expanding world that map publishing soon became a lucrative calling. By the latter half of the sixteenth century the profession was generally in good standing and well supported, albeit its products were still far from being first-class examples of objective scientific thought.

One of the circumstances that contributed greatly to the rapid advance of cartogra-

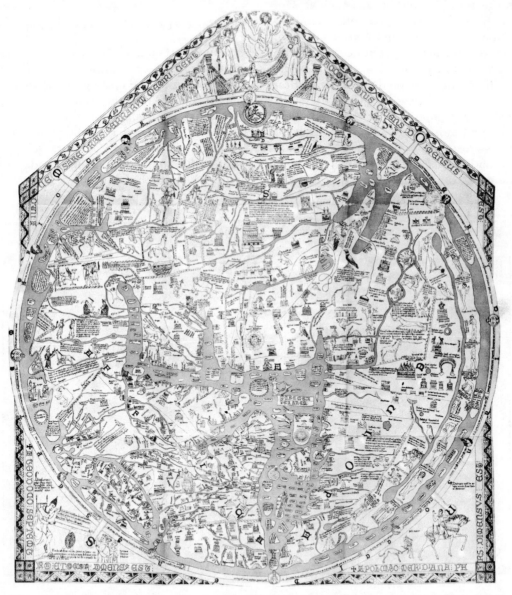

Figure 1.4   The Hereford world map. This map, made in the thirteenth century, illustrates the degree to which cartography had degenerated from the time of Ptolemy, a thousand years earlier. The map is oriented with east at the top and Jerusalem at the center. (From the Library of Congress collection.)

phy was the invention in Europe shortly after 1450 of printing and engraving, which soon made possible the reproduction of maps in numerous copies. Previously, each map had had to be laboriously hand drawn. Great map-publishing houses such as Mercator, Blaeu, Hondius, and others in Holland and France rose and flourished. Their maps were primarily reference maps containing not much more than coastlines, rivers, cities, and occasional crude indications of mountains. Fancy and intricate craftsmanship was popular, and the maps were richly embellished with ornate scrolls, compass roses, and drawings of men, animals, and ships. Except for religious and some navigational data, the mapping of any geographic information beyond what we call today base data was unknown. The mapping of most other kinds of geographical distributions had to wait for the inquiring minds who would want such maps.

The Early Period of Modern Cartography

Between 1600 and 1650 a new and fresh attitude toward cartography arose. For the first time since the ending of the classical era, accuracy and the scientific method again became fashionable. This attitude, replacing the dogmatic and unscientific outlook that was more or less dominant during the long Dark Ages, made itself evident in a number of ways in cartography.

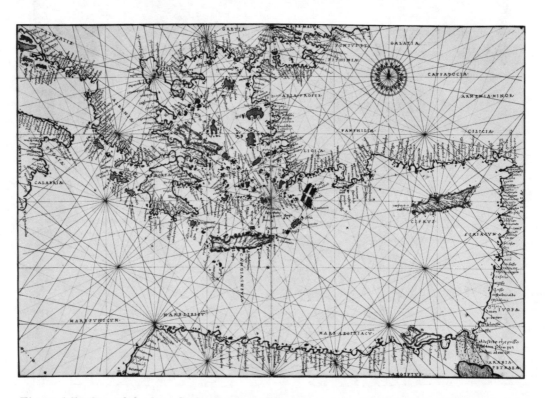

Figure 1.5 One of the later harbor-finding charts or *portolani*. Note the detail and relative accuracy of the delineation of the coasts. From Plate 12 of the manuscript atlas prepared by Battista Agnese of Venice about 1543, now in the Library of Congress. The maps are drawn on vellum and are decorated in colors and gold. (From the Library of Congress collection.)

In the second half of the seventeenth century the French Academy was founded with the avowed purpose of improving charts and navigation. This, of course, included cartography among its important concerns. Precise navigation was becoming a serious problem and its solution depended upon an accurate determination of the size and shape of the earth and upon the development of a practical method for determining the longitude. The necessity for increased mobility in military actions also made desirable the development of land survey methods.

The French Academy began by measuring accurately the arc along a meridian, and, with the new technique of triangulation, it began to position accurately the outlines of France (Fig. 1.6). Because of differences noted in the lengths of degrees along the meridian and the behaviour of the pendulum at various latitudes, the question arose as to the precise shape of the earth. During the first half of the eighteenth century French expeditions were sent to Peru and Lapland to measure other arcs along meridians. Their determinations settled once and for all that the polar radius was shorter than an equatorial radius. Of particular interest in the history of cartography was Edmund Halley's publication in 1701 of what we would now call a thematic map (Fig. 1.7). On it he showed the distribution, as known at that time, of the declination of the compass by means of lines passing through points of equal declination. Now known as isogonic lines, these Halleyan lines or curves led to the widespread use of this kind of line symbol in the following century.

In the middle of the eighteenth century the French initiated a detailed topographic survey of their country at a scale approximately 1 1/4 mi to the inch and almost completed it prior to the end of the century. Harrison's chronometer for longitude de-

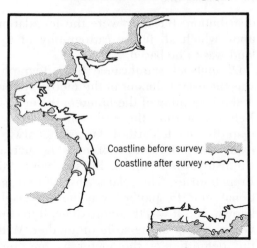

Coastline before survey
Coastline after survey

Figure 1.6 The outlines of France before and after the first accurate triangulation (essentially completed by 1740). Similar corrections of other areas could be made after they were accurately triangulated.

termination was perfected in England in 1765; and there were many other evidences of curiosity about the earth. But perhaps the most notable and significant cartographic trend was the realization by many that their fund of knowledge about the land behind the coastlines was quite erroneous. Even the administrators and rulers of countries, particularly in Europe, became aware that it was impossible to govern (or fight wars) without adequate maps of the land. Soon other great national topographic surveys of Europe were established, such as that of England in 1791, and the relatively rapid production of topographic maps followed.

The problem of representing the land-surface form arose, and, almost as quickly, devices such as the hachure and contour were developed. By the last half of the nineteenth century a large portion of Europe had been covered by topographic maps. These maps were expensive to make, however, and did not have a wide distribution. But they were the foundation upon which all future cartography of the land was to be based.

Of unusual significance to cartography was the establishment of the metric system at the beginning of the nineteenth century. Before that time the scale, or relation of map distance to earth distance, was always expressed in local units of measure such as English yards or miles, Russian versts, or French toises. The relationship of one national unit to another was not precisely known, and it was therefore difficult to convert one map to the scale of another. With the definition of the meter as one tenmillionth part of the arc distance from the equator to the pole, as then calculated, an international unit of measure became available. Soon thereafter the scales of maps began to be expressed as fractions or proportions; with them conversions are

easy to make since such a proportion is independent of any one kind of unit.

The Rise of Thematic Cartography

In addition to the nautical chart and the topographic map, a third great class, the thematic map, was added to the repertoire of cartography by the early nineteenth century. Built upon sketchy beginnings in the eighteenth century, thematic cartography had to wait for the broadening of science and the study of man and his institutions that grew rapidly in the nineteenth century. The thematic map is quite different from the navigational tool or the reference topographic map. It has also been called in the past "geographic map" or "special purpose map." Its main objective is specifically to communicate geographical concepts such as the distribution of densi-

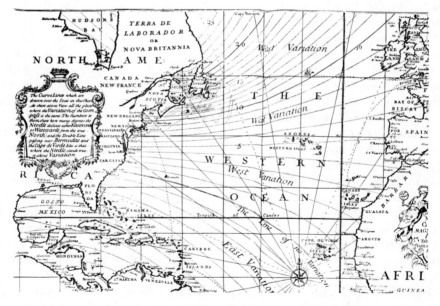

Figure 1.7 A portion of Halley's isogonic chart of 1701 showing "curve lines" supposedly connecting points with the same declination of the compass. Note the compass rose in the lower right with its radiating rhumb lines. (From the Library of Congress collection.)

ties, relative magnitudes, gradients, spatial relationships, movements, and all the myriad interrelationships and aspects among the distributional characteristics of the earth's phenomena.

Most important to the development of thematic cartography was the branching out of science into a number of separate fields, in contrast to its previous state which was a kind of all-inclusive complex of physical science, philosophy, and general geography. To be sure, the more exact studies such as physics, chemistry, mathematics, and astronomy had progressed far; but earth and life scientists who were concerned with certain classes of earth distributions, such as the physical geographer, geologist, meteorologist, and biologist, as well as the host of investigators whom we now call by the general term "social scientist," were just getting under way. Also the taking of censuses and the recording of all sorts of geographical data from weather observations to crime rates was generally initiated during the early part of the nineteenth century, and this too had a significant effect upon the development of thematic cartography. Most of the rapidly growing earth, life, and behavioral sciences needed maps; and in general their needs were for the smaller-scale, compiled, thematic maps.

In addition to the proliferation of scholarly study and the increase of mappable data many other factors helped to promote a rapid change in cartography during the nineteenth century. One can name only a few: the development of lithography, which made possible the easy and inexpensive duplication of drawings; the invention of photography; the development of color printing; the rise of the technique of statistics; the growth of mass transport, such as the railroad; and the rise of professional scholarly societies; the list is endless.

By the beginning of the twentieth century thirst for knowledge about the earth had led to remarkable strides in all aspects of cartography. Many investigations had been made into projection problems; the colored lithographic map was fairly common; a serious proposal to map the earth at the comparatively large scale of 1:1,000,000 had been made; some great map-publishing houses such as Bartholomew in Great Britain and Justus Perthes in Germany had come into being; and of special importance, the discipline most concerned with the description and analysis of spatial distributions, geography, was beginning to grow rapidly.

Twentieth Century Cartography

Cartography has probably advanced more, technically, since 1900 than during any other period of comparable length. It is probably correct to say that the number of maps made since the turn of the century is greater than the production during all previous time, even if we do not count the millions made for military purposes. Almost everyone in the United States has handled numerous maps; maps are frequently in the newspapers; atlases are sold by the hundreds of thousands; and a comforting proportion of the population now knows the meaning of the words "cartography" and "cartographic." The profession is again attaining a scholarly and professional status comparable to that which it held during the period of Dutch and Flemish dominance in the sixteenth and seventeenth centuries.

Many factors have combined to promote this phenomenal growth, and although this is not the place to list them all, a few can be mentioned. One of the most important is the fact that two world wars have occurred. Both wars have required vast numbers of maps for military purposes. Particularly

World War II, with its requirements of rapid movement and air activity, made necessary literally millions of maps and led to the growth of great mapping agencies. War-time travels and military activities all over the globe created a demand for information from the general public which was supplied by a flood of small atlases, separate maps, and newspaper and magazine maps.

The very rapid growth of population during the present century has had a powerful effect on cartography. The demand for environmental information by planners and administrators has greatly promoted all classes of the cartography of the land. The urban, rural, resource, industrial, or other kind of planner-engineer is usually concerned with relations in the spatial dimension; for this he must have maps, and he needs topographic as well as thematic.

Of tremendous significance to modern cartography was the development of the airplane. It has operated as a catalyst in bringing about the demand for more mapping, and at the same time the airplane has made possible photogrammetry (mapping from photographs). The need for smaller-scale coverage of larger areas, such as the aeronautical chart, promoted larger-scale mapping of the unknown areas. Furthermore, the earth seen from the vantage point of an airplane in flight is somewhat like a map, and the air traveller frequently develops an interest in maps. The development of techniques connected with space exploration is having a similar effect, but at a different level. Remote sensing, moon mapping, precise measurement of the earth, and of relative position on it from orbital observation, are just a few of the consequences.

Photography, in which rapid advancements were made during the middle of the nineteenth century, has also had enormous significance for mapping. The potential of photographs for photogrammetric uses was evident almost immediately upon the invention of the camera, and the advantages of aerial over terrestial photographs was obvious. At first, kites, balloons, and even carrier pigeons were used in efforts to provide suitable air stations, and some remarkably good photographs were produced; but they were not adequate for extensive mapping.

The invention of the airplane finally made available a vehicle for the camera that provided mobility and navigability. During World War I aerial cameras were used, and the post-war years saw more stable aircraft, which provided more rigid air stations. The ability to obtain large numbers of photographs at a rather consistent scale permitted the employment of the photogrammetric principles that had been established in the early 1900's. By the 1930's government mapping agencies were beginning to use photographs in stereo plotters for the preparation of large-scale topographic maps. Today, photogrammetric methods are extensively used in the preparation of nearly all published large-scale topographic maps.

The union of photography with the printing processes in the latter part of the nineteenth century provided at last a relatively inexpensive means of reproducing an image on almost any substance. From then on technical developments in the photolithographic and photoengraving fields have been rapid and continuous, and they have become inextricably woven into the methods of cartography. Today, although costs are considerable, high-speed, multicolor presses are capable of handling any kind of cartographic problem. Modern map construction with its scribing, open window negatives, photo lettering, etc., is as different from the map making of fifty years ago as is the automobile from the horse and buggy.

One other factor that bears mention in

this short list is the development of the electronic computer and all the attendant devices and accessories such as the XY digitizer and XY plotter. These have opened the way to innumerable methods that bid fair soon to revolutionize cartography yet again. One may now use automatically plotted coastlines and boundaries on any selected projection entirely from magnetic tape (Fig. 1.8) or have automatic print-outs of statistical data in map form (Fig. 1.9). It is well to observe, however, that the mechanical and technical advances simply make the solutions to problems easier. Car-

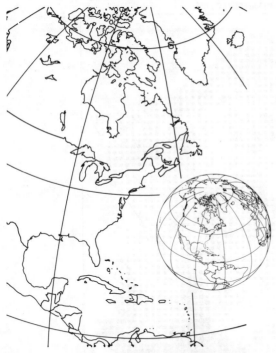

Figure 1.8 An example of computer cartography. The coastlines and boundaries are on magnetic tape in a "data bank." A computer is programmed to use these data and to prepare this map at a given scale on this projection (orthographic) centered at any place — in this case, Boston. It directs the automatic plotter that drew the entire map in about 1 1/2 hours. Shown here as an "inset" is the much reduced full drawing surrounded by a detail.

tography is a body of theory and method for dealing with problems of recording and communicating geographical information graphically; the computer, the air photograph, the scribe coat, and the other innovations are only aids to better, more economical, or faster map making.

The Field of Cartography

Since the beginning of the nineteenth century, the field of cartography has expanded greatly, and while doing so, it has branched out into several somewhat specialized areas.

The entire field is usually thought of as consisting of two more or less distinct phases. The first is concerned with the preparation of a variety of maps used for basic reference and operational purposes. This category includes, for example, large-scale topographic maps of the land, hydrographic charts, and aeronautical charts. The second major division has to do with the preparation of an even larger variety of maps used for general reference and educational (in the broad sense) purposes. This category includes the usually small-scale thematic maps of all kinds, as well as atlas maps, road maps, maps to be used in company with written text in books, planning maps, and so on. Within each of these two broad categories there is also considerable specialization such as may occur among the survey, design, drafting, and reproduction phases of making a topographic map. All such divisions and activities shade one into another, however, and it is to be expected that sharp compartmentalization rarely occurs.

The first-mentioned category of cartography works primarily from data obtained by field or hydrographic survey or by photogrammetric methods. Of fundamental concern are such things as the shape of the earth, height of sea level, land elevations, precise distances, and detailed locational information. Complex electronic and photogrammetric instruments are an integral part of this sort of cartography. Generally speaking, this group includes the great national survey organizations, the oceanographic and aeronautical charting agencies and most military mapping organizations. From their work come the fundamental bases for all mapping.

The other category, which includes thematic cartography, draws upon the basic work of the first group but is mostly concerned with the communication of general information and with the effective graphic delineation of relationships, generalizations, and a host of geographical concepts. The specific subject matter may be drawn from such fields as history, economics, urban planning, rural sociology, engineering, and many others of the physical and social sciences—there is no limit. Uppermost, however, is the concern for a properly designed, clear, legible, graphic communica-

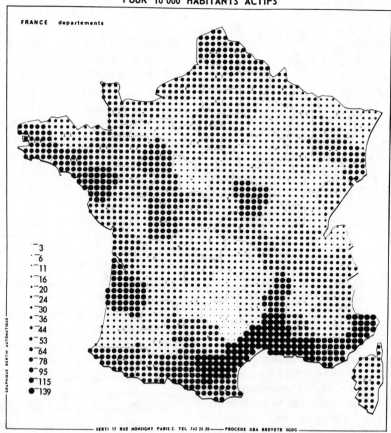

Figure 1.9 An automatically produced statistical map. The simple choropleth map was produced by punch cards directing the operation of a specially keyed electric typewriter. (From Jacques Bertin, *Semiologie Graphique*, Paris, 1967, courtesy the author.)

tion to assist in understanding and interpreting the social and physical complex on the earth's surface. The data for such maps, ranging from the thematic to the small-scale reference maps in an atlas, are usually compiled.

It should not be inferred that the practical considerations that tend to separate the cartographers basically concerned with large-scale maps and survey from those concerned with small-scale maps and compilation necessarily create a well-defined void between the two groups. Quite the contrary is the case. The fundamental problems of each group are conceptually similar. For example, the cartographic concepts upon which the delineation of landforms by contours depends are the same as those upon which the delineation of an abstract statistical surface by isopleths depends. Most of the principles upon which cartographic techniques are based are equally applicable in either division of the field. Rather, the major distinction is in the methods of acquiring the data to be mapped; these are commonly different. On the other hand, the cartography, that is, the conception, the designing, and the execution of the map, as distinct from the gathering of the data, is fundamentally the same in both divisions.

The Essential Cartographic Process

In view of the infinite variety of maps it may seem impossible to summarize what they all have in common. Yet cartography, being a distinct body of theory and practice, does include a series of processes that are peculiarly cartographic and common, in the broad sense, to all maps.

All maps are reductions. This means that the first decision of the cartographer must have to do with the scale of the map he will produce. The choice of a scale is of primary importance because it sets a limit on the information that can be included in the map and on the degree of reality with which it can be delineated. In addition, such matters as the expected use of the map affect the decision as to scale (for example, page size in a book or a term paper, folded in an automobile, handling in the field in military operations, in the cockpit, or on the bridge of a ship), as do such other factors as the economical reproduction size and relation with similar series.

By their very nature all maps are the presentation of spatial relationships. Therefore, another of the cartographer's major problems is to transform the surface of a sphere that curves away in every direction from every point (called the allside curved surface) into a surface that does not curve in any direction at any point (a plane). Such a radical transformation introduces some unavoidable changes in the directions, distances, areas, and shapes from the way they appear on the spherical surface. A system of transformation from the spherical to the plane surface is called a map projection, and the choice, utilization, and construction of projections is of perennial concern to the cartographer. The choice of a projection is an interesting problem for it involves the weighing and balancing of the assets and liabilities of a surprisingly large number of possibilities. The theoretical and actual derivation of many systems of projection are undertakings that require considerable mathematical competence, as would be expected. On the other hand, their utilization and very often their construction require little more than arithmetic, some very elementary geometry, and, most important, some clear thinking.

Every map has an objective in terms of communication. It is made for a general or specific purpose, and it is to be read by a given group of map users. It is, therefore, a complex form of communication and like any form of communication must be carefully planned and executed. Because a map is a reduction and because it must serve its purpose, another major task of the cartographer is to generalize. He must select the proper information, he must simplify where necessary, and he must plan to make either more or less graphically prominent (according to his purpose) those categories of data he intends to include.

Generalization is one of the most difficult of the cartographer's tasks, for, as was pointed out by Max Eckert, one of the great German cartographers, "Only a master of the subject can generalize well." If we are to represent a coastline at a reduced scale we must know the characteristics of that coastline or at least of its type. Similarly, if we are to generalize a river we must know whether it is a dry-land or a humid-land stream, something of its meanderings, its volume, and other important factors that might make it distinctive from or in character with others of its nature. The cartographer's portrayal and the selection of the important and the subordination or elimination of the nonessential factors of the map data require that the map maker be well acquainted with the subject matter of the maps he intends to make.

A fourth major task of the cartographer is the planning of the graphic characteristics of the map. The map must be legible, the symbolism or notation must be suited to the objective of the map, and the whole must be fitted together to make an efficient communication. The preparation of a graphic communication is like that of a written communication; they must each be planned carefully so that the reader will be able, as easily as possible, to obtain the information being conveyed. This phase of the cartographic process involves a variety of operations: the settling upon the methods of portrayal, the choosing of lettering sizes and styles, the specifying of widths of line, the selection of colors and shadings,

the arranging of the various elements within the map, the designing of a legend, and so on. Cartographic design is complex, but like writing with words, communication with the graphic language of maps can be learned.

The fifth major aspect of the cartographic process is the actual construction or drawing of the map and its reproduction (usually). Formerly, one drew a map with pen and ink and then "sent it out" for reproduction by a printer. During the past several decades a major revolution has also occurred in this phase of cartography. In some instances we still follow the old practice, but this is becoming increasingly rare. The development of scribing, of peel coat materials, and other such materials and processes has drawn the two aspects of construction and reproduction so closely together that it is impossible to think of them separately anymore. The cartographer must be familiar with all the processes involved in order to design his map efficiently and have it constructed and reproduced properly and economically.

Cartographic Drafting

The term "drafting" brings to mind pen, ink, T-squares, and the like; but today's cartographic draftsman is likely to be working much of the time with scribing instruments, films, opaque, stick-up lettering, and so on. Drafting is an old and honorable calling, and a good, versatile cartographic draftsman is rare enough to command great respect. There is no other craft quite comparable to that of cartographic drafting. The judgment and skills required make it always interesting, and there is considerable creative pleasure in completing a well-drawn map. On the other hand, the act of drawing a map or preparing the artwork for one that is to be reproduced is not all there is to cartography.

It is important to understand the relation between drafting and cartography. There has been and still is a tendency to think of the two as synonymous. Many draftsmen think of themselves as cartographers, and many cartographers, who are themselves good draftsmen, feel that drafting ability is indispensable to the cartographer. There is no question that drafting is a desirable and useful ability, but it is by no means indispensable. A lack of manual skill need deter no one from entering the field of cartography and, particularly, it should not deter any physical and social scientists from learning the principles of graphic expression. Since learning by doing is one of the better ways of gaining a well-founded understanding of any endeavor, it is to be expected that someone who would learn the principles of cartography would also at least be exposed to the elements of map construction. If he finds that he is deficient in the manual skills he should in no way be discouraged. Trying the techniques and studying their difficulties and possibilities will better prepare him to direct a more skillful craftsman.

Science and Art in Cartography

Cartography has been often described as a meeting place of science and art. There seems no question that as we learn more and more about communication that more of the principles and precepts of cartography are being based on understanding and less on individual aesthetic intuition.

The cartographic process involves many intellectual activities that are scientifically based. We must apply logic in our approach to projections, generalizations, line characterizations, and so on, and in this respect the cartographer is a kind of practical scientist, much like an engineer. He must study the characteristics of his building materials and know the ways and

means of fitting them together so that the end product will convey the correct intellectual meaning to the reader.

Of equal, and sometimes greater, importance are the visual relationships inherent in this form of expression. People, or map readers, think and react in certain ways to visual stimuli. With knowledge of the principles and laws governing these reactions, the cartographer can design his product to fit these perceptual patterns. The selection of lettering sizes and styles, of circle sizes for representing statistics and the colors and tones to be used in representing gradations of amount, are examples of questions to be answered by reference to the principles of visual perception as applied to symbols. They constitute problems of visual logic.

The cartographer is scientific in other ways. One of the largest categories of information with which he works is that contained in other maps, and since he can hardly have first-hand familiarity with all places in the world, he must be able to evaluate his source materials. He must have a good geographical background, and he must be familiar with the state of topographic mapping and its geodetic foundations. To supplement his map information he needs to be able to evaluate and rectify census data, air photographs, documentary materials, and a host of other kinds of sources.

Cartography is neither an experimental science in the sense that chemistry or physics is, nor is it searching for truth in the manner of the social sciences. Nevertheless, it employs the scientific method in the form of reason and logic in constructing its products. Its principles are derived through the analysis of scientific data. It has its foundations in the sciences of geodesy, geography, and psychology. In the sense that it is based on sound principles and seeks to accomplish its ends by way of intellectual and visual logic, it is scientific in nature.

Before the last century the question of whether cartography was an art never arose, for it very definitely was. This is evident when we view the products of the earlier cartographers who embellished their maps with all sorts of imaginative things, together with fancy scroll work, ornate lettering, and intricate compass roses. Special coloring methods and ingredients were carefully guarded secrets Even as late as the middle of the nineteenth century the coloring of maps in one of Germany's greatest map houses was done by the society ladies of the town. Throughout the history of cartography, great emphasis has been laid on fine pen and brush skill, and the aim has been to make something good to look at and perhaps even to hang on a wall as a decoration. Today the con-

cern is less for aesthetic appeal than it is for the efficiency with which the communicative objective is attained.

Maps today are strongly functional in that they are designed, like a bridge or a house, for a purpose. Their primary purpose is to convey information or to "get across" a geographical concept or relationship; it is not to serve as an adornment for a wall. On the other hand, one of the cartographer's concerns may be to keep from producing an ugly map; in this respect he is definitely an artist, albeit in a somewhat negative sense. Cartography is certainly a creative art in the way that careful, creative literary expression is an art. Every map is a different problem requiring a new solution in the field of graphic design. It requires a command of the principles of graphic communication to build a map successfully.

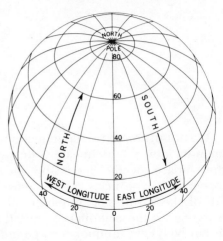

THE EARTH, COORDINATE SYSTEMS, AND SCALE

THE EARTH

ALTHOUGH there is a large mapping program underway to delineate systematically the surface of the moon by means of topographic-type maps, as yet most maps are of the earth. The problems of moon mapping (or the mapping of any other celestial body) are uniquely compounded by its distance, movements, and the difficulty of obtaining data, but the basic cartographic objectives and the means to attain them are not unique. The problems involved in locating and representing on a plane the relative planimetric (horizontal) and hypsometric (vertical) positions of places at a reduced scale are essentially the same regardless of the object being mapped because celestial bodies are spherical, are large, and require a coordinate system. The earth is a good example.

The Shape of the Earth

If we disregard the relatively minor irregularities of the earth's outer shell, such as its continents with their irregular land-surface form, the shape of the earth may be defined roughly as that of the surface that would be assumed by the average level of the sea and a system of sea-level canals crisscrossing the land. Even with that great simplification, the earth is a complex geometric figure. The shape assumed by our plastic planet, spinning on its axis through space, is the result of the interaction of several internal and external forces, such as gravity, the centrifugal force of rotation, and variations in the density of its rock constituents.

The interaction of tectonic and gradational forces has produced other irregularities such as mountains, plains, and ocean basins. This class of irregularities, so noticeable to the human eye, is relatively so small, however, that they are significant only to the cartographic problem of the delineation of the land form. For example, on a globe with a diameter of 1 ft, mountains and ocean basins would scarcely be noticeable; the maximum deviations from average sea level would be less than 1/100 of an in.

Maps are representations of the complex shape of the earth on a plane, and accordingly it is necessary to transfer systematically the geometric relations from the one shape to the other.* If this is to be done accurately it is apparent that the characteristics of both shapes must be known. Furthermore, in order that the transformation may be done systematically it is necessary to assume the earth shape to be a simple solid form. It is important to understand that this part of the mapping process involves three steps: (1) the determination of the regular geometric figure that most closely approximates the actual form of the earth, (2) the transfer, for mapping purposes, of the earth positions to that form, and (3) the transformation of that form to a plane. The determination of the precise figure of the earth is part of the responsibility of the science of geodesy.

Man's first ideas of the earth around him probably included little beyond that which he could see, since his view was limited by the horizon; consequently, the surface appeared flat. The idea of sphericity was

*Strictly speaking, we may map on surfaces other than a plane. The representation on a reduced globe and that of a three-dimensional terrain model are both maps. Ninety-nine percent of all maps are on a plane.

not generated until the philosophers of the pre-Christian era applied reason to the problem, and some 500 years or more before the birth of Christ the earth was theorized to be spherical. By the time of Claudius Ptolemy (second century A.D.) the spherical earth was generally recognized. Although the idea of sphericity certainly did not die out during the Dark Ages, it languished. After the reissuance of Ptolemy's *Geographia* and the subsequent Age of Discovery following the fifteenth century, most doubts about the spherical shape of the earth died forever.* In the late seventeenth century the idea of oblateness because of rotation was advanced by Newton and eagerly pursued. During the last century or so several determinations of the amount of bulging and flattening have been made, and there are several oblate spheroids now in use by mapping organizations.

The oblateness, the largest of the earth's deformations from a true sphere, must be acknowledged, even occasionally in small-scale mapping. On account of its spinning on an axis, the earth ball is bulged somewhat in the area midway between the poles and consequently flattened a bit in the polar regions. The amount of the polar flattening (f) is usually simply expressed as the ratio $f = a - b/a$ where a is the equatorial semiaxis and b is the polar semiaxis. For most spheroids f is very close to 1/297. The actual amount of flattening is of the order of some 13.5 miles difference between the polar and equatorial radii, the equatorial, of course, being the larger.

Because of the bulging and flattening, a line extending around the earth that passes through the poles will not be a circle but will be slightly oval in shape. The flatter portion is in the polar regions and the more rapid curvature is in the equatorial areas. Since a considerable amount of navigation is based upon observations aimed at finding the angle between some celestial

body and the horizon plane (or a perpendicular to it—the vertical) at a point, it is apparent that complications result from this departure from a true sphere. Consequently, whenever maps are being prepared for navigation or for plotting exact courses and distances from one place to another, it is necessary to take the oblateness into account. In most cases of very small-scale mapping it may safely be ignored.

As a result of measurements made for topographic mapping, astronomic observations, and observations of orbiting satellites, it is clear that the generally oblate spheroidal form is also deformed, primarily because of variations in the character of the materials of which the earth is composed. Accumulation of observational data from measurements of gravity and orbits, which will finally reveal the world-wide nature of these irregularities of the spheroid, is underway, and in the forseeable future the precise shape of the earth, called the *geoid*, may finally be ascertained. When this is known then the regular geometric shape that most closely approximates it can be chosen for the mapping spheroid.

The Size of the Earth

Before the beginning of the Christian era both Eratosthenes (about 250 B.C.) and Posidonius (about 100 B.C.) calculated the size of the earth and apparently came close to the figures we now accept, but their closeness was partly the result of fortunate compensation of observational errors rather than precision. These early estimates were reported by others and recorded by Ptolemy, who, recognizing the errors, accepted "corrected" values that reduced the earth's circumference by nearly a fourth. Unfor-

*There are always some diehards, and occasionally we still run across "philosophers" who insist the earth is flat. Astronauts and artificial satellites have made the logic even more difficult.

tunately, or fortunately, depending upon one's viewpoint, Ptolemy's convictions were generally accepted. If Columbus had known the true length of a degree of arc on the earth he probably would not have dared set sail to find the Indies by going west.*

Since 1850 the dimensions of the earth have been calculated with great care from various astro-geodetic arcs measured with precision. Because the earth is not a regular shape the results differ slightly. International standardization has been proposed but is not completely accepted as yet. The values in Table 2.1 are those generally used in the United States and are those of the Clarke spheroid of 1866.

Areas on the Earth

As the spherical earth complicates the cartographic problem of representing on a plane such concepts as distance and direction, the allside curved surface likewise makes difficult the reckoning and representing of areas. To arrive at the area of a polygonal segment of the surface of a sphere is relatively easy, but the earth is not a sphere, and for various other reasons, the establishment of exact position is difficult. If positions are doubtful, then the shape of the spherical segment is in doubt and thus the area of it is open to question. Besides, most of the areas in which we would be interested, aside from small land holdings, are extremely irregular, for instance, continents or countries with complex boundaries or coastlines. Consequently, the only way to determine the actual size of such areas is to map them first and then measure or calculate in some manner the area enclosed. Of course, such a map must be one in which the transformation from the spherical surface to the plane surface has been made so that areas are

uniformly represented as to size anywhere within the map.

The Great Circle

The shortest distance between two points is a straight line; however, on the earth it is obviously impractical to follow this straight line through the solid portion of the planet. The shortest course along the surface between two points on a sphere is the arc on the surface directly above the straight line. This arc is formed by the intersection of the spherical surface with the plane passing through the two points and the center of the earth. The circle established by the intersection of such an extended plane with the surface divides the earth equally into hemispheres and is termed a *great circle*.

Great circles bear a number of geometrical relationships with the spherical earth that are of considerable significance in cartography and map use.

1. A great circle always bisects another great circle.
2. An arc of a great circle is the shortest course between two points on the spherical earth.
3. The plane in which any great circle lies always bisects the earth and hence always includes the center of the earth.

Because a great circle is the shortest distance between two points on a spherical

*When Columbus added the (overestimated) "known" distances eastward from the Azores to the Orient and then divided the sum by his too-small degree, the result when subtracted from 360° meant that the remaining distance westward to the Orient from the Azores was hardly more than the length of the Mediterranean Sea. To the end of his days Columbus believed he had reached the Indies and the Orient —hence the *West Indies* and the term *Indians* applied to the natives!

Table 2.1 Dimensions of the Earth

Equatorial radius	6,378,206 m	3963.3 statute mi
Polar semiaxis	6,356,584 m	3949.7 statute mi
Radius of sphere of equal area	6,370,997 m	3958.7 statute mi
Area of earth (approx)	510,900,000 sq km	197,260,000. sq mi
Equatorial circumference	40,075 km	24,902 mi

surface, air and sea travel, insofar as is possible or desirable, move along such routes. Radio signals and certain other electronic impulses tend to travel along great circles, and for this reason many maps must be made on which great circles are shown by straight lines or simple curves.

COORDINATE SYSTEMS

To locate points relative to one another it is necessary to employ the concepts of direction and distance. These can only be specified in terms of some system; primitive man probably did so in relative terms using such aids as the directions of the rising and setting sun, forward and backward, left and right, and so on and he probably expressed distance in terms of travel time—all of these being reckoned with respect to his location. Any universal or general system must, however, be established in relation to some unique reference or starting point. If such a point is designated, then the location of every other point can be stated in terms of a defined direction and distance with it.

Plane Coordinates

On a limitless plane surface there is no natural reference point; that is, every point is like every other point. An arbitrary system of location on a plane surface has long been used by establishing a "point of origin" at the intersection of two conveniently located, perpendicular "axes." The plane is then divided into a grid by an infinite number of equally spaced lines parallel to each axis. The position of any point on the plane with reference to the point of origin may then be stated by indicating the distance from each axis to the point, measured in each case parallel to the other axis, and expressed to any desired precision. In the familiar rectangular coordinate system (for example, cross-section paper) the "horizontal" distance is called the X value or *abscissa*, and the distance perpendicular to it is called the Y value of the *ordinate* (Fig. 2.1).

Spherical Coordinates

On a sphere a similar (but much older) coordinate system is used. A spherical surface is an allside curved one (that is, it slopes away equally in every direction from every point), and the use of parallel straight lines is impossible. On a motionless spherical surface there is also no natural reference point, but celestial bodies are not motionless with respect to other bodies. On the earth, for example, two convenient reference points are established by the poles, the two points where the axis of rotation intersects the spherical surface. If we imagine the earth before us with the axis vertical, the ordinate values, corresponding to the Y values in the rectangular coordinate system, are called latitude and the abscissa values are called longitude. The arrangement anywhere away from the poles of these two sets of perpendicular coordinates establishes the cardinal directions on the earth. A rectangular and the earth's spherical coordinate system have much in common; both systems may be represented by two sets of grid lines (called the graticule on the earth) that are perpendicular to one another, but only in one of the sets making up the graticule are the lines parallel. On a spherical surface we may conveniently specify distance by degrees, minutes, and seconds of arc.

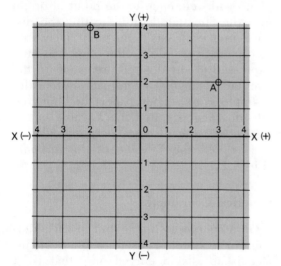

Figure 2.1 A plane coordinate system. The origin is O. XOX constitute the abscissa values and YOY, the ordinate values. The position of point A is 3, 2; the position of point B is −2, 4. In the earth's coordinate system, Y values correspond to latitude north and south and X values to longitude east and west of the origin (0° latitude and 0° longitude). In designating position in a plane coordinate system, the X value is always given first; on the earth's coordinate system, latitude is usually given first.

Latitude

In classical times the system of locating oneself between the two poles was devised. Imagine a line joining the poles on the earth's surface, i.e., a half-circle containing 180°. When someone stands anywhere on this line, his horizon seems to him to bound a flat (or nearly flat) circular plane. If he can visualize himself as being in a space capsule and observing this little horizon plane, he will note that the horizon plane is tangent to the circle, and that if he were to slide the plane north or south along the line it would always be tangent. If the star Polaris is assumed to lie on an extension of the earth's axis, then an observer at the north pole would see Polaris at a 90° angle with his horizon plane. If it were possible for him to see Polaris through the great thickness of the hazy atmosphere at the equator, he would find the direction to Polaris to be tangent to his horizon plane, that is, at an angle of 0°. Movement along the line toward a pole is accompanied by a change in the angular elevation of celestial bodies in relation to the horizon plane on the earth in a one-to-one relationship; that is, for each degree of arc distance traveled, the elevation above the horizon of a celestial body will change by 1°. Any star or the sun can be observed, and the result will be the same.

The foregoing simplifies the problem somewhat, for the earth rotates on its axis and most of the celestial bodies therefore seem also to move while the observer is moving from one place to another. The essential information to correct for apparent celestial motion is, however, readily available from an ephemeris. The fundamental fact remains that position north-south can be determined by measuring the angle between the horizon and a celestial body.

To utilize this relationship in a spherical

coordinate system was natural. The ancients imagined an infinite number of circles around the earth parallel to one another (Fig. 2.2). The one dividing the earth in half, equidistant between the poles, was named, as might be expected, the equator. The series north of the equator was called north latitude and the series south, was called south latitude. To determine which circle one was on, and hence his arc distance north or south of the equator, required only the observation of the angle between the horizon and some known celestial body such as the sun, Polaris, or some other star.

No change has been made in the system since it was first devised nearly 2200 years ago.

The Length of a Degree of Latitude. In the usual system of angular measurement, a circle contains 360°; a half-circle, 180°. Consequently, there are 180° of latitude

from pole to pole. The quadrant of the circle from the equator to each pole is divided into 90°, and the numbering starts from 0° at the equator and goes by degrees, minutes, and seconds to 90° at each pole. Latitude is always designated as north or south.

Because the earth approximates an oblate spheroid, a north-south line (a meridian) has more curvature near the equator and less near the poles. Therefore, to observe a 1° difference in the altitude of a celestial body requires a lesser traverse along the meridian in the equatorial regions and a greater in the polar. Consequently, degrees of north-south arc on the earth are not quite the same lengths in units of uniform surface distance but vary from a little less than 69 statute miles (68.7) near the equator to a little more than 69 (69.4) near the poles. It is apparent that this difference of less than 1 mile in 69 is of little significance in small-scale maps, but is important on large maps of small areas. For ordinary purposes it is well to keep in mind that degrees of latitude are very nearly the same length. Table 2.2 shows a selection of the lengths of the degrees of the meridian, that is, latitude. A more complete table is included in Appendix D.

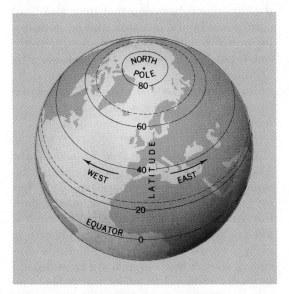

Figure 2.2 The parallels of latitude (distance north-south) coincide with the directions east-west. (From Trewartha, Robinson, and Hammond, *Elements of Geography*, 5th ed., McGraw-Hill Book Company, 1967.)

Table 2.2   *Lengths of Degrees of the Meridian*

Latitude	Statute Miles	Kilometers
0–1°	68.703	110.567
9–10°	68.722	110.598
19–20°	68.781	110.692
29–30°	68.873	110.840
39–40°	68.986	111.023
49–50°	69.108	111.220
59–60°	69.224	111.406
69–70°	69.320	111.560
79–80°	69.383	111.661
89–90°	69.407	111.699

Longitude

Only position north-south on the earth is established by the latitude component of the spherical coordinate system. The transverse component, longitude, or distance east-west is provided by an infinite set of meridians, arranged perpendicular to the parallels. Unlike the equator in the latitude system, there is no meridian with a natural basis for being the starting line from which to reckon distance east-west in degrees, minutes, and seconds of longitude. From a given meridian, selected as a starting line, east-west position is designated by the angular distance along the parallel circle in the latitude system (Fig. 2.3).

Prior to the middle of the eighteenth century only latitude could be easily reckoned with any precision. Distance east-west depends upon time differences and, for easy reckoning, requires that one know the times of day at the two places at the same instant. Without accurate time pieces that can be carried or instantaneous communication, this can only be done by very elaborate astronomical observation and cal-culation. Through the years this led to considerable error in east-west location, and it was one of the contributing factors to the glorious error of the fifteenth century, the idea that the distance separating Europe from Asia westward was less than half its actual value.

When the determination of longitude became critical for navigation, generous prizes for its solution were offered, and a variety of suggestions were put forward, ranging from the observations of a celestial timepiece, such as the behavior of the satellites of Jupiter, to the employment of the variations (declination) of the magnetic compass. When the chronometer (a very accurate clock) was developed by Harrison and others in the middle of the eighteenth century, the problem was solved. Because all parallels are concentric circles, they all rotate at the same angular speed—360° per day or 15° per hour. By carrying a clock showing the accurate time somewhere else, the difference between that time and local suntime in hours, minutes, and seconds can be converted to the longitude difference between the two places merely by arithmetic. Today this is accomplished by radio time signals broadcast at regular intervals, as well as with the aid of chronometers.

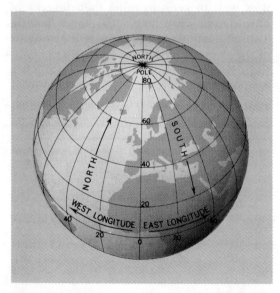

Figure 2.3 The meridians of longitude (distance east-west) coincide with the directions north-south. (From Trewartha, Robinson, and Hammond, *Elements of Geography*, 5th ed., McGraw-Hill Book Company, 1967.)

The Length of a Degree of Longitude. The length of the equator is very nearly the same as the length of a meridian circle, but as we go toward the poles all other parallels become smaller and smaller circles; yet each is divided into 360°. Therefore, each east-west degree of longitude becomes shorter with increasing latitude and is finally reduced to nil at the poles. The relationship between the length of a parallel (the circumference of a small circle) and the circumference of a great circle (for example, the equator or a meridian circle) is the circumference of the great circle multiplied by the cosine of the latitude of the parallel; or stated another way, the length of a degree of longitude = cosine of the latitude × length of a degree of latitude.

A table of cosines (see Appendix B) will show that

$$\cos\ \ 0° = 1.00$$
$$\cos 30° = 0.87$$
$$\cos 60° = 0.50$$
$$\cos 90° = 0.00$$

Thus, at 60° north and south latitude the length of a degree of longitude is half that of a degree of latitude. Table 2.3 is included here as an illustration of the decreasing lengths of the degrees of longitude from the equator toward the pole. A more complete table is included in Appendix D.

Table 2.3   Lengths of Degrees of the Parallel

Latitude	Statute Miles	Kilometers
0°	69.172	111.321
10°	68.129	109.641
20°	65.026	104.649
30°	59.956	96.448
40°	53.063	85.396
50°	44.552	71.698
60°	34.674	55.802
70°	23.729	38.188
80°	12.051	19.394
90°	0	0

The Prime Meridian. The meridians are all alike, and any one can be chosen as the meridian of origin from which to start the numbering for longitude. The choice became, as might be expected, a problem of international consequence. Numerous countries, each with characteristic national ambition, wished to have 0° longitude within its borders or as the meridian of its capital. For many years each nation published its own maps and charts with longitude reckoned from its own meridian of origin. This, of course, made for much confusion.

During the last century many nations began to accept the meridian of the observatory at Greenwich near London, England, as 0°, and in 1884 it was agreed upon at an international conference. Today this is almost universally accepted as the prime meridian, but some maps still show two sets of meridians, one based on a local prime meridian and the other on the Greenwich system. Since longitude is reckoned as either east or west from Greenwich (to 180°), the prime meridian is somewhat troublesome because it divides both Europe and Africa into east and west longitude. The choice of the meridian of Greenwich as the prime meridian establishes the "point of origin" of the earth's coordinate system in the Gulf of Guinea. The opposite of the prime meridian, the 180° meridian, is more fortunately located, for its position in the Pacific provides a convenient place for the international date line. Days on earth must begin and end somewhere and only a few deviations from 180° are needed to keep from separating inhabited areas into different-day time zones.

Rectangular Coordinates

The spherical coordinate system is useful for large areas, and the measurement of distances and directions in angular measure in degrees, minutes, and seconds can hardly be improved upon. But for small areas it is cumbersome. With the increasing range of artillery in World War I it became more and more difficult to arrive at accurate azimuth (bearing or direction) and range (distance). In order to simplify the problem, the French constructed a series of local plane, rectangular coordinate grids on maps. Since the formulas of plane geometry are far simpler than those of spherical geometry, other nations quickly followed suit and between World Wars I and II a great many systems of plane rectangular coordinates were devised and put into use.

By now the use of rectangular grid systems is "standard procedure."

The procedure is as follows: first, a map is made by transforming the spherical surface to a plane (by a system of projection), preparing the map on the plane, and then placing a rectangular plane coordinate grid over the map. The two sets of straight, parallel grid lines are equally spaced and perpendicular to one another. To locate a position we need only specify the X and Y coordinates to whatever degree of precision we desire in whatever earth distance units are used, usually in a decimal system. This is much simpler than degrees, minutes, and seconds of latitude and longitude.

To simplify the reckoning of position only the upper right-hand part of a plane coordinate system is used (Fig. 2.1) so that both sets of coordinates are positive, and therefore there will be no repetition of numbers east and west or north and south of the axes. Normally, the origin of the numbering is assumed to be outside the map area to the lower left.

The reading of a grid reference proceeds in the normal way a point is located on cross-section paper. In rectangular map coordinates the X value is always given first and is called an *easting*; the Y value is called a *northing*. A rule of thumb is that when using grid references we must always "read right up." Reference to Fig. 2.4 will show that point P can be given an easting value of 145 and a northing value of 201 by decimal subdivision of the squares. The grid reference would simply be 145201. With lesser precision it would be 1420 and with greater precision it would be 14562011. Grid references always contain an even number of digits and the first half refers to the easting and the second to the northing. If each square in Fig. 2.4 represented 1 sq km (1000 m on a side) then the reference 145201 would be a statement of location to within a 100 m square, that

being the location of the southwest corner of the square. Adding an additional digit would narrow the position to a 10 m square.

Normally, rectangular coordinates are only used on large-scale maps since the distortions that result from the transformation of the spherical surface to the plane make small-scale maps undesirable for detailed reference and calculation. Small-scale reference maps sometimes employ a rectangular letter-number indexing system to help locate map data, but that is not a rectangular coordinate system.

The UTM and the UPS Military Grids.

The United States has adopted projection systems for large-scale military mapping called the transverse Mercator and the polar stereographic. (These will be described in Chapter 10.) On all such maps there is superimposed a rectangular grid called the Universal Transverse Mercator (UTM) or the Universal Polar Stereographic grid (UPS). The UTM grid is employed between 80°N and S lat., and the UPS poleward of 80°.

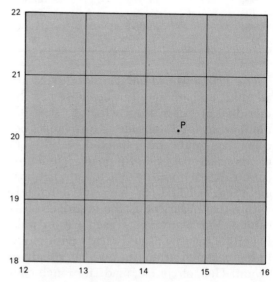

Figure 2.4 A portion of a plane rectangular grid. Point P is located at 14562011.

In the UTM grid system the area of the earth between 80° N and S lat. is divided into 60 north-south columns 6° of longitude wide called zones. Within each zone the meridian in the center of the zone is given an easting value of 500,000 meters. The equator is designated as having a northing value of 0 for the northern hemisphere coordinates and an arbitrary northing value of 10,000,000 meters for the southern hemisphere.

The UPS grid system superimposes a rectangular grid over the polar areas with 90°E and W being the "horizontal" central line and 0°–180° being the "vertical". The 0° meridian is uppermost in the south polar grid and 180° in the north polar grid. Each pole is given an arbitrary easting and northing of 2,000,000 meters.[*]

Figure 2.5 illustrates the use of the UTM grid without the identification of the particular 6° × 8° quadrilateral and the 100,000 square number-letter designations.

State Plane Coordinates. In order to provide the convenience of plane coordinates and a way to ensure the permanent recording of the location of original land survey monuments, the U.S. Coast and Geodetic Survey has worked out a system of plane coordinates for each of the states (on particular projections—transverse Mercator and

[*]Each zone in the UTM is divided into 6° × 8° quadrilaterals, called grid zones, which are identified by reference numbers and letters. These are further subdivided into 100,000 meter squares which also have identifying letter combinations. The UPS is similarly subdivided. A complete description is given in Departments of the Army and Air Force, *The Universal Grid Systems*, *TM* 5-241/*TO* 16-1-233, U.S. Government Printing Office, Washington, D.C., 1951.

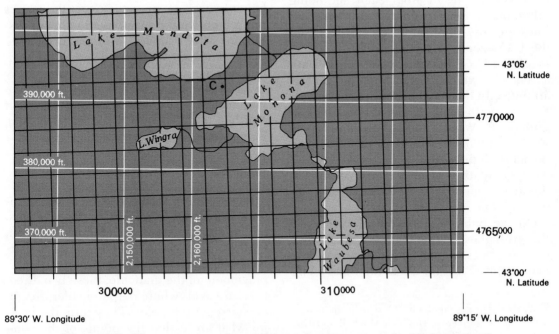

Figure 2.5   Three systems of coordinates that appear on the Madison Quadrangle, U.S.G.S., 1:62,500. The bordering ticks show tha graticule; the white lines show the Wisconsin Coordinate System, South Zone; and the interior black lines show the UTM grid.

Lambert's conic) which are specifically tied to the geodetic locations in the national triangulation system. To keep the inevitable scale variation to a reasonable minimum, the states usually have two or more overlapping zones each of which has its own projection system and grid. The units used are feet. The large-scale topographic maps published by the U.S. Geological Survey now carry tick marks showing the locations of the 10,000 ft grid.*

Fig. 2.5 shows both the 1000 m UTM grid lines of Zone 16 and the 10,000 ft grid lines of the Wisconsin Coordinate System, South Zone extended over the southern third of the area included on the Madison Quadrangle, 1:62,500, U.S. Geological Survey topographic map. The three different orientations (the graticule, i.e. the net of parallels and meridians, the UTM grid, and the grid of the Wisconsin Coordinate System) result from each set being differently arranged. In the graticule all lines are true north-south or east-west. In the UTM zone 16 the meridian of 87° W longitude is central and is the north-south axis about which the rectangular grid is arranged. In the Wisconsin Coordinate System the north-south axis of the rectangular grid is 90° W longitude. Accordingly, on Fig. 2.5 it will be seen that the UTM grid deviates slightly from a north-south, east-west orientation in a fashion opposite to the arrangement of the Wisconsin Coordinate System.

On the published map the positions of the grid lines in the two sets are shown only by ticks around the margins of the map. Point C in Fig. 2.5 is the State Capi-

tol. Grid references of C would be read as follows.

1. Normally the small initial digits in the UTM grid shown on a map (3 in the eastings and 47 in the northings on Fig. 2.5) are not used. In the UTM Grid System the reference 0571 would locate Point C within a 1000 m square and 058716 would locate it within a 100 m square, these designations being the locations of the southwest corners of the respective squares. Adding yet another digit to each set would locate it within a 10 m square, and adding a fifth digit to the easting and the northing would locate it within a 1 m square. Naturally, accurate, larger-scale maps would be needed to obtain that much precision.

2. Wisconsin Coordinate System. On the published topographic map 1/100 in. represents 52 ft so that the reading of a coordinate position to that precise value would be quite useless, since the paper the map is printed on would be subject to greater distortion with changes in humidity. State plane coordinates are surveyed to the fraction of a foot in the field, and then are fully given. Point C in Fig. 2.5 is located at approximately 2,164,600 ft East and 392,300 ft North in the Wisconsin Coordinate System, South Zone.

Direction

Directions on the earth are entirely arbitrary since a spherical surface has no edges, beginning or end. By definition, then, north-south is along any meridian and east-west along any parallel; because of the arrangement of the graticule, these two directions are everywhere perpendicular except, of course, at the poles. They form the basis for what are called the points of the compass, collectively called the compass rose when graphically portrayed. Such diagrams were formerly common items on maps, and

---

*A complete description of the state plane coordinate systems is given in Hugh C. Mitchell and Lansing G. Simmons, *The State Coordinate Systems*, U.S. Coast and Geodetic Survey, Special Publication No. 235, U.S. Government Printing Office, Washington, D.C., 1945.

often they were embellished and made quite ornate. Today the compass rose appears less frequently except that, of course, a circle divided into degrees is often printed on charts intended for navigational purposes.

The needle of the magnetic compass aligns itself with the total magnetic force, and usually this results in its not paralleling the meridian, that is, in its having some declination. Consequently, most areas will have a different "magnetic north" from "true north." When we add to these the "grid north" of a local rectangular grid system, considerable confusion can result (Fig. 2.5). We must always make clear which "north" is being specified.

The direction of a line on the earth is called many things: bearing, course, heading, or azimuth. Their meanings are essentially the same, differing largely in the context in which they are used. The two of importance in cartography are azimuth and bearing.

The Azimuth. As is apparent from studying a globe, the directions on the earth, established by the graticule, are likely to be constantly different if we move along the arc of a great circle. Only on a meridian or on the equator does direction remain constant along a great circle; but since arcs of a great circle represent the shortest distances between any two points, it is convenient to be able to designate the "direction" the great circle has at any starting point toward a destination. This direction is reckoned by observing the angle the arc of the great circle makes with the meridian of the starting point. The angle is described by the number of degrees (0° to 360°), reading clockwise from either north or south.

In these days of radio waves and air transport directions, routes of travel along great circles are of major importance. Hence many maps are constructed so that

directional relations are maintained as far as possible.

The computation of azimuths in the earth's spherical coordinate system is quite involved and is ordinarily not needed except in geodetic work. In plane rectangular coordinates the grid azimuth (Az_G) from A to B is

$$\tan Az_G = \frac{E_2 - E_1}{N_2 - N_1}$$

in which E_1 is the easting value for A, and E_2 is the easting value for B; N_1 is the northing value for A, and N_2 the northing value for B. Az_G will be found in a table of tangents such as in Appendix B, and must then be converted to an angular measure clockwise from grid north. If B lies in the northeast quadrant *from A*, then the tangent value = Az_G = the grid azimuth; if B is in the southeast quadrant, then $Az_g = 180° - \tan$; if B is in the southwest quadrant, then $Az_G = 180° + \tan$; and if B lies in the northwest quadrant, then $Az_G = 360° - \tan$.

The Loxodrome or Rhumb Line. A bearing is the direction from one point to another, usually expressed in relation to the compass rose either in such a fashion as northeast or as north 45° east. A great circle, being the shortest route from one point to another on the sphere, is the most economical route to follow when traveling on the earth. A pilot may do this by following a radio beam, but it is practically impossible otherwise, except when travel is along a meridian or the equator. The difficulty arises from the fact that, except for those particular great circles, directional relations constantly change along all other great-circle routes. This is illustrated in Fig. 2.6. Because a course of travel must be directed in some manner, such as by the compass, it is not only inconvenient but impracticable

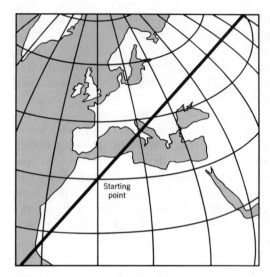

Starting
point

Figure 2.6   How azimuth (direction) is read. The drawings show a great circle
on the earth's graticule. The drawing on the right is an enlarged view of the
center section of the drawing on the left. The azimuth, from the starting point,
of any place along the great circle to the northeast is the angle between the
meridian and the great circle, reckoned clockwise from the meridian. Notice
that the great circle intersects each meridian at a different angle.

to try to change course at, so to speak, each
step.

A line of constant bearing is called a loxodrome or rhumb line. Meridians and the
equator are loxodromes as well as great circles, but all other lines of constant compass
direction are not great circles. As a matter
of fact, loxodromes are complicated curves,
and if one were to continue along an oblique loxodrome or rhumb line (other than
the meridians or the equator), he would
spiral toward the pole but in theory never
reach it.

In order for ships and aircraft to approximate as closely as possible the most direct
route between two points, great-circle
movement is directed along a line of constant bearing in relation to a (corrected)
compass. Such a course is planned to begin
on the great circle and shortly return to it,
then depart and return again as shown in
Fig. 2.7. This procedure is similar to fol-

lowing the inside of the circumference of a
circle by a series of short straight-line
chords. It is not the same, of course, since
the great circle is the "straight line" and
the rhumb lines of constant bearing are
curved lines and are actually longer routes.

Orientation. Many conventions exist in
cartography, and one of the stronger is that
of orientation or the way the directions on
the mapped earth segment are arranged on
the sheet. Naturally, on a spherical surface
there is no up or down along the surface.
But a sheet when looked at or held in the
hand has a top and a bottom. The top
seems to be the direction in which the
reader is looking.

Several centuries ago the convention of
placing north at the top became the practice and has become so strongly established that we think of "up north" and
"down south"—Australia and New Zealand

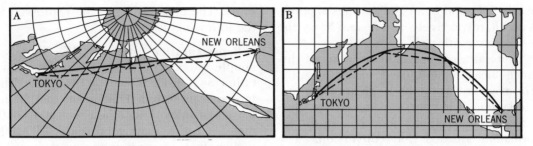

Figure 2.7 Two maps showing the *same* great circle arcs, and rhumb lines. Map A is constructed so that the great circle arc appears as the shortest distance between the two points, that is, a straight line, while the rhumbs appear as somewhat longer "loops." This is their correct relationship. In map B the representation has been reversed by constructing the map in such a way as to make the rhumbs appear as straight lines, which "deforms" the great circle arc into a curve.

are "down under," and upper Michigan and lower California are examples of the unconscious adjustment to this convention. Needless to say, save for the convention, there is no reason why a map cannot be oriented any way the cartographer pleases. Since we think of the top of a sheet as "away" from us, it is apparent that orienting the maps in the direction of interest or movement, if any, may well promote the purpose of the map.

Distance

Distances on the earth's surface are always reckoned along arcs of great circles unless otherwise qualified. Because no map, except one on a globe, can represent the distances between all points correctly, it is frequently necessary to refer to a globe, to a table of distances, or to calculate the length of the great-circle arc between two places. A piece of string or the edge of a piece of paper can be employed to establish the great circle on a globe. If the scale of the globe is not readily available, the string or paper may be transferred to a meridian and its length in degrees of latitude

ascertained. Since all degrees of latitude are nearly equal and approximately 69 miles, the length of the arc in miles can be determined.

The arc distance D on the sphere between two points A and B, the positions of which are known, can be calculated by the formula

$$cos\ D = (\sin a \sin b) + (\cos a \cos b \cos P)$$

in which
D = arc distance between A and B
a = latitude of A
b = latitude of B
p = degrees of longitude between A and B
When D is determined in arc distance, it may be converted to any other convenient unit of measure.

Note. If A and B are on opposite sides of the equator, then the product of the sines will be negative. If P is greater than 90°, the product of the cosines will be negative. Solve algebraically.

The plane grid distance D_G, between two points A and B, on a rectangular coordinate system can be calculated by the formula

$$D_G = \sqrt{(E_2 - E_1)^2 + (N_2 - N_1)^2}$$

in which E_1 and E_2 are the eastings of points A and B, respectively, and N_1 and N_2 are the respective northings.

There are, of course, many units of distance measurement used in cartography. For maps not using the English or the metric system it is necessary to refer to glossaries or some other source having the information needed for conversion. The common English and metric units of length are shown below.

	Feet	*Meters*
Statute mile	5280	1609.35
Nautical mile (Int.)	6076.10	1852.00
Kilometer	3280.83	1000.00
Foot	–	0.3048
Meter	3.2808	–

SCALE

Since maps must necessarily be smaller than the areas mapped, their use requires that the ratio or proportion between comparable measurements be expressed on the map. This is called the map scale and should be the first thing of which the map user becomes aware. The scale of a map may be shown in many ways: it can be specifically indicated by some statement or graphic device, and it may be shown indirectly by the spacing of the graticule and even subtly by the size and character of the marks on the map.

Scale is an elusive thing in maps because by the very nature of the necessary transformation from the sphere to the plane the scale on a map must vary from place to place and will commonly also vary even in different directions at a point.

Statements of Scale

The scale is commonly thought of as being an expression of a distance on the map to distance on the earth ratio with the distance on the map always expressed as unity. The map scale may be expressed in the following ways.

Representative Fraction (RF). This is a simple fraction or ratio. It may be shown either as 1:1,000,000 or 1/1,000,000. The former is preferred. This means that (along particular lines) 1 inch or 1 foot or 1 centimeter on the map represents 1,000,000 inches, feet, or centimeters on the earth's surface. It is usually referred to as the "RF" for short. The unit of distance on both sides of the ratio must be in the same units.*

Verbal Statement. This is a statement of map distance in relation to earth distance. For example, the RF 1:1,000,000 works out to be approximately 1 in. to 16 mi. Many map series are commonly referred to by this type of scale, for example, 1-in. or 6-in. maps of the British Ordnance Survey (1 in. to 1 mi, 6 in. to 1 mi).

*It is probably preferable (although not common) to define the RF merely as the ratio of reduction from the earth to the map, and to think of it only as a statement of relative magnitude, i.e., simply a numerical relationship since it exists at a point that has no dimension.

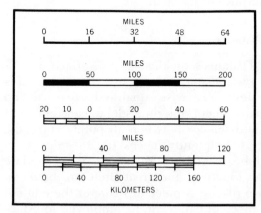

Figure 2.8 Examples of graphic or bar scales. Often, the left end of the bar is subdivided in smaller units in order to provide easier estimation of precise distances.

Graphic or Bar Scale. This is a line placed on the map that has been subdivided to show the lengths of units of earth distance. One end of the bar scale is usually subdivided further in order that the map reader may measure distances more precisely (Fig. 2.8).

Area Scale. This refers to the ratio of areas on the map to those on the earth. When the transformation from the sphere to the plane has been made so that all area proportions on the earth are correctly represented, the stated scale is one in which 1 unit of area (square inches, square centimeters) is proportional to a particular number of the same square units on the earth. This may be expressed, for example, either as $1:1,000,000^2$ or as 1 to the square of 1,000,000. Usually, however, the fact that the number is squared is assumed and not shown.

Scale Factor

It is not possible to transform the spherical surface to a plane without "stretching" or "shrinking" differentially the spherical surface in the process. This means that the stated scale, the RF, will be a correct statement of the scale only at selected points or along particular lines; elsewhere the actual scale will be either larger or smaller than the given RF. This is true to some degree in *all* flat maps. The statement of the relation between the given RF and the actual scale value is called the *scale factor.*

Perhaps the simplest way to appreciate the concept of the scale factor is to imagine the necessary reduction and transformation of the spherical surface as being accomplished in two stages: (1) the reduction of the earth to a globe of a selected scale, and (2) the transformation of that spherical globe to the plane of the map. The stated RF of the map will then be the RF of the globe and is called the *principal* (or nominal) RF. The actual RF will be the real scale on the map and will, of course, vary from place to place.

The scale factor (SF) may be computed by the following formula:

$$SF = \frac{\text{denominator of principal scale}}{\text{denominator of actual scale}}$$

This expresses the SF as a ratio related to the principal scale as unity. A scale factor of 2.000 would mean that the actual scale was twice the principal scale, which would be the case if, for example, the actual scale were 1:15,000,000 and the principal scale were 1:30,000,000. (The student must remember that the larger the denominator of the RF the smaller the scale.) Similarly, a SF of 0.5000 would show that the actual scale was half that of the principal scale, as would be the case if, for example, the actual scale were 1:60,000,000 and the principal scale were 1:30,000,000.

Scale factors of the magnitudes used for the foregoing illustrations are common only on small-scale world maps. On large-scale maps they will vary only slightly from unity. For example, on the transverse Mercator projection used for the UTM grid system, the scale factors within the 6° longitude zones vary only from 0.99960 to 1.00158. On the Wisconsin Coordinate System, the South Zone part of which is illustrated in Fig. 2.5, the scale factors vary only from 0.9999326 to 1.0002282.

Determining the Scale of a Map

Sometimes maps are made that do not include a statement of scale. This is poor practice, to say the least, but nevertheless it occurs. More often it is necessary to determine the scale for a particular part of the map, since, as was observed previously, the scale can never be the same all over a flat map. The map scale along a particular line may be approximated by measuring the map distance between two points that are a known earth distance apart and then computing the scale. Certain known distances of the graticule are easy to use, such as the distance between parallels (average of 69 miles) or the distance between meridians (see Appendix D). Care should be exercised that the measurement is taken in the direction the scale is to be used, for frequently the distance scale of the map will not be the same in all directions from a point.

If the area scale is desired, a known area on the earth (see Appendix D) may be measured on the map with a planimeter and the proportion thus determined. It should be remembered that area scales are conventionally expressed as the square root of the number of units on the right of the ratio. Thus, if the measurement shows that 1 square unit on the map represents 25,000,000,000,000 of the same units on the earth, it would not be recorded that way but as $1:5,000,000^2$ or merely by the square root, 1:5,000,000, which approximates the linear scale.

Transforming the Map Scale

Frequently the cartographer is called upon to change the size of a map, that is, to reduce or enlarge it. The mechanical means of accomplishing this are dealt with in a later chapter, but the problem of determining how to change it in terms of scale is similar to the problem of transforming one type of scale to another. If the cartographer can develop a facility with this sort of scale transformation, he will experience no difficulty in enlarging or reducing maps.

There is, of course, no difficulty in transforming decimal scales commonly used by countries employing the metric system. The English system is more bothersome. The essential information is that 1 mi (statute) = 63,360 in. With this information we can change each of the linear scales (RF, graphic, inch to mile) previously described to the others. Examples follow.

If the RF of the map is shown as 1:75,000:

Example 1. The inch/mile scale will be:

(1) 1 in. (map) represents 75,000 in. (earth),
and
(2) $1/75,000 = x/63,360$,
and
(3) $x = .844$.
Therefore,
(4) .84 in. represents 1.0 mi.

Example 2. To construct the graphic scale, a proportion is established as:

(1) .84 in./1.0 mi. = x in./10 mi
(2) $x = 8.44$ in.
(3) 8.44 in. represents 10 mi, which may be easily plotted and subdivided.

If the graphic scale shows by measurement that 1 in. represents 35 mi:

Example 3. The RF may be determined:

(1) 1 in. represents $35 \times 63,360$ in., or
(2) 1 in. to 2,217,600 in., and therefore
(3) RF : 1:2,217,600

The number of miles to the inch may be read directly from the graphic scale.

If the inch to the mile scale is stated as 1 in. to 26 mi the graphic scale may be constructed as in Example 2. The RF can be determined as in Example 3.

Changing the scale of a map that has an area scale is accomplished by converting the known area scale and the desired area scale to a linear proportion.

Common Map Scales

Maps are made with an infinite variety of scales. An experienced map reader learns to associate a given level of generalization and accuracy with particular scales, and the RF then becomes a kind of index of precision and content. It is helpful to translate the RF mentally into common units of measure. For example, on a map at a scale of 1:1,000,000, 1/8 in. represents about 2 mi and 1 mm represents one km. Table 2.4 below contains a listing of some of the more common map scales.

Table 2.4   Common Map Scales and Their Equivalents

Map scale	One inch represents	One centimeter represents	One mile is represented by	One kilometer is represented by
1:2,000	56 yd	20 meters	31.68 in.	50 cm
1:5,000	139 yd	50 meters	12.67 in.	20 cm
1:10,000	0.158 mi	0.1 km	6.34 in.	10 cm
1:20,000	0.316 mi	0.2 km	3.17 in.	5 cm
1:24,000	0.379 mi	0.24 km	2.64 in.	4.17 cm
1:25,000	0.395 mi	0.25 km	2.53 in.	4.0 cm
1:31,680	0.500 mi	0.317 km	2.00 in.	3.16 cm
1:50,000	0.789 mi	0.5 km	1.27 in.	2.0 cm
1:62,500	0.986 mi	0.625 km	1.014 in.	1.6 cm
1:63,360	1.00 mi	0.634 km	1.00 in.	1.58 cm
1:75,000	1.18 mi	0.75 km	0.845 in.	1.33 cm
1:80,000	1.26 mi	0.80 km	0.792 in.	1.25 cm
1:100,000	1.58 mi	1.0 km	0.634 in.	1.0 cm
1:125,000	1.97 mi	1.25 km	0.507 in.	8.0 cm
1:250,000	3.95 mi	2.5 km	0.253 in.	4.0 mm
1:500,000	7.89 mi	5.0 km	0.127 in.	2.0 mm
1:1,000,000	15.78 mi	10.0 km	0.063 in.	1.0 mm

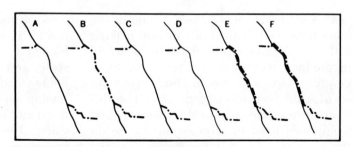

COMPILATION AND GENERALIZATION

As suggested in Chapter 1, one of the more important distinctions in map making is the one between the processes employed to produce large-scale reference or topographic maps and the methods used for small-scale thematic cartography. The difference lies primarily in the methods and techniques used to acquire and locate on the maps the data to be portrayed. On the other hand, the fundamental cartographic process, namely, the designing of any kind of map, ranging from the choice of the symbology, the graphic organization, the selection and positioning of the lettering, even to the adoption of the projection system and the generalization plan, is totally concerned with but one set of theoretical and practical principles, regardless of scale. The employment of color, the representation of the third dimension, and the use of patterns, lines, and geometrical symbols are based on such matter as perception, reproduction techniques, costs, objectives, and so on, and are, therefore, essentially independent of scale. Only in the sense that convention may decree, for example, that the cartographer should use a particular symbol or color on topographic maps or that physical dimension may place a limitation on symbols does scale enter into this part of the cartographic process. Clearly, however, there is a vast difference between topographic mapping (large scale) and thematic mapping (small scale) in obtaining, locating, and processing the map data.

Large-scale maps, generally considered to include scales larger than about 1:125,000, are ordinarily made by way of photogrammetric operations on the one hand to field survey on the other. Their planimetric accuracy, that is, the correctness of geographical position in the horizontal, is controlled as carefully as possible, and within the limits of definition, scale, and human error, they are correct. Some medium-scale maps are made by tracing selected data from the larger-scale maps and reducing the result photographically. These are essentially mechanical processes, and relatively little interpretation and generalization take place.

The small-scale thematic map is quite a different operation. The primary subject matter of the map is presented against a background of locational information, which is called the base data. This base material is ordinarily compiled first, and the accuracy with which it is done determines in large part the accuracy of the final map. This is because of the practical requirement that the thematic cartographer must compile much of the subject-matter data by using the base material as a skeleton on which to hang it. These data, usually consisting of coasts, rivers, lakes, and political boundaries, are generally available from larger-scale, generally accurate, survey or reference maps.

The methods employed in precise field survey and large-scale photogrammetry are essentially noncartographic except in the sense that the end product is a map. The disciplines involved are surveying and civil engineering, and, except for an occasional reference where the principles overlap or impinge on the cartographic process, they will not be treated in this book. On the other hand, compilation, whether for thematic or smaller-scale reference maps, and whether from air photographs or other sources, is generally considered to be part of the cartographic process.

Similarly, cartographic generalization increases in importance as scale decreases so that it is intimately involved with compilation.

COMPILATION

The Compilation Process

The compiling of data requires using many maps (as well as other sources) from which to gain the desired information. The maps may be on different projections; they may differ markedly in level of accuracy; the dates of publication may vary; and their scales will probably be different. The cartographer must pick and choose, discard this, and modify that, and all the while he must place the selected data on the new map, locating each item precisely.

The first rule of compilation is to work from larger to smaller scales. The reason for this is that all but the largest scale maps show data that have been simplified from reality. Although these data (for example, linear, such as coasts, rivers, roads, or boundaries) may be "accurate" for the scale at which they are presented, the generalization made necessary because of scale and purpose would usually not be "accurate" for a larger scale. If we compile from smaller to larger scale, then, we will be building inappropriate error into the compilation.

The techniques of compilation, that is, the means employed to position the data on the new map, range from those that are largely mechanical to those that consist of the transfer of data by eye. It is possible by photography or with the help of a projector to transform a scale difference of as much as four or five times, but much larger changes are difficult. Part of the reason for this is found in the relation between linear and areal scale: the size (area) of maps of the same region will vary as the square of the ratio of their linear scales. For example,

the region shown in 1 sq. in. at a scale of 1:62,500 will occupy only 0.01 sq. in. at a scale of 1:625,500. Therefore, although positions of individual points or simple lines could be easily determined, anything that involves complications in areal spread would be reduced to an almost indecipherable complexity.

The process of compilation often requires that much of the data be transferred by eye. The graticule shown on the projection in each case constitutes the guide lines and all positions must be estimated. The eye is remarkably discriminating and with practice can position data with all the precision that is ordinarily necessary. Very often the use of mechanical means for reduction is precluded by a change in projection. Lest the reader be concerned about the accuracy of such a process he should remember that 90% of all small-scale maps have been compiled in this manner.

When the projections differ between those of the sources being used and those of the map being compiled, it is necessary for the cartographer to become adept at imagining the shearing and twisting of the graticule from one projection to another and to modify his positioning accordingly. He must continuously generalize and simplify. The difficulties occasioned by projection differences between the sources and the compilation can be largely eliminated by making the graticules comparable. That is to say, the same interval on each will greatly facilitate the work (Fig. 3.1).

Compilation is most easily undertaken by first outlining on the new projection the areas covered by the source maps. This outlining is similar to the index map of a

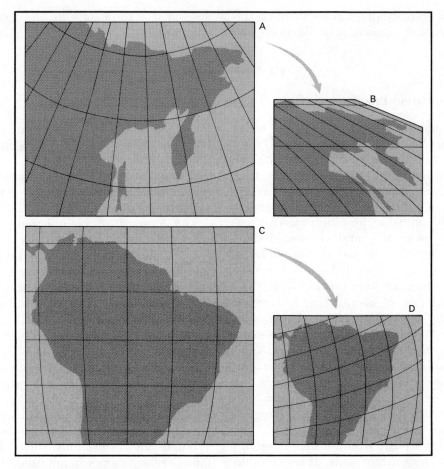

Figure 3.1 Shearing and shape changes in the compilation procedure. Maps *B* and *D* are derived from *A* and *C*.

map series. The sheet outlines may be drawn, and the special spacing of the graticule on each source (5°, 2°, etc.) may be lightly indicated.

The Significance of Base Data to Thematic Maps

The importance of including on the finished map an adequate amount of base data cannot be overemphasized. Nothing is so disconcerting to a map reader as to see a large amount of detail presented on a map and then be confronted with the realization that there is no "frame" of basic geographic information to which he can relate the distributions. The most important objective of thematic mapping is to communicate geographical relationships. Since few people are able to produce an acceptable "mental map" and project it, so to speak, on the thematic map, it is incumbent upon the cartographer to provide it.

The amount and detail with which the base data are shown will, of course, vary from map to map. The usual thematic map must have on it the coastlines, the major rivers and lakes, and at least the basic civil divisions. The graticule, in most cases, should also be indicated in some fashion. The objectives of the map will dictate the degree of detail required, but it is a rare map that can be made without these kinds of information to aid the reader in obtaining the relationships presented.

Compiling Base Data. It is important to appreciate that there are different kinds of surveys according to the definitions and assumptions made by the survey organization. Cadastral survey is commonly done for a limited area, and because the curvature of the earth's surface is relatively insignificant over a small area, it may not be taken into account. The lines of such a plane survey are determined from ground observations and are normally mapped as observed, rather than first being referred to a geodetic or three dimensional base for mapping purposes. Topographic maps, on the other hand, are normally made by referring the ground observations to a geodetic base and, consequently, the two kinds of surveys usually do not match.

Most large-scale plane survey and cadastral maps do not show much physical data. If compilation requires the union of the two kinds of data, the cartographer may be hard put to resolve the differences. For example, if we wish to make a map showing up-to-date information concerning (1) the streams, lakes, and swamps, and (2) the roads of a region, we may find the first category on topographic maps (of different dates!), but not the second; the roads will probably be available from county or state road maps, but these may not show the drainage details. The two sources will be essentially "accurate" according to the definitions used for their mapping, but they will not match one another. In general, the practical significance of these kinds of problems varies according to the scale of

the map being compiled; the smaller the scale the less the difficulties, since positional discrepancies diminish and the desirability of generalization increases with reduction in scale.

Determining the Scale

All maps are, of course, constructed to a scale. In actual practice many thematic maps are prepared in a size to fit a prescribed format, the format being the size and shape of the sheet on which the map will appear. The format may be a whole page in a book or atlas, a part of a page, a separate map requiring a fold, a wall map, or a map of almost any conceivable shape and size. Whatever the format may be, the map must fit within it.

The shapes of earth areas may vary considerably when mapped, depending on the projections upon which they are plotted. Hence, an important concern of the cartographer regarding scale may be the projection on which the map will be made (see Chapter 10). When the projection choice has been narrowed to those that are suitable, then the variations in the shapes of the mapped area on the different projections can be matched against the format to see which will provide the best fit and maximum scale.

The easiest way to do this is to establish the vertical and horizontal relationship of the format shape on a proportion basis and then compare the proportion against representations on the various projections. The proportion can be set on a slide rule or plotted on graph paper, so that for any one dimension the other may be readily determined (Fig. 3.2). We can usually find examples of projections or can easily calculate critical dimensions from the tables or directions for their construction. In this instance the actual dimensions as to length and breadth are not the critical matter; it is the

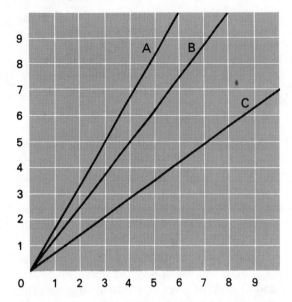

Figure 3.2 Dimension proportion graph. The ratio is constant along each line. A defines a rectangle with sides in a 5:3 ratio; B, a 5:4 ratio; and C, a 7:10 ratio. Whatever pair of values is obtained by reading the ordinate abscissa of any point on one line, the defined shape will always be the same.

proportion between them in relation to the format that is important. When the projection has been selected that best fits the purpose and format of the proposed map, the scale of the finished map may be determined. Normally it is good practice to use either a round number RF or a simple stated scale (map distance to earth distance).

The cartographer must also decide at this stage whether he is going to compile on an already drawn projection or base map, or construct his own projection and compile the entire map. Most projections may be used more than once by simply tracing the graticule and renumbering the longitude. We must be careful, however, not to use a projection that has been copyrighted or patented unless permission is obtained. As a general rule, it is far better to construct

the projection anew to fit precisely the objective of the map. Most projections are not difficult to construct except in special phases, and it is poor practice to produce an inferior map "projection-wise" in order to save a few hours' time. Frequently mapped areas such as continents or countries have, however, appeared on most of the appropriate standard projections, and if they are available there is no reason why a good projection already at hand should not be used. We should always, however, carefully check such a projection to be certain that it has been properly constructed.

Compilation Scale in Relation to Finished Scale. If the map is to be scribed it is normal practice to compile at the finished scale. If carefully done the compilation itself will be the basis for the image that can be made to appear in the scribe coat.* Similarly, if the map is to be reproduced from an original drawing on plastic or paper by a process that does not allow changing scale, then compilation will have to be done at the finished scale. On the other hand, if reproduction is to be by a process that allows reduction, then it is desirable to compile and draft at a larger scale and then to reduce the artwork to the finished scale.

Sometimes the complexity of the map or other characteristics makes it desirable to compile at a larger scale even for a map to be scribed. In this case reduction would then be used to place the image on the scribe coat.

In general, reduction of an ink drawing has certain advantages: at the larger scale drafting is easier, reduction "sharpens up" the lines, etc. The amount of reduction will depend upon the process of reproduction

---

*Scribing is a drafting process by which an image may be fixed in the coating (scribe coat) on a sheet of plastic. The draftsman, being guided by the image, then removes the coating with special tools where lines, etc., are to appear (see Chapter 14).

and upon the complexity of the map. It will also depend upon whether definite specifications have been determined for the reproduction, as may be the case with maps of a series. In general, ink-drafted maps are made for from one-quarter to one-half reduction in the linear dimensions. It is unwise to make a greater reduction because the design problem then becomes difficult.

Changing Scale

The scale of a base map or a source map may be changed in a number of ways, namely, by optical projection, by photography, by pantography, or by similar squares.

A variety of projection devices are available and most cartographic establishments have one. They may operate by projection from overhead onto an opaque drawing surface or from underneath onto a translucent tracing surface. A source map is placed in the projector and, by adjusting the enlargement or reduction, the projection of the image to the drawing surface may be changed in scale to that of the map being compiled. When properly aligned, the desired information may then be traced. Needless to say, if the projections of the source map and the map being compiled are much different, such a projector cannot be used.

Changing scale by photography is also a very easy process, but it is expensive. It may be accomplished either by photographic enlargement or reduction through the use of the conventional film process or by several kinds of photocopy processes, for example, Photostat. It is necessary to specify the reduction or enlargement for some types of photocopy work by a percentage ratio, set on the machine, such as a 50 or 75% reduction. Care should be exercised in specifying the percentage change, since 50% reduction or enlargement means one-half in the linear dimension and, in some processes, is the most that can occur in one "shot." In such instances a change of more than 50% would require repeating the process; to reduce something by 75% in the linear dimension would then require two exposures of 50% each, not one at 50% and another at 25%. Available photocopy paper is often limited to 18 × 24 inches, and anything larger may need to be done in sections. It is difficult to match the sections since the paper ordinarily changes shape unevenly in the developing and drying.

Photographic enlargement or reduction using film negatives may be made somewhat more precisely, since the image may be either projected or viewed in the camera and thus dimensions can be scaled more exactly. All the cartographer need do is to specify a line on the piece to be photographed and then request that it be reduced to a specific length. The ratio can be worked out exactly, and the photographer needs only a scale to check his setting. Any clearly defined line or border will serve as the guide. If none is available he may place one (with light blue pencil) on his drawing.

The pantograph is an ancient device for enlarging and reducing, the common form of which is illustrated in Fig. 3.3. They are

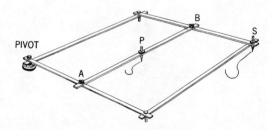

Figure 3.3 A pantograph. Stylus (S) and pencil (P) are interchangeable for enlargement or reduction. Adjustment to the desired ratio is made by moving arm *AB* parallel to itself while the pencil or stylus on *AB* is adjusted so that it, the pivot point, and *S* are kept in a straight line.

easy to operate for reduction, but enlargement is relatively difficult and accuracy is hard to obtain.

In some instances we are forced to change scale by a method called "similar squares." This involves drawing a grid of squares on the original and drawing the "same" squares, only larger or smaller, on the compilation. The lines and positions may then be transferred by eye from one grid system to the other (see Fig. 3.4). With care it is quite an accurate process, for it is the same as compilation.

Occasionally, a cartographer is called upon to produce an over-sized chart or map that involves great enlargement. In most cases, extreme accuracy is not required. If the outlines cannot be sketched satisfactorily because of their intricacies, it is possible to accomplish an adequate solution by projection onto a wall. A slide or film positive (in some cases, a negative will do) may be projected to paper affixed to the wall and the image traced thereon. If an opaque classroom projector with a large projecting surface (that is sufficiently cool in operation) is available, it may be used directly so that the necessity of making a slide is eliminated.

Compilation Procedure

The compilation process has as its objective the preparation of a composite that contains all the base data, the lettering, the geographical distribution(s) being mapped, and so on. This becomes the guide for the construction of the map, either by scribing or by ink drawing. The mechanical processing of the compilation is not of concern in this chapter (see Chapter 14), but it is important that the cartographer proceed at this stage in a manner that will make the ultimate map construction as easy as possible.

Perhaps nothing helps the compiling procedure so much as a transparent or translucent material with which to work. A tracing medium of some sort (for example,

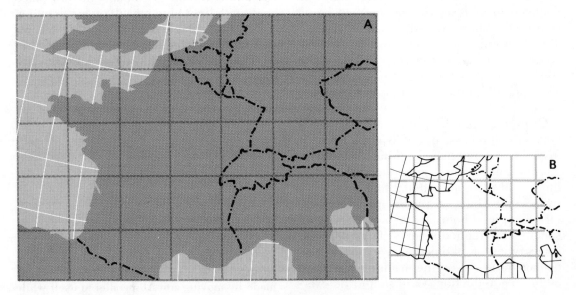

Figure 3.4   Changing scale by similar squares. Map *B* has been compiled from Map *A*.

plastics of various kinds and tracing papers) enables the compiler to accomplish a number of things in addition to the convenience of being able to trace some data. He may lay out lettering for titles, etc., and move the layout around under his compilation worksheet. If he wishes to draw a series of parallel lines, letter at an angle, or place dots regularly, he need only place some cross-section paper under the tracing paper. The use of tracing paper, however, occasions the problem of maintaining size, since paper contracts and expands with changes in humidity and temperature. Plastic materials are more stable. If registry, that is, the matching of several compilation or final drawings, is a problem, a dimensionally stable material should be used.

The Worksheet. The composite that results from the compilation process is called the worksheet. For a simple map the worksheet contains everything, but for complex maps there may be need for more than one worksheet, each being carefully registered with its companions. When completed, a worksheet is rather comparable to a corrected, "rough draft" manuscript: all that is then necessary is for the final flaps (the artwork) to be prepared by scribing or tracing.

The various marks on the worksheet may be done by pencil, ball-point pens, or with any satisfactory medium. It will be of great help in compilation if each kind of line that is to be shown differently on the final map is put in a different color. Lettering, if its positioning is no problem, may be roughly done; if the positioning is important, the lettering should be laid out with approximate size and spacing. If the first try does not work, then it may be erased and done over. Borders and obvious line work need only be suggested by ticks.

When the worksheet and the compilation have been completed, the image may be transferred to a scribe sheet for processing or ink drafting may be done on translucent material directly over the worksheet. If the map is simple and the artwork is to be prepared by the cartographer, he will probably have in mind such things as the character of the lines; but if it is to be drafted by someone else he must prepare a sample sheet of specifications to guide the draftsman. Even for the cartographer-produced map this is wise. This is simple to do if each category is in a different color or otherwise clearly distinguished. Separation drawings for small maps may easily be made from a single worksheet and will register.

It is a common experience for a cartographer to find that many of the elements included in one map might easily be used for another. Base materials such as boundaries, hydrography, and even lettering may not vary much, if at all, from one to another in a series of maps. Consequently, the cartographer can save himself considerable future effort if he tries to anticipate possible subsequent use of his compilation efforts, and prepares his worksheet and plans with these possibilities in mind. Modern reproduction methods make it relatively easy to combine different separation drawings even when printing in one color (see Chapter 12).

Coastlines

The compiling of coasts for very small-scale maps is not much of a problem, for they usually require so much simplification that detail is of little consequence. This is not the case when compiling at medium scales where considerable accuracy of detail is necessary.

Perhaps the major problem facing the cartographer is the matter of source material. It is well to bear in mind that some coasts will be shown quite differently on different maps, yet both may be correct. Hydrographic charts are made with a datum, or plane of reference of mean low water, whereas topographic maps are usually made with a datum of mean sea level. The two are not the same elevation, and it is to be expected that there will therefore be a difference in the resulting outline of the land. In parts of the world that experience high tidal ranges or where special planes of reference are used, the differences will be greater. Another difficult aspect of dealing with source materials is that the coloring of the various charts and maps of the same area may be inconsistent. Marsh land, definitely not navigable, is likely to be colored as land on a chart, and a compiler would assume it to be land by its appearance. On the other hand, low-lying swamp on a topographic map is likely to be colored blue as water, and only a small area may be shown as land. On many low-lying coasts the cartographer may be faced with a decision as to what is land; the charts and maps do not tell him.

Through the years some coasts change outline sufficiently so that it makes a difference even on medium-scale maps. Figure 3.5 shows the north coast of the Persian Gulf in the past and at present, and Figure 3.6, a portion of the Atlantic coast. If we were making maps of an historical period we would endeavor to recreate the conditions at the period of the map. This problem is particularly evident on coastal areas of rapid silting, which in many parts of the world seem to be important areas of occupancy.

Another problem of considerable concern to the cartographer is the representation of coasts on maps wherein the scale varies considerably over the map. As is pointed out later in this chapter in connection with generalization, it is not uncommon on some conventional projections that there is much more detail possible in the

higher latitudes than in the lower. Areas such as bays, inlets, and fjords, located in the higher latitudes, may therefore, take on great apparent significance on such maps, and they often look more detailed and complex than they should if the coast is not a focus of interest. For some purposes it may be necessary to vary the simplification and generalization according to the scale variations of the map.

In a number of areas of the world, notably in the polar regions, the coastlines, like many other elements of the map base, are not well known, and they vary surprisingly from one source to another. On some maps a particular region may appear as an island; on others, as a series of islands; and on still others, as a peninsula. On simple line maps a broken or dashed line suffices for un-

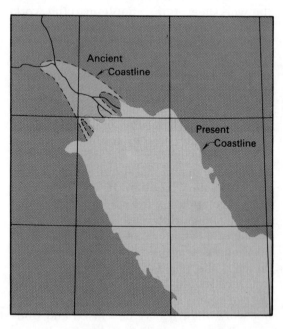

Figure 3.5 Major changes in coastlines occur over long periods of time, and these changes may be significant even on small-scale maps. A portion of the Persian Gulf.

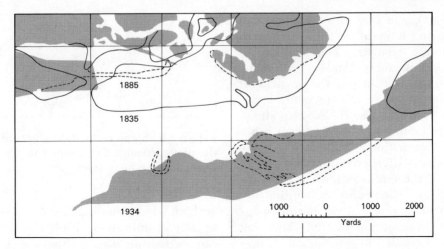

Figure 3.6   Frequent changes have occurred in some areas. The various lines show the positions of the shoreline of Rockaway Inlet, Long Island, at several periods. (Modified from Deetz, *Cartography*, Special Publication 205, U.S. Coast and Geodetic Survey.)

known positions of coastlines, but it becomes a larger problem when the water is to be shaded or colored, for no matter what type of line is used to delineate the coast, the tonal or color change outlines it clearly.

The Design of Coastlines. Generally, coastlines should be uniform, relatively thin lines. Occasionally it is desirable to have a thinner or lighter line in a complex area and a slightly heavier line in an adjacent simple area in order that the two coastlines may appear more nearly uniform as shown in A in Fig. 3.7. Likewise, embayments, estuaries, etc., may sometimes need to be altered (generalized) slightly to allow for the reproduction process or to be delineated with a line lighter than the main trend of the coast, as shown in B in Fig. 3.7. This is especially desirable for the thick coasts of wall maps. Coasts produced by thick uniform lines, as in C, cannot show points and sharp changes in direction.

Administrative Boundaries

Compiling political boundaries is sometimes a remarkably complex problem, since the boundaries must be chosen for the purpose and date of the map. The problem becomes more difficult as the area covered by the map increases. Almost all man-made

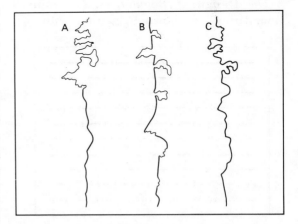

Figure 3.7   Various kinds of coastlines. Letters refer to the text.

boundaries change from time to time, and it is surprising how difficult it is to search out the minor changes. The problem is compounded in some kinds of thematic maps. For example, a population map of the distribution of people in central Europe prior to World War II, but also showing present boundaries, raises these problems: (a) international boundaries today and (b) census division boundaries as of the dates of the enumerations in the various countries. The major difficulties are twofold. The first is that of finding maps that show enumeration districts and that also show latitude, longitude, and other base data so that the boundaries may be transferred to the worksheet in proper planimetric position. The second is that of placing later international boundaries in correct relation to the earlier enumeration boundaries.

It is not uncommon for official civil division boundary maps to be without any other base data, even projection lines. Such a condition should impress on the cartographer who uses such deficient maps the need for him to provide base data for the users of his maps.

The Design of Boundaries. Generally, boundaries are shown by a line with or

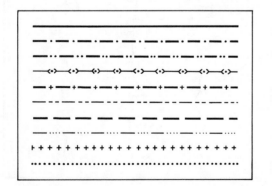

Figure 3.8   Some examples of kinds of boundary symbols.

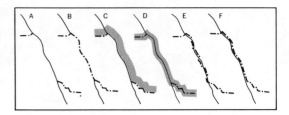

Figure 3.9   Some methods of showing boundaries along streams. The shading shown in *C* and *D* would often be in color.

without added shading or color. The line is usually of uniform width but is commonly broken in one way or another (Fig. 3.8). There is no generally accepted design for boundary symbols, but many agencies and governments have standardized their form for their own official maps. Although there are exceptions, generally speaking, decreasing significance of boundaries on a map is accompanied by a decrease in width of line and commonly by a progressive change in complexity of line.

One difficulty the cartographer regularly encounters is the problem of symbolizing the boundary along a water course, also shown as base data. Some possible solutions are shown in Fig. 3.9. *A, B, C,* and *D* show only that the boundary follows the river; it may be on either bank or in midstream; *E* shows that it is in midstream; *F* shows that it is on the east bank.

Hydrography

The compiling of rivers, lakes, and other hydrographic features as part of base data is very important. These elements of the physical landscape are in some instances the only relatively permanent interior geographical features on many maps, and they provide helpful "anchor points" both for the compilation of other data and for the map reader's understanding of the "place correlations" being communicated.

The selection of the rivers and lakes depends, of course, upon their significance to the objective of the map. On some maps the inclusion of well-known state or lower order administrative boundaries makes it unnecessary to include any but the larger rivers. Maps of less well-known areas require more hydrography, since the drainage lines, which indicate the major landform structure of an area, are sometimes a better known phenomenon than the internal boundaries. Care must be exercised to choose the "main stream" of rivers and major tributaries. Often this depends not upon the width, depth, or volume of the stream but upon some economic, historical, or other element of significance. Often it is necessary to eliminate relatively important rivers or lakes because they will interfere with the planned use of the maps. For example, Lake Winnebago in Wisconsin becomes a visual focus on many maps to the detriment of the communication of the geographical data that prompted the making of the map.

Just as coastlines have characteristic shapes so do rivers, and these shapes help considerably on the larger-scale maps to identify the feature. The braided streams of dry lands, intermittent streams, or the meandering streams on flood plains are examples. On small-scale thematic maps it frequently is not possible to include enough detail to thus differentiate between stream types, but the larger sweeps, angles, and curves of the stream's course should be faithfully delineated. Likewise, the manner in which a stream enters the sea is important. Some enter at a particular angle, some enter into bays, and some break into a characteristic set of distributaries. Swamps, marshes, and mud flats are also commonly important locational elements on the map base. Some examples of symbols for hydro-

graphic features are shown in Fig. 3.10.

The Design of Rivers. It is usual when representing streams to show them as growing from thin to thicker lines near their mouths. It should be pointed out, however, that streams do not always become larger or wider downstream. In dry land areas of the world the reverse is ordinarily true, with the streams carrying more water near their sources. This is also occasionally the case even in humid regions; for example, the Congo River is wider in its middle course than near its mouth.

On all but the largest-scale maps it is impossible to represent the width of a river truly and, consequently, the width of line chosen is an important consideration. When the existence of a river is known but not its precise location, its unknown portion may be represented by a dashed line of the same width as the known course.

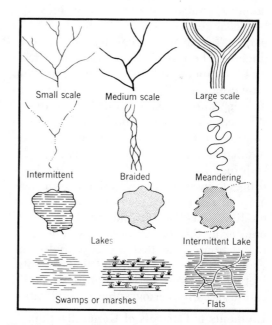

Figure 3.10   Some examples of kinds of symbols used for hydrography.

GENERALIZATION

The earth is so large, relative to diminutive man and his ordinary vision, that for geographical relationships to be comprehensible they must be reduced. As observed earlier, the objective of this essential scientific process is closely analagous to the employment of magnification to aid in the study of those phenomena that are too small to be directly observed. Both reduction and enlargement, in the senses these terms are used here, are means to the same end—comprehensibility; but the consequences of these processes are quite different.

Unmodified magnification, as by microscope, is accompanied by a general loss of clarity and intensity of coloring, by an emphasis of the relative visual significance of minor features, that is, by an increase in the relative visual importance of the specific compared to the general, and so on. Reduction, when applied to earth phenomena, is also accompanied by inescapable changes: widths and lengths of the earth's linear and areal features are reduced in the ratio of the reduction; intricacies are increased in similar proportion; crowding of adjacent discrete items increases; and clarity in general is reduced. Whereas there is not much one can do to counteract the undesirable consequences of enlargement (beyond such techniques as staining, for example), we can do many things to increase the effectiveness of cartographic reduction. There are a variety of modifications that can, and must, be carried out during the reduction process, and these range from some that are essentially mechanical processes to others that are intellectual exercises. The sum of these modifications is called cartographic generalization.

Elements & Controls of Generalization

Cartographic generalization is born of the necessity to communicate. It is quite impossible to portray everything, even theoretically, at a 1:1 scale, to say nothing of the large reductions required for most maps. In order to portray the important aspects of reality, various manipulations of the data to be mapped are necessary; these may be grouped into four major categories which we shall term the *elements* of cartographic generalization. These are as follows.

1. *Simplification*, that is, the selection of characteristics of the data, possibly its geographical modification, and the elimination of detail.

2. *Symbolization*, that is, the graphic summarizing and coding of essential characteristics, comparative significance, and relative position.

3. *Classification*, that is, the ordering or scaling and grouping of geographical phenomena.

4. *Induction*, that is, the application in cartography of the logical process of inference.

These fundamental and complex processes are all consciously accomplished by the cartographer in the practice of map making, but each map provides a different set of requirements. The "mix" of the elements of generalization as they are combined will, therefore, vary from map to map, and the manner in which each will be applied depends upon the dictates of the *controls* of cartographic generalization. These are as follows.

1. *The objective*, that is, the purpose of the map.

2. *The scale*, that is, the ratio of earth/map reduction.

3. *The graphic limits*, that is, the capability of the systems employed for the communication.

4. *The quality of data*, that is, the reliability and precision of the various kinds of data being mapped.

It is only fair to observe that the separation of cartographic generalization into four elements and four controls is in itself a generalization about generalization. The aim in doing so, like the aim of all generalizing, is to simplify to mangeable proportions what is in reality a very complex intellectual-visual process. To examine it in great detail is not possible in a book of this nature, any more than it would be possible to map on a page this size all that could be mapped on a sheet twenty times this size. Generalizations are always needed—even about generalization.

The four elements of generalization are not clearly separable in many cartographic situations. The selection and elimination of factors in simplification grades imperceptibly into characterizing by symbolization. For example, we wish to smooth a coastline by ignoring some details, but at the same time, we wish to convey by the character of the line symbol the fact that the coast is typically estuarine. The term "typically estuarine" implies classification of types of coasts. (Keep in mind that the subsequent discussions of each of the elements of generalization cannot be as simply focused as would normally be possible if they were less complex concepts.)

One other introductory observation is appropriate here: cartographic generalization is essentially a creative act. Some of the elements and some of the controls can be treated systematically, by quantitative methods, for example, and a kind of automatic generalization can be specified or even programmed for a computer. Not all can be treated this way to best advantage, however, and it takes long experience, good judgment, and familiarity with the data to manipulate the elements properly, according to the dictates of the controls.

Simplification

The process of simplification is the most universal of the four basic elements of cartographic generalization. If examined in detail it includes the selection of the charac-

teristics of the particular phenomena being mapped, the elimination of unwanted details, and in some instances even the rearrangement, reshaping, or modification of geographical features so as to attain a "truthful" representation.* It needs to be observed here that the simple process of "selection" can also occur at a first order stage in the cartographic procedure. For example, we may decide that we wish to make a "railroad" map, which will presumably show railroads and associated features of importance to the objective of the map. This primary decision making to "select" this category of transportation is at a different stage than is the decision making involved in the simplification of the railroad data whose inclusion has already been decided upon.

Any map is a reduction and because such elements as line widths, type sizes, and symbol forms must be kept to a visible and legible size, it follows that as reduction occurs in the scale of a map, each map item will occupy a proportionately larger amount of space. Consequently, selective simplification must be practiced in order to insure legibility and truthful portrayal. Since the reduction of area on a map takes place as the square of the ratio of the difference in linear scales, the amount of information that can be shown per unit area decreases in geometrical progression. Töpfer and Pillewizer have developed a law or principle of generalization which states in general terms the amount of detail that can be shown at different scales.‡ Called the *Radical Law* or the *Principle of Selection* by its authors, it is expressed in its simplest form as

$$n_f = n_a \ \sqrt{M_a/M_f}$$

*The observation, "To tell the truth one often has to lie a little," is not as contradictory as it first seems, especially in cartography.

‡F. Töpfer and W. Pillewizer, "The Principles of Selection," (with introduction by D. H. Maling), *The Cartographic Journal*, **3**, 10–16 (1966).

where

n_f is the number (n) of items on the newly compiled map (f)

n_a is the number of items on a source map (a)

M_a is the scale denominator of the source map

M_f is the scale denominator of the newly compiled map

The basic equation of the Radical Law must be modified depending upon the nature of the phenomena being compiled. The modification is made by introducing two constants, C_b and C_z, in the righthand side of the basic equation.

Constant C_b is called the *constant of symbolic exaggeration* and takes three forms.

$C_{b1} = 1.0$ for normal symbolization, that is, for elements appearing without exaggeration.

$C_{b2} = \sqrt{M_f/M_a}$ for features of areal extent shown in outline, without exaggeration, such as lakes and islands.

$C_{b3} = \sqrt{M_a/M_f}$ for symbolization involving great exaggeration of the area required on a compiled map, such as a settlement symbol, with its associated name.

Constant C_z is called the *constant of symbolic form* and also takes three forms.

$C_{z1} = 1.0$ for symbols compiled without essential change.

$C_{z2} = (S_a/S_f) \sqrt{M_a/M_f}$ for linear symbols in which the widths of the lines on the source map (S_a) and the newly compiled map (S_f) are the important items in generalization.

$C_{z3} = (f_a/f_f) \sqrt{M_a/M_f}$ for area symbols in which the areas of the symbols on the source map (f_a) and the newly compiled map (f_f) are the important items in the generalization.

It will be observed that the basic equation, with or without modification by the introduction of the constants, is solely a statement of the number of items that can be expected on the newly compiled map as compared to their existence on the source map. The value of the equations is in their expression of the basic relationship among scale and the nature of the mapped phenomena in terms of the number of items that will normally appear on a newly compiled map. When compiling from larger to smaller scales, the number of items that can be shown on the smaller scales will diminish according to the Radical Law, that is, in geometric progression.

Although the law specifies, with high probability, how many items or how much detail can be retained when working from a larger to a smaller scale it cannot, of course, specify which of the items should be selected and which should be discarded in order to characterize, that is, generalize, a distribution. One cannot, for most thematic maps, make a selection of the features within a class, for example, rivers or cities, on purely objective grounds such as size. Importance is a subjective quality; a selection of cities on a map of the United States which included only those of more than 100,000 inhabitants would eliminate many in the western United States that are far more "important" in their region than many of those that would have been included in the more populous eastern United States. For a consideration of this aspect of generalization we must turn to the second major element.

Symbolization

All the marks on a map are symbols, from the lines to the dots representing cities and from the blue ocean to the hummocks showing marshlands: by their employment the cartographer is generalizing a concept, a series of facts, or the character of a geographical distribution. The very fact of symbolization is generalization, sometimes to a rather small degree, as when we characterize the internal areal administrative hierarchy of townships and counties by a set of boundary symbols, or sometimes to a high degree, as when we symbolize an entire complex urban area with a simple dot.

Such generalization by symbolization ordinarily increases with decreasing scale, that is, the least amount occurs on the largest scale maps and as the scale of the map decreases the degree of the generalization increases.* It is important to observe, furthermore, that the degree of this kind of generalization commonly varies within a map. Some symbols, such as the lines of the graticule indicating latitude and longitude, involve almost no generalization while on the same map we may include a large smooth arrow symbolizing the migration of population from one region to another, the arrow taking no note whatever of the many different routes taken by the peoples who moved.

The symbolizing process does not stop with the simple act of employing marks to represent selected discrete items or smooth lines to represent linear elements; it also involves the capturing of the essential character of a distribution, a line, or a concept. The symbolizing of the general character of the population distribution in the United States by a judicious selection of municipalities is a case in point. The best representative of this element of generalization by symbolization is the line, and the symbolization of coasts and rivers provides a good example.

It must be strongly emphasized that we cannot merely smooth out coasts and ignore islands. The basic or fundamental character of the shapes and outlines must be retained and emphasized in their simplicity, since the eye will not quibble with the representation if the general shapes are as expected. On the other hand, the cartog-

rapher must be careful not to overdo the simplification, for, as is abundantly shown in many newspaper maps, too much simplification can make the representation of a known shape appear ridiculous.

Although the matter of retaining the basic characteristics of shapes and distributions in the compilation process through judicious selection and simplification appears complicated, it is not so difficult in practice as it may seem. Pannekoek gives several excellent illustrations of the principles involved.† Two that are reproduced here exemplify the care this kind of symbolization requires.

Figures 3.11 and 3.12 show a well-known area, the delta region of southwestern Netherlands. In Figure 3.11 the region is shown in A at relatively large scale. Professor Pannekoek points out that, when generalized as in B and D, the representation by only three rivers ignores the insularity of the region and "pushes simplification too far, unless it is meant as a prophecy." He observes that it would be

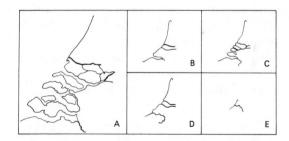

Figure 3.11 Generalizations of the complex of islands and inlets in southwestern Netherlands as they appear on various published atlases. In this series the land area is emphasized. See text for explanation. (From Pannekoek.)

*This opposing use of the terms large and small scale in connection with (a) ratio of earth to map reduction and (b) generalization is very confusing. The student must be wary indeed.

†A. J. Pannekoek, "Generalization of Coastlines and Contours," *International Yearbook of Cartography,* **2**, 55–74 (1962).

better to proceed as in *C*, by adding a few lines to retain the insular character. At a very small scale, as in *E*, it is proper to omit all rivers except the Rhine and symbolize the coast as a single line.

In Figure 3.12 the region shown in *A* is a poor symbolization which, Professor Pannekoek observes, "suffers from a fear of smooth lines." In *B* the estuaries become too much a part of the sea. In *C* and *D* "the whole thing goes wrong," being shown as a large inlet with a few small islands. In *E* "only a yawning gulf remains, which shows rather the situation which existed in about 1200 A.D."

The series shown in Figures 3.11 and 3.12 illustrates also how necessary it is to refer to basic sources when the compiler is not fully familiar with an area. To generalize uncritically from the generalizations of others is likely to lead to ludicrous representations.

In generalizing lines the cartographer must also be concerned with the effect on the viewer of the graphic character of the line. It is particularly important in symbolizing coasts and boundaries when the reference value of the line itself is slight (in terms of the objective of the map), as is the case with a great many thematic maps on which boundaries and coasts appear. As a general rule we may say that any graphic

form that is more complicated than the surrounding forms will draw attention to itself, because it is more interesting to the eye. The fjorded coast of Norway is an example. If the reader will refer to Figure 3.13 he will at once see that *A*, the more complicated representation of the coast, catches his eye. This is partly because the degree of complexity also makes the coastal region a relatively dark tonal area. By careful symbolization the map maker can help the reader refrain from giving undue attention to details that are extraneous to the purpose of the map.

In the total process of symbolization the cartographer must positively assign relative prominence to the features he is including in his map. Ordinarily this aspect of symbolization is dealt with in the design of the graphic presentation (see Chapter 11) where we are concerned with the levels of visual significance and the total visual hierarchy of the graphic elements of the communication. On the other hand, we cannot separate these elements that clearly in practice, and generalization during compilation is made easier if some of these decisions have been made prior to beginning the map. In the compilation stage it shows up primarily in the character of the line work: thin, weak lines for items of lesser importance; strong, thicker, smoother lines for more important features.

Classification

Classification is a standard intellectual process of generalization that seeks to sort phenomena into classes in order to bring relative order and simplicity out of the complexity of imcomprehensible differences or the unmanageable magnitudes of information. It is difficult to imagine any intellectual understanding, beyond the very elementary, that does not involve classification. We classify without thinking about it:

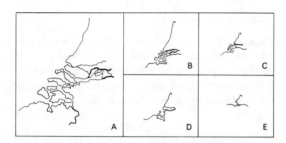

Figure 3.12 Generalizations of the southwestern Netherlands in various published atlases in which the water area has been overemphasized. See text for explanation. (From Pannekoek.)

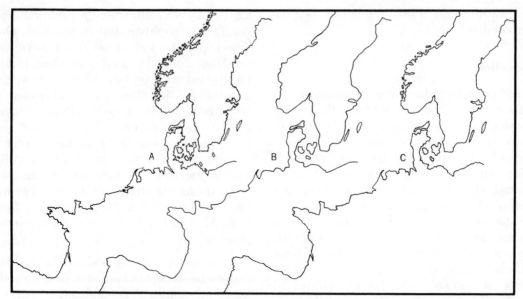

Figure 3.13 Generalization of coasts. A and C are least generalized; B the most. All, however, are greatly simplified from reality.

we sort numerical data into averages, above and below average, extremes, and so on; or we condense all the varieties of phenomena into simple classes such as "roads," "rivers," or "coastlines." Generalization by classification takes two basic forms in cartography, each of which is explicitly dealt with in another section of this book. They will merely be mentioned here.

In Chapter 5 the general concepts of symbolization in cartography are discussed at some length, and there it is observed that all data can be differentiated or grouped, that is, classified, according to several kinds of scaling systems: simply by class, called *nominal*; by class and rank, called *ordinal*; and by class with stated numerical limits, called *interval*.* The representation of geographical phenomena on maps by appropriate marks generally conforms to this system, and consequently the very act of employing symbols is one of the ways in which cartography inherently involves generalization by classification.

A good share of thematic cartography

concerns what are called distribution maps, and a large proportion of these portray quantitative data. The majority of these maps must separate the data into a given number of classes so as to be able to communicate it by various patterns, tones, etc. The selection of class intervals for use on such maps is a second way that cartography involves a fundamental kind of generalization. As Professor Jenks has so clearly observed:

The map-maker can present to the map reader for interpretation only one of the multitude of different versions of a statistical distribution. In the creative process of developing this one map, the cartographer makes a series of simultaneous judgments involving . . . the most desirable degree of generalization . . . and . . . a mathematical process for classing the data. These . . . control and shape a generalized statistical sur-

*In cartography the term *scale* is rather overworked. Do not confuse the use of the term scaling system, as here employed, with map scale or with scale (degree) of generalization.

face which is then symbolized to represent the abstract data.*

Induction

Induction or inductive generalization is a rather specialized element in the cartographic process. It applies particularly to those kinds of map making where the cartographer is performing operations that will allow him to end up with more than he started with. For example, if we have the average January temperatures for a series of stations, we can, by suitable "logical contouring," construct a set of isotherms. The set of isotherms has resulted from making logical inferences about the occurrence of average January temperatures in the areas between the data points. By analogy, any such logical extension of data founded upon accepted associations is inductive. Such would be the mapping of a classification system, such as climates or soils, where data are available only for a relatively few points. Any overall extension of a system in mapping is inductive generalization, since the cartographer ends up with more than the information with which he started.

The Operation of the Controls of Generalization

The elements of generalization are what the cartographer does, and the controls are those factors that influence his actions. It is quite impossible for a textbook to do more than suggest the operations of the controls, because there is no limit to the number of possible combinations of objectives, scales, symbolisms, and qualities of data. Nevertheless a few observations can be made to suggest their functioning.

Objective. The purpose of a map clearly has a great deal to do with its design. The kind of audience to which it is aimed, geographically sophisticated or ignorant, children or adults, and so on, is one basic factor. How it is to be used is another. Is it to be studied with no time limit, so to speak, as might be the case with an atlas map, or is it to be used as an illustration to be shown briefly during a lecture? Is it primarily to communicate a theme or is it to present evidence to support a thesis? Before he begins, the cartographer must answer these questions; otherwise he will have no basis for making the many decisions that are a part of the generalizing process.

Scale. Obviously the scale of the finished map-to-be is a fundamental factor in the kind and degree of generalization that will be employed. As a general rule, of course, the smaller the scale the greater will be the degree of generalization required. When the cartographer has complete freedom to select the scale of the map, that decision may be greatly influenced by the generalization needed to fit the objective. There may well be, as has been suggested, an optimum scale for every objective in thematic cartography† The influences of scale in cartography are so ubiquitous and pervasive, however, that any such simplification of what is a most complex problem must be applied with care.

Probably the most important task of the cartographer when generalizing the various categories of geographical data within the context of scale is to keep the various degrees of generalization "in tune" with one another and with the chosen scale. At each scale there is a range of generalization that will "fit" the scale, that is, the treatment

*George F. Jenks, "Generalization in Statistical Mapping," *Annals of the Association of American Geographers,* **53,** 15–26 (1963).

†O.M. Miller and Robert J. Voskuil, "Thematic Map Generalization," *Geographical Review,* **54,** 13–19 (1964).

will be neither too detailed nor too general: the difficulty comes in the attempt to maintain a balance with a series of categories such as coasts, boundaries, or roads. The combinations of scale, data, and objective are legion, and in practice it is probably wise to leave it as a matter of "good judgment."

The cartographer should remember that when he is working at very small scales there is usually a great scale range within the map. For example, on the conventional Mercator's projection the values of S (see Chapter 9) in the following table show the relative exaggeration of areas at the given

Latitude	S
0°	1.0
20°	1.1
40°	1.7
50°	2.4
60°	4.0

latitudes. This means that at latitude 60° on that projection there is four times as much map area available within which to represent phenomena as there is at the equator. The tendency is to use all the space available, so there is commonly a great difference in the amount of generalization. Whether one should be consistent in the degree of simplification and other factors or not depends upon the objective of the map.

Graphic Limits. The kinds of symbolism to be used on a map function as an important control that operates in many ways. For example, if we are making a map on which color can be used we may include more detail (assuming such is desirable) because color can be used as a simplifying agent, an example being where color is used to symbolize the sea. Much more detail can be introduced into a coastline if the land-water separation is to be enhanced by color on the sea (or on the land).

Another example of the control of graphic limits upon generalization is given by the requirement of using enumeration units to portray distributional data, such as in a map utilizing either states or counties in the United States to portray distributions, such as power plants, which have locations in those areal units but which are not further localized on the map. The use of such a graphic system involves a loss of locational information by generalization. The near-distance formula, as employed by McCarty and Lindberg, when applied to the areas of (a) the states and (b) the counties of the United States, works out as follows: the average loss of "locational accuracy" as determined from theoretical squares the average size of the states and counties would be 118 miles and 16 miles, respectively.[*] As a general rule, of course, the smaller the areal unit used to portray such data, the less the degree of locational generalization. Furthermore, as Das Gupta has pointed out, the smaller the size of the areal unit the greater the probability that its average value will be representative of the values for a variable distribution spread throughout it.[†]

Maps symbolizing quantitative distributions by choropleth methods (area units) must ordinarily employ a system of grouping the values, since it is all but impossible to symbolize each individual value that may occur. As was pointed out in connection with the element of classification, the selection of the class intervals required by such maps is a form of generalization. There are many other ways that the control of graphic limits operates. To the extent that they can be explained or illustrated in

[*]Harold H. McCarty and James B. Lindberg, *A Preface to Economic Geography*, Prentice–Hall, Inc., Englewood Cliffs, New Jersey, 1966, pp. 23–28.

[†]Sivaprasad Das Gupta, "Some Measures of Generalization on Thematic Maps," *Geographical Review of India*, **26**, 73–78 (1964).

a book like this, it will be done when the particular kind of symbolism is treated in later chapters.

Quality of Data. It is readily apparent that the more reliable and precise the data to be mapped are, the more detail is potentially available for presentation. In cartographic generalization the converse is unusually important, that is, the cartographer must sometimes go to considerable pains in order not to let the map give an impression of an accuracy greater than the source material warrants.

Just as basic as the quality of the data is to proper generalization is the scholarly competence and intellectual honesty of the cartographer. A thorough knowledge of the data, which, of course, includes the area being mapped, is indispensable. A Swedish cartographer, Gösta Lundquist, has observed that it always seems easier to generalize faraway places; and he stated a familiar and revealing reaction when he pointed out, "I always find . . . that other peoples' maps are extremely good—except for their treatment of Sweden"! Intellectual honesty is particularly important in cartography because a well-designed map has about it an authoritative appearance of truth and exactness. The cartographer must, therefore, take unusual pains to ensure that the data are correct and that their presentation on the map does not convey a greater impression of completeness and reliability than is warranted. For example, there is no theoretical limit to the number of contours that can be interpolated from a set of spot elevations; yet if the set is small, a large number of contours would give an impression of detail and accuracy quite out of line with the quality of the data.

In commenting on the attributes necessary in a map maker, a distinguished cartographer and geographer, John K. Wright, puts it as follows.*

Fundamental among these qualities is scientific integrity: devotion to the truth and a will to record it as accurately as possible. The strength of this devotion varies with the individual. Not all cartographers are above attempting to make their maps seem more accurate than they actually are by drawing rivers, coasts, form lines, and so on, with an intricacy of detail derived largely from the imagination. This may be done to cover up the use of inadequate source materials or, what is worse, to mask carelessness in the use of adequate sources. Indifference to the truth may also show itself in failure to counteract, where it would be feasible and desirable to do so, the exaggerated impression of accuracy often due to the clean-cut appearance of a map.

One of the most difficult tasks of the cartographer is to convey to the map reader a clear indication of the quality of the data employed in the map. When writing or speaking, words such as "almost," "nearly," and "approximately" can be included to indicate the desired degree of precision of the subject matter. It is not easy to do this with map data.

There are several ways the cartographer may proceed. One is to include in the legend a statement, when necessary, concerning the accuracy of any item. Another and more common method on larger-scale maps, is to include a reliability diagram (Fig. 3.14), which shows the relative accuracy of various parts of the map.

It is also good practice to include in the legend, if warranted, such terms as "position approximate," "generalized roads," or "selected railroads," in order that an idea of the completeness and accuracy may be given to the reader.

Sources and Copyright

It is beyond the purpose of this book to suggest sources of compilation materials, but the cartographer, regardless of what sources he uses, must always give proper

*From "Map Makers Are Human . . .," *The Geographical Review*, **32**, 528 (1942).

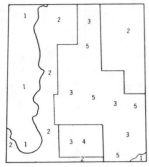

1. USC&GS Charts, (Reliable survey)
2. Luzon, 1:63,360 (Trigonometric
    survey - reliability fair)
3. PIC&GS Charts
    (Compiled map - reliability fair)
4. Luzon Island, 1:200,000
    (Reconnaissance map - reliability fair)
5. Sectional Aeronautical Chart, 1:600,000
    (Compiled map - reliability fair)

Figure 3.14 Coverage diagram from a map of the Lingayen Gulf area, Philippine Islands, giving an annotated list of sources together with an index. (Courtesy of the Army Map Service.)

credit for materials gathered and presented by others. It is necessary to do this because it is the honest way of doing things. But this does not mean that he must identify the source of every single item on a reference map, even though it is obvious that he could not have gathered any of the material himself in the field by original survey. It does mean that he should identify the source of any material that is not generally common knowledge or does not obviously come from good public authority. He may also, however, wish to identify such

sources of well-known information in order to justify the quality of his map.

Equally important for the map maker is the problem of copyright protection and the use of materials that have been copyrighted. This is a particularly difficult problem in cartography, since there has never been a clear and complete definition of the way in which the copyright laws apply to maps. Furthermore, the general copyright law is now under revision. As a rule, no copyrighted map may be copied in any way without permission because the precise manner in which the geographical material is cartographically arrayed and generalized is protected. Most United States government maps and publications may be used, but in the case of specialized materials containing judgments and opinions of named authors, it is not only courteous but wise to request permission, since the material may have been copyrighted by the authors separately or some of it may have come from copyrighted sources.

Generally speaking, governmentally produced survey maps in the United States of the topographic variety, census materials, and the like may be used freely as sources of data, but actual reproduction, tracing, and copying of all privately produced and many publicly produced materials must depend upon written permission from the holder or holders of the copyright.

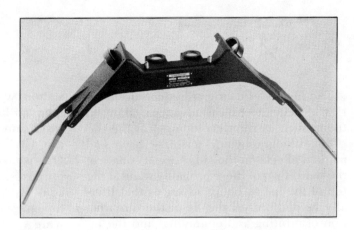

COMPILATION FROM AIR PHOTOGRAPHS

The history of cartography includes a series of major technical advances. Many of them, such as the development of the isarithm, lithography, or scribing, have had marked effects on the subsequent course of the field. Two of these probably stand at the top of the list as being advances that literally revolutionized the field: the introduction of printing and engraving, and the invention of photography.

The mechanical reproduction of maps enabled man to disseminate widely the geographical information carried on maps and, in a sense, increased his ability to communicate a thousandfold, at least, to numbers of people. In quite a different way the process of photography has also increased a thousandfold man's capabilities for making available geographical information, in this case by expanding his observing and recording powers, which has led to a fantastic increase in the number of maps made, to say nothing of its effect upon map reproduction processes. Instead of having to go into the field for all the information needed for mapping, the cartographer can bring the field into the laboratory. Now air photographs are a standard part of the mapping process, and both major topographic mapping and much survey and thematic mapping are unthinkable without them.

The History of Aerial Photography

The application of photographs to mapping became apparent almost immediately upon the production of the first daguerreotype before the middle of the nineteenth century. The first attempts at aerial photography were crude, as would be expected, because the technique of photography itself was in its infancy and a satisfactory air station was difficult to obtain. Kites and then balloons were successfully used to carry cameras aloft, and during the American Civil War, the military used photographs taken from balloons for intelligence purposes. By the end of the nineteenth century, methods for the compilation of topographic maps from aerial photographs were well established. Still lacking, however, was a suitable means for carrying the camera aloft so that photographs could be produced as desired.

The development of heavier-than-air craft enabled man to move through the atmosphere as he wished, and eventually they provided the capability of transporting the necessary equipment to produce excellent aerial photographs. During the early part of the twentieth century a few attempts were made to use photographs taken from airplanes for topographic mapping, and in World War I, pilots, using hand cameras, obtained pictures that were used for intelligence purposes. After the war, great and rapid strides were made in the production of better photographic equipment as well as in improvements in aircraft.

The production and improvement of the quality of photographs tended to expand their application. Better control of scales and improved definition of the photographic images provided data that became much more useful in fields other than mapping, such as geology, forestry, agriculture, and planning. During the 1930's progress was made in the development of techniques and new equipment for better utilization of photography; and at the beginning of World War II, the potential for intelligence was obvious to most military leaders.

Activities during the second World War contributed much in the way of methods and procedures in air photo interpretation. The mass of photography that had to be handled required that a great number of interpreters be trained, and many of these people retained an interest after the war and contributed to research in the field. Likewise, the urgent need for maps during the war resulted in greater use of photogrammetric methods, both because of the speed with which maps could be made and because the photographs provided a good source of information for inaccessible areas.

Increasing civilian (as well as military) needs since 1950 have given even greater impetus to the use of photogrammetric methods in the preparation of maps. Currently, attempts are being made to automate as many of the steps as is possible; for example, equipment is being developed that automatically determines heights by sensing the photographic images. Work is also being carried on in the use of radar and infrared imagery both for photogrammetric and interpretation purposes. There seems no end in sight to ways in which photography can assist in the major tasks of cartography.

The Employment of Aerial Photographs

Aerial photographs have a great variety of applications that overlap more often than not, and it will be useful at this point to categorize briefly some of these and describe them in general terms.

Photogrammetry. In a rather broad sense, photogrammetry is the art and science of making maps from aerial photographs, but more specifically it involves using photographs for making earth measurements. Photogrammetry is not merely tracing information directly from a single photograph, but includes the ways of using the photographs to produce accurate planimetric and hypsometric control, that is, the establishment of precise relative horizontal and vertical locations. Generally speaking, photogrammetry contributes the positional element to the compilation process. Very sophisticated equipment is now commonly used to solve the geometric problems of aerial photographs in order to produce accurate large-scale maps; we can, however, engage in photogrammetric work with quite limited equipment. The radial line plot, considered later in this chapter, can be accomplished with merely overlay material, a straight edge, stereoscope, and pencil.

Photo Interpretation. Photo interpretation or photo identification is not primarily concerned with measurement of horizontal and vertical location from the photographs but with the identification of the images. The photo interpreter learns to recognize the various kinds of land uses, forest types, crops, industries, transportation facilities, and so on from the manner in which they appear in the photographic emulsion. We may utilize the information in noncartographic ways, but, needless to say, the information available through photo interpretation methods is invaluable in cartographic compilation.

Because most maps of medium and small scales are relatively generalized, we can often obtain information directly from the photographs without great concern about the comparatively minor scale variations that occur within the photographs. Photographs are especially useful as source materials, since they often provide the only means for obtaining recent information for map revision. They also are sometimes the only source available for areas poorly mapped or difficult of access.

A Base for Mapping. In the field the photograph is an excellent base on which to map such things as soils, geology, and vegetation. Much preliminary work can be done using the photographs before entering the field: routes for the work can be planned and tentative, rather generalized, mapping can be accomplished, such as delineating areas of topographic differences for soils mapping. In the field the detailed mapping can be done efficiently and accurately because many of the boundaries to be mapped are represented on the photographs by the natural breaks or change in topography or vegetation (Fig. 4.1).

Mosaics and Photomaps. Photographs can be assembled into mosaics which, in turn, can be overprinted with selected map symbols to produce photomaps. Because they can be prepared quickly, photomaps are often used during military operations as an expediency when large-scale topographic maps are not available. Mosaics also serve well as a base upon which information about soils, geology, or vegetation

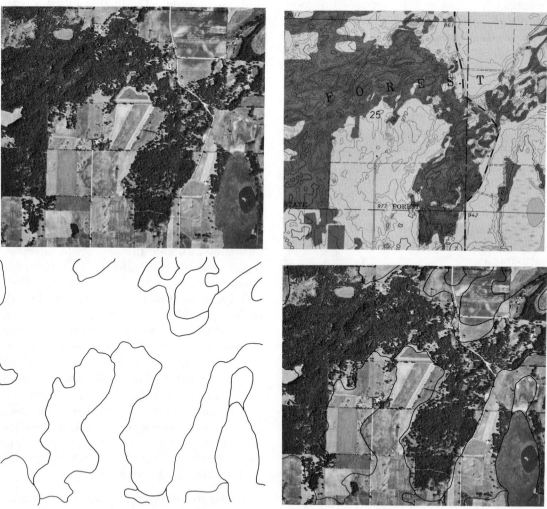

Figure 4.1   Changes in vegetation and elevations are obvious on the photograph and the portion of a USGS topographic map covering the same area (above). Below, soil boundaries (left) obtained in the field are shown superimposed on the aerial photograph (right).

may be overprinted. This technique provides a map with a great deal of base reference material for the map reader.

Mosaics are of two kinds, controlled and uncontrolled. Usually only the central parts of photographs are used, thereby considerably reducing the amount of error due to relief displacement.*

The central areas are carefully trimmed along features in such a way that the joints between pieces can be easily camouflaged upon assembly. If the sections of the photographs are laid in place only by matching images, the mosaic is referred to as being uncontrolled. If precise horizontal locations on the ground are first plotted on a map base (thereby providing control) and if the photographs are positioned so that the photo images of the control points coincide with the plotted control, the mosaic is said to be controlled. Although the overall scale of a controlled mosaic is likely to be better, it must be remembered that none of the perspective displacement has been removed from the individual photographs. Consequently, the scale relationships of the mosaic are usually not comparable to that of a map.

*Since a photograph is a perspective view (projection), objects often do not appear in their true horizontal position.

THE CHARACTERISTICS OF AERIAL PHOTOGRAPHS

Throughout the world, the greatest portion of aerial photography is produced by or for civilian government agencies and the military establishments, which are commonly the primary map-making organizations. Availability of photographs for private use varies depending on national regulations. In the United States anyone can purchase photography from government agencies. Other countries limit the sale to their citizens; and some countries prohibit sale to or even its use by private individuals. Existing photography can usually be purchased for only a fraction of the cost of contracting with a commercial organization for coverage. Whether coverage exists and is available can best be determined by communicating with the government concerned.

The United States has been completely photographed, and most parts more than once. The Map Information Office of the U.S. Geological Survey regularly publishes an index map of the United States showing air photo coverage produced by the various government agencies. Specific information may be secured by contacting the holding agency. We can learn which agencies hold cover for a given area by writing to the Map Information Office, U.S. Geological Survey, Washington, D.C. Contact prints can usually be purchased for the cost of materials and handling.

Production of Photography

Producing high quality photographic coverage requires a good-quality areal camera, a stable aircraft, and fine weather.* As the aircraft follows a predetermined path at a constant speed and elevation, the camera automatically exposes the film at regular timed intervals. The time between exposures determines the amount of overlap along the flight line. At least 50% overlap is necessary for complete stereoscopic coverage, therefore 60% is usually specified. Flight lines are usually arranged so that adjacent photos have a side lap of about 35%

*Further information concerning the photographic mission can be found in Robert N. Colwell, "Procurement of Aerial Photography," *Manual of Photographic Interpretation*, American Society of Photogrammetry, 27–38 (1960).

Although flight lines are usually oriented north-south, the shape of an area may dictate more economical flight lines in some other direction; therefore, we must not assume that north is always at the top of the photograph relative to the marginal information. In some cases an arrow on the edge of the print may be included to indicate at least general direction. Other marginal information may include such things as the date, film role and print identification, time of the day, altitude of the aircraft, and focal length of the camera (Fig. 4.2).

Ideally, the edges of prints along a flight line should match exactly and be parallel to the flight line. Wind may cause the aircraft to drift off course, and the resulting photographs will then be successively offset in an arrangement called "drift" (Fig. 4.3). The pilot, however, usually will try to compensate for the wind by turning his craft slightly into it. Photographs produced under this condition will be twisted into an arrangement called "crab" (Fig. 4.3).

The plan for the photography of an area designates a particular altitude for the aircraft and a specific focal length for the camera. This combination determines the general scale, which can be calculated using the formula

Representative Fraction (RF)

$$= \frac{\text{Camera Focal length (Cf)}}{\text{Elevation of Camera (H)}}$$

Example: RF $= \dfrac{.5}{10,000}$ or $\dfrac{1}{20,000}$

or 1:20,000

The elevation of the camera and the focal length must, of course, be in the same units. The elevation is relative to a specific horizontal known as the datum. In undulating terrain, the datum selected will, naturally, be only one of an infinite number possible. The general scale, then, occurs only at the particular datum selected. Images of the areas lying above this datum are represented at a larger scale and those below at a smaller scale.

The general scale is also affected by variations in the height of the aircraft. If, in the above example, the airplane was sometimes as low as 9800 ft and as high as 10,200 ft, the general scale would vary on a specific datum from 1:19,600 to 1:20,400. Since the general scale applies only to points at a particular elevation on the landscape, and since the general scale changes with changes in altitude of the aircraft, it should not be used in calculations for precise distances.

Oblique versus Vertical Photography

Although the vertical photograph is most common, the oblique photograph, taken with the camera axis intentionally directed between the vertical and horizontal, has some advantages. The appearance of objects is much more like the familiar horizontal view, and the use of oblique views along with verticals has proven to be a useful technique in training photo interpreters. A second advantage is that coverage of a much greater area may be obtained on one photograph.

Figure 4.2 The date the photography was produced is in the upper left corner of the photograph. In the upper right corner, the letters are the Department of Agriculture's code identifying the county and the following two numbers are the role number of the film and the picture number.

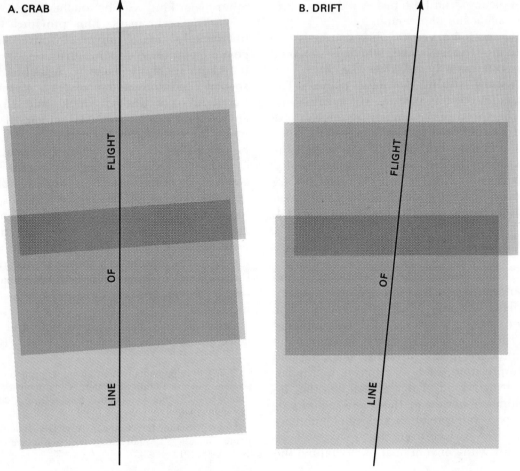

Figure 4.3   A. Crab. B. Drift. In A the aircraft was turned slightly into the wind and has followed the planned flight line. In B the wind has pushed the aircraft off the proposed flight line.

There are two types of oblique photographs that differ only as a consequence of the angle of the camera axis. A high oblique shows the apparent horizon, and the low oblique does not. The high oblique has greater use because the horizon line reveals information relative to tilt and swing that is necessary for photogrammetric purposes. Because obliques have more complicated scale relationships and are often more difficult to view stereoscopically, most photo interpreters prefer the vertical. Both vertical and oblique photographs are used for photogrammetric mapping, the latter mainly for small-scale reconnaissance maps. Trimetrogon photography, using three cameras, one vertical and two obliques, exposed simultaneously, produces a photograph of a strip of the earth from horizon to horizon. The resulting three photographs produced by this method can be treated as one photograph for mapping purposes. Multiple-lens systems, which use from two to nine lenses in a single camera, are used to obtain a wide-angle field of coverage. With proper rectification the pictures can be transformed into a single composite photograph.

A variation of vertical photography is continuous strip coverage, which is obtained by continuously passing the film over a narrow slit in the focal plane.* The speed of the film past the slit is calibrated to the speed of the aircraft in order to provide a continuous sharp image. Two such cameras mounted laterally will permit the two strips to be viewed stereoscopically. Continuous strip is especially useful for producing large-scale photography of linear features such as roads or railroads (Fig. 4.4)

Kinds of Images

Until recently, photo interpreters and photogrammetrists could use only conventional black and white photography. The past few years has seen the development and use of a variety of sensing equipment to produce images useful in mapping and interpretation, such as infrared, and radar. Each new kind of image requires the development of new methods and techniques for both interpretation and photogrammetric uses. Concern in this book will be mainly with conventional photography; the

other major kinds can be touched upon.

While panchromatic film provides the photography most universally used for general purposes, certain attributes of infrared photography make it desirable for special conditions and uses (Fig. 4.5) Infrared rays cut through smoke and haze, and pictures can be produced on hazy days when the use of panchromatic film would be unsatisfactory. More contrast in images on infrared makes it especially useful for some kinds of interpretation. For example, it is a favorite of foresters since varieties of forest cover can be more easily distinguished by variations in tone. Because water registers as a very dark grey (or black) it is desirable as a source for the delineation of hydrographic features. Shadows, however, are very dark and can easily hide detail.

Only recently has the real potential of color photography been recognized.* Be-

*Early developments in continuous strip photography are given in Phillip S. Kistler, "Continuous Strip Photography," *Photogrammetric Engineering*, **12**(2), 219–223 (1946).

*John T. Smith, Jr., "Color, A New Dimension in Photogrammetry," *Photogrammetric Engineering*, **29**(6), 999–1013 (1963).

Figure 4.4 Because of the low altitude at which this continuous strip photography was made, displacement is enough to be quite evident.

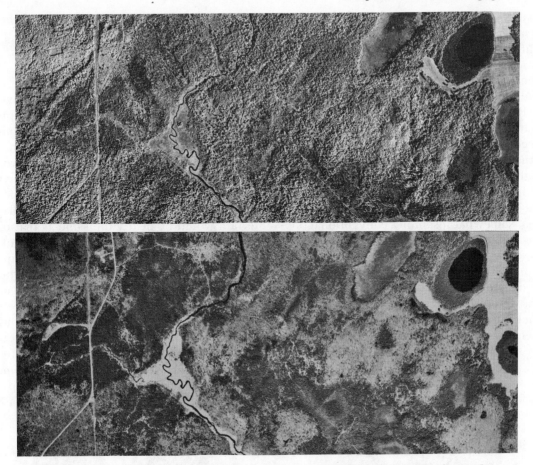

Figure 4.5 The infrared photograph (above) was taken during September and the panchromatic (below) was taken some years later, during May.

cause of rather poor resolution, supposed high cost, and the inability to use color positives in stereoplotting machines, color has been employed only for special purposes or problems. It has, for example, been used for the detection of disease in vegetation. Recent improvements in resolution suggest an increased use of color in the future. Although the cost of processing the film is somewhat more than for black and white, the generally high cost of operating an aircraft plus the other fixed costs in air photo production suggest that, with more extensive application, the extra cost of color will have less effect in curtailing production and use.

Devices that record the characteristics of the landscape by means other than conventional photographic methods provide a product called by the general term imagery. Although sometimes its form can resemble a conventional aerial photograph, the details of tone, texture, and scale relationships usually differ.* The tone on infrared imagery, for example, depends on the objects infrared radiation emission. Objects that may reflect equally in the visual portion of the spectrum and therefore not be distinguishable on conventional panchro-

*George L. Laprode, "An Analytical and Experimental Study of Stereo for Radar," *Photogrammetric Engineering*, **29** (2), 294–300 (1963).

matic photography may, because of their temperatures, be easily identified by different tones on infrared imagery (Fig. 4.6). Another obvious advantage of infrared is the ability to produce the imagery during the hours of darkness.

Radar imagery, made by a device that transmits pulses of microwave energy and receives reflections from objects on the landscape, can also be produced without light. The images are recorded in relationship to the time required for the pulse to be transmitted and reflected. Although it, too, somewhat resembles conventional photography, identification on the radar imagery is complicated. Problems such as reflection characteristics of surface materials and resolution raise new problems for the interpreter. In the field of photogrammetry the radar image presents difficulties because of its complicated geometry.* Pro-

*Earl S. Leonardo, "Comparison of Imagery Geometry for Radar and Camera Photographs," *Photogrammetric Engineering*, 29 (2), 287–293 (1963).

gress is, however, being made toward the use of radar imagery in stereo-plotters.

Geometry of Aerial Photographs

The perspective projection of the image of the land surface on an aerial photograph causes the scale relationships to differ from those that would occur on an orthogonal projection of the land to a map. Disregarding such things as the distortion produced by the lens, paper, and film, the scale of a truly vertical photograph of perfectly flat terrain would be nearly the same as that of an accurate map. The occurrence of relief, however, causes variations in scale to appear because of the perspective view of the camera. Relative to one particular level of the terrain, higher points will be displaced away from the center of the photograph and lower points displaced toward the center. These differential variations in scale preclude our merely tracing information from photographs directly to large-scale maps. The amount of displacement can be meas-

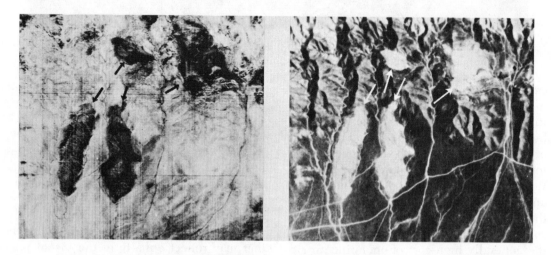

Figure 4.6   Infrared imagery, left, obtained at night, and a conventional aerial photograph, right, display an almost complete reversal in tone. The arrows point to outcrops of a Pliocene shale formation. (Images courtesy of U.S. Geological Survey and National Aeronautics and Space Administration.)

ured, and the disadvantage of not being able to trace information directly for large scale map production is far outweighted by our ability to use displacement to determine distances above or below a chosen datum.

Consideration of the complex geometry of an aerial photograph will be limited here to the aspects of vertical photographs with no tilt. The optical or lense distortion, which on a typical 9 x 9 photograph might be in the magnitude of a displacement of a few thousandths of an inch, can usually be ignored except for determination of elevations with stereoscopic plotting instruments. Distortion resulting from changes in the dimensions of the film base are usually very slight, but paper distortion of prints can be considerably more bothersome.

Scale. The precise scale ratio between two points on a vertical aerial photograph usually differs from that of the general or average scale. As we observed earlier, the general scale is the ratio of the focal length of the camera to the elevation of the camera with respect to some specific elevation on the landscape; it follows that this ratio will not be correct for any other elevation or datum. Each of the infinite number of horizontal planes has its own specific ratio or scale. Figure 4.7 illustrates how scale varies with differences in the vertical positions of points on the landscape. The true map locations of two towers on the land surface at the same elevation is at points A and B. On the photograph, at a scale of 1:20,000 these locations will be at a and b. Because of the perspective in the photograph, the tops of the towers will appear in the photograph at a' and b', which are clearly farther apart and therefore at a larger scale than a and b, for example, a scale of 1:18,000. In this particular case, we could plot the positions of the towers at a scale of 1:20,000 because the bases as well as the tops of the towers would be visible

on the photograph. On the other hand, if the problem involved a hill, the base

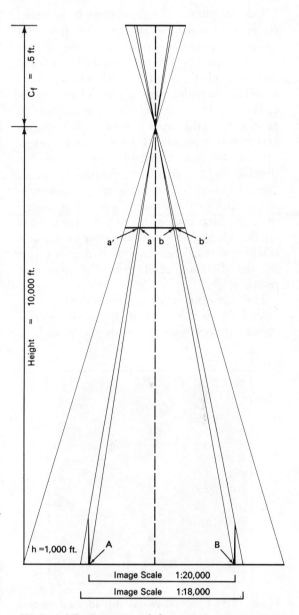

Figure 4.7 Because of the perspective projection of an aerial photograph, the locations of images on the film (and print) are determined by their vertical positions and their distances from the point directly beneath the camera.

would not be visible and, therefore, we could not plot the position of a hill directly from the photograph at a scale of 1:20,000.

Displacement. Displacement because of relief occurs at a radial direction from the nadir, the point on the plane of the photograph located by the extension of a vertical line through the center of the camera lens which, on a truly vertical photograph, coincides with the principal point or geometric center. Straight lines across a photograph connecting opposite fiducial marks intersect at this principal point (Fig. 4.8). No displacement occurs at the principal point, then, when it is aligned with the perspective center of the camera lense, but image distance from the center to any other point on the photograph will depend on (a) the relative vertical location of that point and (b) the distance of it from the principal point.

The amount of displacement changes directly with the vertical departure from a chosen datum and the distance from the

principal point, and inversely with the height of the camera. A comparison of similar triangles in Figure 4.9 indicates this relationship; for as either h or d increases, rd also increases; and as H increases, rd decreases. Photographs made at low altitudes, then, display more displacement than those made at high altitudes. Although the latter have scale relationships that more closely approximate a map, the low altitude photographs with the greater displacements turn out to be more useful for the

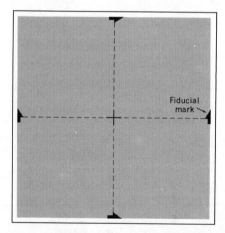

Figure 4.8 Fiducial marks are located in the focal plane of the camera. They are etched on the surface of the glass, which is in contact with the film; or, in the case of the open type focal plane, they are projections of metal that extend into the negative area.

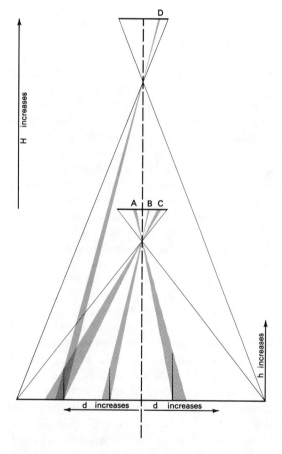

Figure 4.9 As the distance from the center increases and as the height of the object increases, the amount of displacement is greater. As the altitude of the camera increases, displacement is less.

determination of elevations.

We perceive depth in our vision in a variety of ways. With one eye we must rely on sizes of objects, clarity of detail, or whether one object appears in front of another. When using both eyes, each eye sends its own signal to the brain. Because our eyes are separated, each sees an object (if it is within approximately 2000 ft.) from a different angle, and the subsequent signal to the brain causes the sensation of depth.

The displacement of objects on aerial photographs produces parallax, which is the apparent change in position of an object because of a change in the point of observation. This apparent change in position is the principal reason for our being able to view two photographs and produce an illusion of a third dimension. By viewing an object on one photograph with one eye and the same object on an overlapping photograph with the other, we are in effect viewing the object from two points that approximately represent the two camera stations. Each picture shows all objects in perspective from each camera station, and our eyes each send a signal to our brain causing the image to take on an apparent third dimension.

On a photograph with no tilt, the parallax is a linear element used for determination of elevation. It is parallel with the axis of the camera stations and is associated with the height of the object, the focal length of the camera, the distance between the camera stations, and the distances from the camera stations to the object. The algebraic difference of the parallax on two overlapping photographs is used to determine elevations using stereoscopic plotting instruments.

The parallax difference can be measured graphically; however, the error is likely to be somewhat larger than when using plotting instruments. Two overlapping photographs should be carefully aligned, as shown in Figure 4.10. The principal points and conjugate principal points (image of the principal point of the overlapping photograph) of both photographs must fall on a

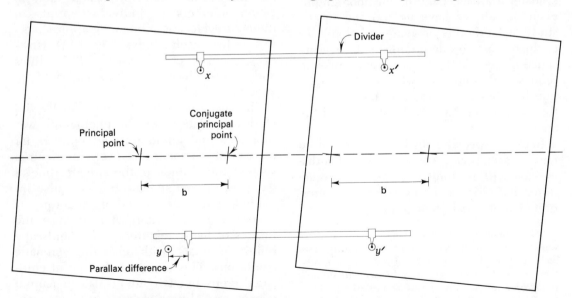

Figure 4.10 Graphically measuring parallax difference.

straight line. The average distance b between principal points and conjugate principal points is then computed. The parallax difference is the difference of the distance xx' and yy' and can be used in the following formula to determine the difference in

elevation between the two points:

$$\triangle e = \frac{H}{b} \times parallax\ difference$$ where H is the average height of the camera above the terrain

PHOTO INTERPRETATION

Although there is no general agreement on a precise definition of what is involved in photo interpretation, most agree that it involves more than just the recognition of features. To determine the significance of images on photographs, the interpreter must evaluate what he recognizes in terms of his own experience and based on his professional understanding. Photo interpretation is carried on in many widely different fields, and since no one can be a true specialist in all subjects, often the interpreting of one individual may be largely limited to the mere recognition of features. We may, for example, learn to identify different kinds of drainage patterns on photographs even though we have no training in the study of landforms or geomorphology. To evaluate or to determine the significance of our identifications may, however, require a thorough knowledge of the relationship between drainage patterns and the form of the land or the geology of the area.

Because special knowledge is required for interpretation in specific fields, this discussion will be limited to general procedures and aids for the handling and interpretation of aerial photographs.*

*Most photo interpretation books are slanted toward a specific field such as geology or forestry. Two books that also have a good bit of general interpretation are Carl H. Strandberg, *Aerial Discovery Manual,* John Wiley and Sons, Inc., New York, 1967; and Donald R. Lueder, *Aerial Photographic Interpretation,* McGraw-Hill Book Company, Inc., New York, 1959.

Handling

Since we usually work with rather large-scale photographs and are often concerned with sizeable areas on the earth's surface, we are likely to be confronted with a large number of prints. Adequate storage such as a file cabinet is required to keep the photos in good condition and readily accessible. Each photo should be identified by developing an overall numbering system or by the systematic use of a number that may be a part of the marginal information. To be able to select individual photographs, some sort of index showing location is necessary. Plotting outlines or even just the identification numbers of individual prints in their correct locations on topographic sheets results in a useful index (Fig. 4.11). However, even a chart showing flight lines and relative locations of prints could be adequate.

The marginal information usually includes the date when the photograph was produced and an identifying number for each print. In some instances, information such as the altitude of the aircraft, time of day, and camera specifications are also included. The date of the photography is important to the interpreter because the images of many features of the landscape reflect changes with changing climatic conditions. This is especially true of cultivated crops but is evident also in natural vegetation. Abrupt changes in images in several prints of a flight line may be ac-

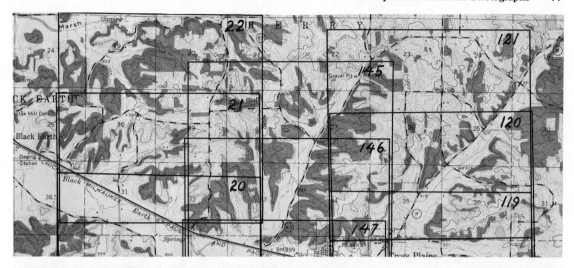

Figure 4.11 Topographic map used as an index for photography compared to a photo mosaic index.

counted for by reference to the date. If, for example, a number of photographs do not conform to the specifications, those portions of a flight line may have been rephotographed at a later time. Consequently, on the later photos, trees may reflect a different amount of light because of changes in leaves or the trees could even have dropped their foliage.

Procedures

Preliminary study of the photography is usually undertaken without the use of the stereoscope, an instrument which necessarily limits your field of vision to a relatively small area. Naked eye study of individual prints and prints laid in a rough mosaic arrangement is often beneficial. Regional relationships, which may easily escape detection in stereoscopic study, may be quite evident in the overall view.

For any photo interpretation project one should gather as much background material as is available and combine field investigation if possible. Reference material in the form of written accounts, maps, and interpretation keys will not only assist in the interpretation but can add precision to the study.

Field study can be carried on at any time; however, it is probably most valuable if accomplished before detailed study of the photographs begins. Using available maps and the photographs, a careful plan should be developed for the conduct of the field work. Preliminary study should reveal problems that are most easily resolved in the field where we can compare the photographic images with ground views of objects and verify identification.

There are numerous ways in which we can record information on the photographic prints. Translucent material fastened to the photograph can be marked with either pencil or ink. If the material is attached to only one edge of the photo, it can be flipped away so a stereoscope can still be used. Marks made on prints with grease pencils can easily be removed with a cloth moistened with trichloroethylene or lighter fluid. For very precise locations, we can prick tiny holes in the photograph and record the information next to the hole on the reverse side.

Stereoscopic study, while it limits the field of vision, permits the interpreter to view features in three dimensions and usually also supplies magnification. Since the vertical view of the landscape is itself somewhat foreign to most people, restricting that view to two dimensions adds a serious limitation. In the study of such things as slope or drainage, it is almost imperative that stereoscopic methods be used.

Several kinds of stereoscopes are available; but the two most commonly used are the lens type (Fig. 4.12) and the mirror type (Fig. 4.13). Without the binoculars attached, the mirror type permits the larger field of vision. The folding lens type is more portable, of course, and for an experienced interpreter is probably more useful. Proper use of a stereoscope will result in a better three-dimensional view and will avoid undue eye strain.

When first attempting to view photographs with a stereoscope we should:

(a) locate and mark the principal point of each photograph. This point is located where lines connecting opposite fiducial marks intersect (Fig. 4.8).

(b) Locate and mark on each photograph the image of the overlapping photograph's principal point (conjugate principal point). A line connecting the principal point and the conjugate principal point on each photograph is the flight line.

(c) On a flat surface overlap the photographs so images on the two photographs match and the flight lines are superim-

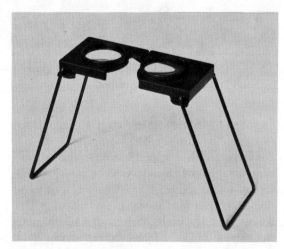

Figure 4.12 Lens type stereoscope. (Courtesy of Gordon Enterprises.)

Figure 4.13   Mirror type stereoscope. (Courtesy of Gordon Enterprises.)

posed. The photographs must be oriented so the flight line extends from left to right relative to the observer. While keeping the flight lines superimposed, separate the photographs approximately the distance between the viewer's eyes (roughly two inches).

(d) Place the stereoscope on the photographs so each lens is directly above the object to be viewed.

(e) While viewing through the stereoscope, each eye must see only that image directly beneath the lens. It might be helpful to imagine that the image is some distance away to assist in keeping each eye looking at its respective image. If the images do not merge and form a good three-dimensional model, we may need to make minor adjustments in the amount of separa-

tion of the photographs, the separation of the lenses of the stereoscope, or in the alignment of the pictures.

After a short period of practice we need no longer follow the foregoing steps since we will be able to judge the alignment and spacing and will be able to achieve the stereoscopic view in a few seconds.

The ability to use a stereoscope requires binocular vision. Some students with a weak eye are unable to experience the three-dimensional view, and others may assume they see it because of apparent relief caused by shadows. The stereogram° (Fig. 4.14) can be used to determine whether or not the student has the three-dimensional view. The two blocks of letters are arranged in such a way that most of the letters will appear on the plane of the paper when viewed through the stereoscope. Some scattered letters, which form three words, appear to float some distance above the plane of the page. The ability to read these three words indicates that the viewer has achieved a stereoscopic view.

Characteristics of Images

Unlike a map in which selected features of the landscape are represented by sym-

°A set of drawings or photos arranged correctly for stereoscopic viewing.

```
E S P B O E Y V E S O D          E S P B O E Y V E S O D
P S A T B L P M G M I D          P S A T B L P M G M I D
F EM L E N MS EN T S             FE ML E NM S E N T S
M T Q H O G I P M M T F          M T Q H O G I P M M T F
Y U E S G J G A Z G J P          Y U E S G J G A Z G J P
B G S O P F S E  F T F M         B G S O P F S E F  T F M
O T H P Z P R P O F P V          O T H P Z P R P O F P V
J T C P F A P R G H G T          J T C P F A  P R G H G T
F L T S F L P T A N T O          F L T S F L P T A N T O
O G M R D  A P P H R Y M         O G M R D A P  P H R Y M
F B E R P F L A E Y J L          F B E R P F L A E Y J L
G O E C P H E A Z G A Z          G O E C P H E A Z G A Z
```

Figure 4.14   When viewed stereoscopically, three words should appear to float some distance above the plane of the page.

bols, an aerial photograph is a record, in various tones of gray, of all objects in the field of view of the camera. Identifying objects on an aerial photograph is usually more difficult than recognizing map symbols. There are, however, a variety of characteristics of the photographic image that can be used to aid in the process of identification.

Tone variations result from differences in the reflective qualities of objects, which are recorded on film in a variety of gray values. As well as defining the shapes of objects, tone is also useful in identification. It is described in terms such as light, medium, or dark; and in interpretation, prepared gray scales can be used for comparison with photographic images. Because of differences in the amount of available light during a photographic mission or because of variations in the developing and printing processes, a given object may have variations in tone on several photographs in one set of coverage. Tone differs even more, depending on the kind of film used. On panchromatic film, for instance, water is usually a medium to medium dark tone; however, if the sun's rays are reflected directly into the camera, an extremely light or almost white tone is the result (Fig. 4.15). On infrared photography, on the other hand, water is recorded in a very dark tone with the sun not affecting it.

Moisture, especially in soils, causes tone to be darker; and marshland tends to be darker than the surrounding areas. Human activity in an area, on the other hand, tends to cause lighter tones. Rural schools are often distinguishable from rural churches because of the lighter tones around a school caused by childrens' play.

Texture, produced when changes of tone in objects are too small to be discernible, may be described with such terms as stippled, granular, or mottled. Texture is commonly used in describing the photographic

Figure 4.15 When the sun's rays are reflected from water directly into the camera (lower left), the tone becomes very light.

appearance of natural vegetation, especially forests. Depending on the scale of the photograph, certain species of trees may have a particular texture resulting from the arrangement and reflective ability of leaves, branches, or crowns. Grass, because of its tiny blades, usually produces a smooth texture on the photograph, whereas brush is somewhat coarser.

Shape or the general form (which includes the three-dimensional stereoscopic view) may be the single most reliable evidence for identification. Buildings often have characteristic shapes because of their use or because of tradition of design. Churches are often in the shape of a cross and usually have a spire. Thermal electric plants are usually housed in structures that have three distinct levels: the lower for administration, the second for generating equipment, and the highest for production of power. In addition, a large smokestack will usually be visible. Shape is also useful for evaluating or determining the significance of features. Distinctive shapes of landforms, for example, often provide clues to the underlying structure or erosional processes.

Shadows are usually most useful when photographs are produced during that part of the day when the shadow is most nearly the size of the object. The shadow, of course, provides a view that is more like the ground view of the object and can, therefore, be an important clue in identification. Lengths of shadows can be used to determine heights of objects.

The heights of some sample objects are first measured on the ground and then their shadows are measured on the photograph. A shadow factor, which is the height in feet for each unit of measurement of the shadow, is then computed and applied to the measured lengths of shadows of objects for which heights are desired.

Occasionally, when viewing the photographs with the naked eye, shadows falling away from the viewer cause an illusion in which the relief on the photograph appears reversed. Streams appear to be on ridges and ridges appear to be valleys. This probably happens because we are accustomed to overhead sources of illumination, in which case shadows generally are beneath the object that intercepts the light. In any event, the reversed effect should not be evident if we orient the picture in such a way that the shadows fall toward the viewer.

Size of images is extremely important because it can be a deciding clue when distinguishing between objects alike in shape. Measuring the object may be necessary before the interpreter can make an accurate identification.

There are a variety of scales available for measuring, and the interpreter should select one that best meets his needs. A scale graduated in thousandths of a foot is useful in that a measurement need only be multiplied by the denominator of the representative fraction to convert the photo distance to ground distance in feet. If we wish to convert to ground distances in the metric system we should, of course, use a scale graduated in metric units. We must keep in mind that when using the general scale assigned to a set of photography, calculations of distance may not be precise.

Pattern, the arrangement on the landscape of both physical and cultural features, is often distinctive and may be useful for recognition and evaluation. Drainage patterns, as mentioned earlier, can be evidence for geological interpretation. Field and road patterns can be useful for work in agriculture and economic studies. Pattern, like texture, has a relationship to scale in that some features may form discernible patterns only at certain scales. For example, at large and medium scales, trees in an orchard may form a pattern; at much smaller scales the reduction in size of images could change the pattern to a texture. Then, at an even smaller scale, it is possible that the distribution of many orchards would form a pattern.

Site, the location on the landscape, can contribute to identification, since many features are found in characteristic places. A particular type of vegetation may, for example, appear only along streams, only on ridges, only on north slopes, or only on south slopes.

Associated features are those we commonly find adjacent to or in conjunction with the object under investigation. We would expect to find heavy industry adjacent to some form of heavy transportation such as rail or water. Thermal electric plants and paper mills need large supplies of water and are usually located on or near lakes or rivers.

Identification should not be made on the

basis of just one of an object's characteristics; all the evidence available should be used. One characteristic may be more obvious or dominant, but consideration of all the characteristics will produce more reliable results.*

Keys as an Aid in Interpretation

A photo interpretation key is an assembly of reference materials prepared to assist the interpreter in identifying and in some cases determining the significance of objects on aerial photographs. A key contains information, much of it usually in graphic form such as stereograms or pictures of ground views, which illustrate the recognition characteristics of particular objects or groups of objects in a prescribed area. Since distinguishing characteristics of objects are not always the same in every part of the world, it usually is not possible to prepare a key that has world-wide application.

There has been disagreement among photo interpreters concerning the value of keys. Keys, of course, cannot replace the interpreter's use of reasoning, but they do have some important uses. An interpreter often finds it necessary to compile information in some unfamiliar field; keys can be used to assist in identification. They are also useful for training photo interpreters or for would-be interpreters to train themselves. An experienced interpreter may find keys helpful to refresh his memory or to introduce him to image characteristics in a new geographical area.

Automation in Interpretation

Experiments in automatic interpretation have met with limited success. Sensing devices must recognize objects in terms of tone, texture, pattern, shape, and size. It is difficult to separate images that represent a single object or to recognize that several separate images represent a larger object. Although individual characteristics of objects can be sensed, the myriad of possible combinations of characteristics for a given feature still presents a problem for automation. Used in conjunction with computers, improved sensors may soon be able at least to make tentative identifications.*

*Useful information about the availability of materials for teaching interpretation is presented in Gene Avery and Dennis Richter, "An Airphoto Index to Physical and Cultural Features in Eastern United States," *Photogrammetric Engineering*, 31 (5), 896-914 (1965).

*Azriel Rosenfield, "Automatic Imagery Interpretation," *Photogrammetric Engineering*, 31 (2), 240-242 (1965).

PHOTOGRAMMETRY

The cartographer increasingly relies on photogrammetric methods to furnish him with planimetric and hypsometric positions for map preparation. Compared with field methods, photographs can provide such information much more rapidly and at a reduced cost. With limited ground control, photogrammetric methods can provide accurate large-scale maps for areas where it is difficult to conduct extensive field surveys because of terrain or climatic conditions. This advantage, of course, is especially important for military operations.

Most detailed large-scale maps made from aerial photographs are produced by government agencies or by commercial organizations working under contract. The high cost of precise stereoplotting instruments prohibits their use by individuals or small commercial firms. Some, however, can be found at educational institutions where they are used mainly for instruction and research.

Even without expensive equipment, cartographers can make use of photographs for extending control. Equipment that is not particularly costly can furnish control that is precise enough for the preparation of smaller-scale maps or for map revision. Space will limit discussion here to a description of a system of extending control using radial line plotting and to a discussion of some plotting instruments.

Radial Line Plotting

On a truly vertical photograph, azimuths from the principal point are correct to any point on the photograph; a condition that permits us to perform triangulation directly from the photographs. Since the direction from the principal point to its conjugate principal point is correct on each of two overlapping photographs, it follows that by superimposing these two lines we produce a base line with ends from which all angles are correct. On pieces of overlay material we can mark the principal points of each photograph, and draw lines through the conjugate principal points and from the principal points through a third point c

(Fig. 4.16). When the overlays are placed so the base lines are superimposed, we have produced a triangle in which the angles at a and a' are correct. If these two angles are correct, the angle at c must, of course, also be correct (Fig. 4.17), the elevation of c having no effect on the horizontal angles. The three points have been located in a true planimetric relationship, and the scale of the three sides of the triangle is the same.

In radial line plotting each point is relocated by the amount of its displacement and all points are then located at a common scale. The actual scale of the plot bears no special relationship to the general scale of the photography and can be either larger or smaller, depending on the distance between the principal points of the two photographs (Fig. 4.17). In this situation scale can be determined if we know the ground (or map) distance between two points in the plot.

Usually a radial line plot is prepared at a predetermined scale by using control points established in the field or from an accurate map. Ideally, we should plot three

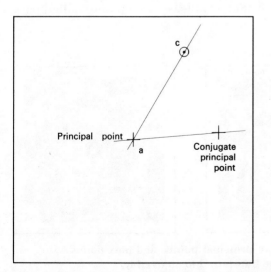
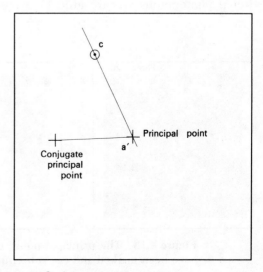

Figure 4.16   The angles at points a and a' are correct.

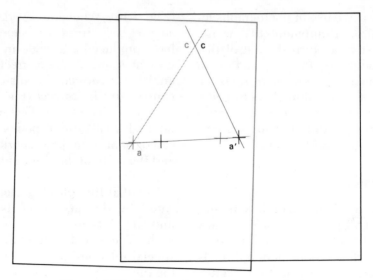

Figure 4.17   The angles at points *a* and *a'* are correct; therefore, the angle at *c* must be correct.

control points at the desired scale in the area of overlap of the first two photographs. The principal points are then either separated or moved closer together to cause radial lines to intersect at the plotted control points.

Hand Template Method. By using additional photographs we are able to extend

control over larger areas by using the following procedure (Fig. 4.18).

1. Locate the principal points on all photographs and mark them by pricking a small hole, point *A* (Fig. 4.18).

2. Locate and mark all conjugate principal points while viewing stereoscopically, point *b* (Fig. 4.18).

3. Select at least three points (called pass

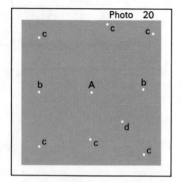
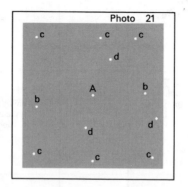
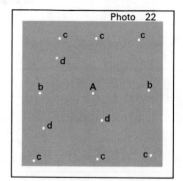

Figure 4.18   The principal point, conjugate principal points, and pass points have been marked and can subsequently be transferred to an overlay.

points) on each photograph in the area of side lap of the adjacent flight line, points *c* (Fig. 4.18). Although it is desirable, pass points need not be features that ordinarily would be symbolized on a map but can be any images that are easily recognized on adjacent photographs.

4. Mark all control points to be used in the plot. (Three control points in the area of overlap of the first two photographs establish the scale, points *d* — Figure 4.18; however, more control points are recommended for other photographs to help hold the entire plot to the predetermined scale.)

5. Templates are prepared on a clear or translucent material that has been cut into pieces slightly larger than the photographs. (Photographs can then be taped face down to the overlay material, thereby preventing the tape from touching the emulsion.)

6. To make templates of the overlay material (Fig. 4.19):

(a) Mark the principal points.
(b) Draw lines from the principal points through the conjugate principal points (base lines).
(c) Draw short lines or rays from each principal point through each pass

point on that photograph extending a half inch or so on either side. If the intended scale of the plot is to depart considerably from the general scale of the photographs, the lines through the pass points must be longer. Extend them away from the center for larger scale and toward the center for smaller scales.

(d) Draw short segments of radials from the principal point of each photograph through features that have been selected for establishing new control points.

7. Identify each template by adding the picture number in any convenient location on the photograph.

8. Separate templates from the photographs and arrange in groups according to flight lines.

9. Prepare a base sheet for template assembly by plotting existing control on a translucent material (Fig. 4.20).

10. Begin laying the templates by selecting two photographs (preferably near the center of the plot) that have the greatest number of control points and placing them on the base sheet with the base lines of the

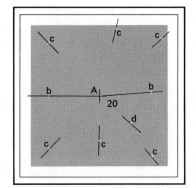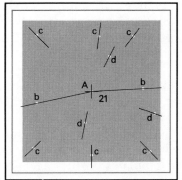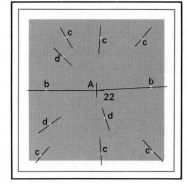

Figure 4.19 Prepared template with principal points marked, base lines plotted, and rays to pass points drawn.

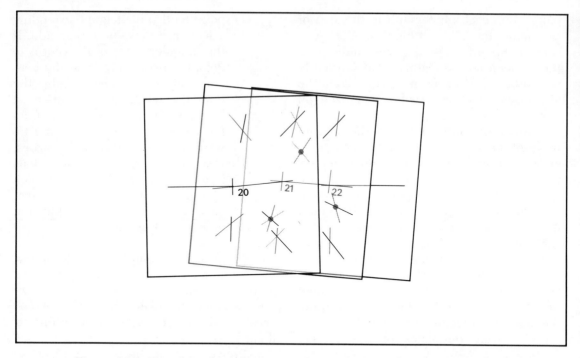

Figure 4.20   Templates have been arranged on the base sheet (on which map or ground control has been plotted) so that appropriate rays intersect at the control points and establish the scale for the plot.

two photographs coincident. Adjust the photographs by separating the principal points or moving them closer together until the radials of the control points intersect at the plotted positions on the base. Tape the template to the bases. The principal points of the first two photographs are now in their correct position relative to all the control plotted on the base. The pass points, likewise, are in their correct positions (Fig. 4.20).

11. Working along the flight line, lay each consecutive template on the plot, keeping the base lines superimposed and moving the template until its radials through the pass points intersect at those points established by the prior templates. As we proceed, the additional plotted control should be used to hold the scale and adjust for accumulated errors.

12. The adjacent flight line is laid in the same manner, using established pass points and control points from the previous flight line.

13. When all the templates are in place, the assembly may be turned over and, since the base sheet is translucent, any points where radials intersect can be selected and marked on the reverse side of the base sheet.

The base sheet with the original control points and those new ones established by the radial line plot can now be used to compile the map features. Photographic images should be adjusted to fit the control by the use of an instrument that will change the scale of the photograph and, if possible, correct for small amounts of tilt.

The accuracy of the radial line plot depends upon several factors. If any of the photographs have considerable tilt, the angles at the principal point will not be correct and rays will not intersect at the correct positions. Sometimes the paper on which the prints are made expands or shrinks in one direction as compared to another causing differences in relative positions. The care with which the plot is prepared is also reflected in the accuracy of positions. Errors in the plot usually appear in the form of triangles where rays should, of course, intersect at a point. Sometimes it is necessary to pick up several templates and adjust them slightly for discrepancies that appear.

Slotted Template Method. Slotted templates, mechanical substitutes for the hand template, are pieces of material (which need not be translucent) in which slots substitute for the lines drawn on the hand templates. A machined round stud, which fits the slot precisely, can be slid within the slot (Fig. 4.21). Since the stud can slide in all the templates having slots for a common point, the templates tend to settle at the correct position. Discrepancies then tend to "shake out" as the plot proceeds.

Since the templates need not be translucent, rather rigid and stable material can be used. Selected points are recorded on the base map by inserting a sharp pin into the hole in the center of the stud and piercing the base material. Special punches are needed to cut the slots, and since the slots are cut for a specific photograph, the templates cannot be used for other plots.

Radial Arm Template Method. Another mechanical solution, using adjustable radial arms, permits reuse of the equipment. Flat metal arms of various lengths with holes at one end and a slot at the other again replace the lines in the hand template. The holes in the arms are placed over a bolt secured at the principal point

Figure 4.21 In the slotted template method, the centers of the slots are the radials from the principal point to the pass points, and to the conjugate principal points.

with the middle of the slots falling on the conjugate principal point and pass point studs. When all radial arms are in place, a nut is tightened to hold the template rigid and the radial arms are used in the same manner as slotted template (Fig. 4.22). When the plot has been completed, the template can be disassembled and the parts used again.

Plotting Instruments

Plotting equipment for map compilation from aerial photographs falls in two general groups, monocular and stereoplotting instruments. Monocular instruments are used chiefly for small-scale mapping or for revising existing mapping. With simple stereoplotting instruments parallax can be measured and at least generalized contour lines or form lines can be drawn. Photogramme-

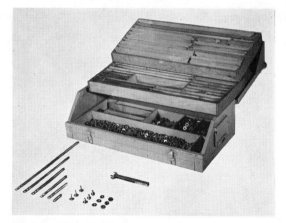

Figure 4.22   Radial arm template set. (Courtesy of Gordon Enterprises.)

tric plotting instruments used in the preparation of large-scale accurate topographic maps are sophisticated machines that provide very precise measurements and plot the map data in orthogonal projection.

Monocular Instruments

Two general types of monocular instruments are those that use the camera lucida principle and those using optical projection. Both types are essentially devices by which photographic detail can be transferred to a base map on which control has been plotted. Changes in scale can be made to facilitate fitting the detail to the control. Some instruments also have the capability of making adjustments for tilt.

The sketchmaster uses the camera lucida principle, and the eye receives two superimposed images, one from the base map and the other from the photograph. The image from the photograph is reflected from a semitransparent mirror, and the base map is visible through the mirror. The instrument is adjustable so that we can change the scale of the photograph to fit the base. Foot screws on each of its three legs can also be adjusted to remove small amounts of tilt from the photograph (Fig. 4.23).

Another machine, useful for medium and small-scale compilation from photographs, is the vertical reflecting projector. The image from opaque copy is reflected from mirrors through a lens to a table on which the map is placed. Usually the projectors are designed so that we are able to change the size of the image up to three or four times, permitting tracing at the desired scale. On some projectors the table on which the base map is placed can also be tilted slightly to allow for tilt in the photograph. Neither an instrument that employs the camera lucida principle or one that uses optical projection has the capability of removing the effect of displacement due to relief.

Stereoscopic Plotting Instruments

Stereoscopic plotting devices differ from monocular instruments in that they employ

Figure 4.23   The sketchmaster can be adjusted for different scales and to compensate for tilt. (Courtesy of Gordon Enterprises)

two overlapping photographs enabling us to make parallax measurements in order to determine elevations.

One of the simpler instruments is the stereocomparagraph, which consists of three major parts: (1) a stereoscope, (2) a device for measuring parallax, and (3) a drawing attachment (Fig. 4.24). In the stereocomparagraph a pair of photographs is viewed stereoscopically so as to produce a three-dimensional image or "model" of the terrain. The measurement of parallax is accomplished by adjustment of a "floating dot." This dot is actually made up of the images of two dots each seen with but one eye. The single dot may be adjusted "vertically" in space by changing the horizontal spacing of the two dots that form its image. The tracing arm is linked to the dot. Changing the apparent vertical position of the dot does not affect the position of the tracing arm, but horizontal movement does.

By viewing the photographs stereoscopically and adjusting the two dots, the floating dot can be brought into apparent contact with the surface of the earth in the three-dimensional stereo model. Then by moving the tracing arm so that the dot remains in contact with the surface of the model, a contour line may be traced. We may move the dot to points of different elevations, and by adjustment, the floating dot can be brought into apparent contact with the surface of the model. Readings from the micrometer showing the position of the dot provide parallax measurements from which differences in elevation can be calculated.

The stereocomparagraph makes no corrections for tilt. Furthermore, the plot produced is still in the perspective projection of the photograph; therefore, each contour line is projected at a different scale. For precise maps we must compensate for tilt and provide means for compiling the information on the map sheet in an orthogonal projection at a consistent scale.

Some photogrammetric plotters or plotting assemblies operate on the basis of systems which, in a sense, reverse the rays that formed the camera picture and, by projection, produce the image of a three-dimensional model of the terrain on a mapping table. When the projectors are properly oriented so as to adjust the projected images to fit already plotted horizontal control, the apparent model may be used to delineate planimetric detail and plot contours.

In such an assembly, two or more projectors are mounted in such a way that they can be moved in any direction and thereby duplicate the relative position and orientation of the camera at the instant the picture was taken (Fig. 4.25). Overlapping photographs in the form of glass positives are placed in alternate projectors, and one photograph is projected through a red filter and the other through a cyan (greenish blue) filter. A cartographer, wearing spectacles with one cyan and one red lens, sees the image of one photograph with one eye and the other photograph with the other eye, resulting in a three-dimensional view. A stereo pair superimposed in red and cyan is called an anaglyph.

A small tracing stand with a white disc

Figure 4.24   With the stereocomparagraph, parallax can be measured for calculating differences in elevation. (Courtesy of Gordon Enterprises.)

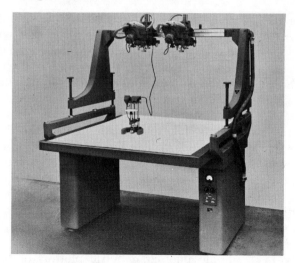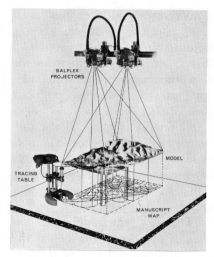

Figure 4.25   Left, a Balplex Plotter used for very precise maps. Right, illustration of the basic operation of the Balplex Plotter. (Courtesy of Bausch and Lomb, Incorporated, Special Products Division.)

or platen has an illuminated aperture in its center (Fig. 4.26). This platen "disappears" in the projected image of the model because it becomes part of it, so to speak, but the tiny illuminated aperture appears to float in space like the dot in the stereocomparagraph. The platen, and thus the dot, can be moved vertically, and its height within the model is indicated on a micrometer scale.

When the floating dot is placed in contact with the surface, the tracing table may be moved, keeping the dot in contact with land. A pencil or scribing instrument built into the tracing stand directly beneath the dot traces the contour in its correct planimetric position. Planimetry, such as the location of a road, may be plotted from the model by tracing the feature and at the same time continually adjusting the floating dot so it is always in contact with the surface.

A recent development by the United States Geological Survey alleviates some of the problems inherent in systems requiring colored filters to produce the three-dimensional model. A new system known

as the Stereo Image Alternator permits an operator to view the two photographs without the need for colored spectacles.

Rotating shutters mounted in front of each projector interrupt the light at a rate of sixty times per second. They are synchronized so as to flash alternating images of the left and right photographs. The operator, viewing through shutters synchronized with those of the projectors, sees the right-hand photograph only with the right eye, and the left with the left eye, thus producing the stereoscopic three-dimensional model.

Advantages of this system over anaglyphic systems are as follows.

1. Much more light reaches the platen.
2. Because the two images may be matched in sharpness, resolution is greatly improved.
3. The operator sees no subimage (intended for the other eye) since the images are completely separated.
4. Color photographs can be used with the system. This capability, along with recent improvements in the resolution quali-

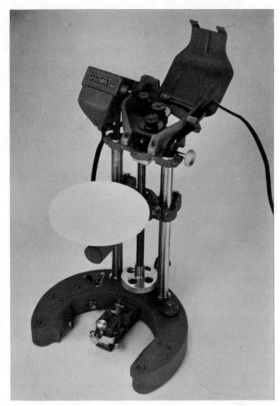

Figure 4.26 Balplex Universal Foot Reading Tracing Table. (Courtesy of Bausch and Lomb, Incorporated, Special Products Division.)

ties of color film, suggest that color photography may be used much more for map production in the future.

Orthophotoscope

The perspective image of a photograph can be changed to an orthogonal* projection by use of a device called an orthophotoscope.† The new picture may be made free of tilt, and thus all points may be placed in their correct locations relative to one another. Scale, then, is constant and all angles are true. The usefulness of such a photograph is quite obvious. Areas measured on it will be correct, and angles on the photograph will match angles in the field.

Because of the great amount of detail on photographs, in some instances the orthophotograph can be more useful than a topographic map for work in the field. For example, an image on a photograph can often be correlated with the corresponding ground feature much more easily than can a map symbol. Furthermore, the correct planimetry of an orthophotograph permits precise location of features that are not and often cannot be symbolized on a map.

Although there are now a variety of instruments that can produce orthophotographs, most operate on the basic principle of the first one developed in 1953. The projection of a three-dimensional stereoscopic image produced by a stereoscopic plotter is exposed to a film through a very small opening moved across the model. As the tiny aperture moves along a narrow strip, the film being exposed remains stationary in its horizontal position but is moved in the vertical dimension to keep the aperture "in contact with the surface" of the three dimensional image. The operation is much the same, then, as tracing correct planimetry from a model by continually adjusting the floating dot to keep in contact with the surface of the stereo model. After the aperture has moved across the model once, it is moved sideways a distance equal to the width of the opening.

Orthophotographs can, of course, be assembled to form a mosaic, which, in turn, can be overprinted with map symbols to produce an orthophotomap. Photomaps, prepared from mosaics of conventional photographs, have long been used as map substitutes, but the displacement inherent in the photos, however, caused scale dis-

---

*The new image is orthogonal in theory; in practice it can only closely approach a truly orthogonal projection.

†Russell K. Bean, "Development of the Orthophotoscope," *Photogrammetric Engineering*, **21** (4), 529-535 (1955).

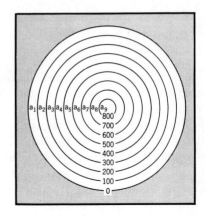

SYMBOLIZING AND PROCESSING DATA

Maps may satisfy many requirements, ranging from serving as useful devices on which spatial information may efficiently be recorded to functioning as scientific tools in assisting in the development of geographical hypotheses. In general terms, however, they always function as media of communication. The cartographer constructs his map for the purpose of communicating to the reader (it may be himself) some geographical concepts, locations, distributions, and relationships. To do so he must encode the facts and concepts he wishes to express, just as he must do when he wishes to convey information by writing, by the media of mathematics, or by the distinctive sound combinations of the spoken word. Cartography employs a graphic language made up of various kinds of marks, hues, lines patterns, tones, and so on, and as an author the cartographer must learn to manipulate these map terms to achieve the best possible communication.

Because of the constraints that result from the everpresent scale reduction and the need for clarity and legibility, the geographical data, concepts, and relationships being mapped cannot be shown as they appear in reality to the eye but, instead, must be symbolized. This is one of the major ways in which a map differs from an air photograph. The variety of data on a map are selected and symbolized (coded) in order to present clearly a chosen set of relationships. On the other hand, an air photograph, although also reduced in scale, is merely a mechanical record, unsymbolized and mechanically selected, of what is "seen" by the camera. One of the major duties of the cartographer, then, is to understand the relations among symbols, their relative effectiveness, and their relative suitability for the purpose for which the map is being constructed.

CARTOGRAPHIC SYMBOLISM

A map is a system for coding a variety of kinds of information so that a trained reader can efficiently retrieve it. In addition to the variety of factual data that is normally included in a map, there is always a considerable amount of derivative information that arises because of the spatial relationships revealed by the locational arrangements of the primary data. This too is communicative information. The basic classes of marks and their numerous variations used for cartographic communication constitute a large body of symbols and method. Because the symbolism of cartography has developed over many centuries, numerous traditions and conventions regarding map representation have grown up, and in a sense, the symbolism of cartography has become somewhat standardized. On the other hand, the great possibilities for variation have effectively prevented any really rigid standardization, although in the case of large-scale reference maps, standardization is more nearly approached. Conversely, the cartographer who works with a thematic map must of necessity be more critical and imaginative so as to adjust the representation and symbolism to the special purpose of his map, since his precise objectives are likely to be unique.

In the following sections of this chapter the fundamental differences in the characteristics of the basic classes of geographical data and the kinds of symbols that may be used for their representation are considered. The various kinds of statistical concepts frequently employed in cartography or that underlie cartographic symbolism are dealt with in the balance of the chapter. The practical application of the principles of cartographic symbolism are dealt with in later chapters.

Geographical Phenomena and Their Scaling

Anything that is anywhere, concrete or abstract, is a geographical phenomenon, since the notion of relative position on earth (or, by extension, any similar body) is central to the concept of geography. Geographical phenomena can be classified in four categories as being *point* (nondimensional), *line* (one-dimensional), *area* (two-dimensional), and *volume* (three-dimensional).[*]

A point is simply a place or location; a line is anything that is elongated, narrow, and rather uniform; an area conceptually is a two-dimensional extent that is nonlinear; and a geographical volume is a three-dimensional phenomenon that occurs over geographical area. The geographical variable in question can be thoroughly tangible: a location (a point), a boundary (a line), a forest (an area), or a depth of precipitation (a volume) can all exist. On the other hand, except for location they may be quite abstract: a line may be the route along which ideas travelled or the average position of a weather front; an area may consist of the two-dimensional extent of a political preference; and a volume can consist of an intangible quantity such as the concept of number of persons per square mile, that is, population density.

We tend not to be very systematic in the way we deal with such matters because we often put the same geographical phenomenon in different categories depending upon how we may be thinking about it. For example, we may conceive of New York City as a place (in contrast to Philadelphia), or we may think of it as an area (as opposed to the surrounding, less populated region), or we may think of it as a "volume" of humanity. Nevertheless, it is not difficult to classify all geographical phenomena in one or more of these categories.

When dealing cartographically with geographical points, lines, areas, and volumes, we must always perform two operations: locate these variables, of course, and, in order to symbolize differences among the elements of a class, such as lines, or the variations of a single variable, classify or order the differences according to some system. The basic methods of differentiating data are to classify them according to *nominal, ordinal,* and *interval* scales.

Nominal ordering occurs when we classify without ranking. Sometimes this is termed qualitative. Nominal classes are mutually exclusive; examples would be the differentiation between forest and grassland (area), places *A* and *B*, (points) railroads and roads (lines).

Ordinal scaling involves not only classifying but ranking as well. For example, we can differentiate major ports from minor ports, intensive from extensive agriculture, among small, medium and large cities, hot and cold temperatures, and so on. It enables the map reader to tell that one is larger or smaller, more or less important, younger or older, and so on, but it does not provide any specific magnitude assignment.

Interval scaling adds the information of distance between ranks to the process of classifying and ranking. For interval scaling we must employ a standard unit and then express the amount of difference in terms of that unit. For example, we differentiate among temperatures by using

[*]John K. Wright, " 'Crossbreeding' Geographical Quantities," *Geographical Review,* **45**, 52-65 (1955).

a standard unit, the degree (F° or C°), among population densities by numbers of persons per square mile, and among differences in elevation by units of linear measure such as the meter or the foot.*

We may cross-classify the classes of geographical phenomena, on the one hand, and the manner in which they can be scaled, on the other, as shown in Figure 5.1. Such a simple cross-classification is useful as a conceptual framework for cartographic symbolization. If we read down under each category of geographical data we will note the change from class, to class and rank, and finally to class and measurable characteristics. The simple diagram shown in Figure 5.1 reduces the complex of geographical phenomena to its basic elements and shows in principle how the cartographer must logically approach the problem of scaling in order to communicate the essential characteristics of the mapped variables.

It would be comparatively simple if we could merely design symbols that could fit into each cell of the above cross-classification, but unfortunately it is not that easy. In the first place, there are only three basic kinds of marks that can be put on flat paper: points (dots, etc.), lines of various sorts, and area symbols (colors or patterns, for example). These three classes

	Point	Line	Area	Volume
Nominal	City	Road	Soil type	Precipitation
Ordinal	Large city	Major road	Good quality	Heavy precipitation
Interval	Total population	Load limit	Exchange capacity	No. of inches per year

Figure 5.1   A cross classification of (1) kinds of geographical phenomena and (2) scaling categories. Under each heading of class of geographical phenomenon, an entry example shows the kind of characteristic that would be identified according to the scaling category.

of symbols must somehow be made to represent the four classes of geographical data—point, line, area, and volume. In the second place, we can make a symbol do double duty and represent, for example, both nominal and interval characteristics at once. Consequently, a simple structuring of the visual "grammar" and graphic "vocabulary" of cartography is not possible.

In the immediately following sections of this chapter, attention will be paid primarily to the classes of marks that constitute cartographic symbols, other than base data. The specific manner in which these marks can be employed to communicate geographical information scaled in different ways will be dealt with in Chapter 6 and 7.

Kinds of Symbols

As was observed above, map symbols may be grouped into three major categories, namely, *point, line,* and *area* symbols, and within each of these classes there may be numerous variations. As was also pointed out, the cartographer is called upon to represent four classes of data, but he has only three classes of marks with which to do it. Needless to say, the discrepancy between the number of kinds of geographical variables and the number of classes of cartographic symbols available to represent them causes confusion to the beginner. This will be considerably clarified if, when thinking about the representation of geographical data, he takes care to maintain a clear distinction between the kind of geo-

*A fourth kind of scaling is generally recognized, called *ratio* scaling. This involves relating the numerical values of an interval scale to absolute magnitudes, such as is done in the Kelvin thermometric scale. For most cartographic purposes this is either not important or the scales are inherently interval-ratio. For a clear, introductory explanation of nominal, ordinal, and interval scaling systems, their analysis and manipulation, see Linton C. Freeman, *Elementary Applied Statistics,* John Wiley and Sons, Inc., New York, 1965.

graphical quantity with which he is concerned (e.g., linear *data*) and the kind of map symbol (e.g., line *symbol*) he may use to represent it: he must understand, for example, that point data need not always be shown by point symbols.

Almost as important as the inherent qualities of a symbol is its design, since, as has often been stated (but less often heeded), the map symbol that cannot be seen or read is wasted and is but a useless encumbrance. The appearance of a visual item may be varied in several ways: by size, shape, and color contrasts. Dots may be large or small, regular or irregular, dark, light, or colored. Lines may vary similarly. As is pointed out in Chapter 11, visual contrast of any of these qualities or combinations of them is the key to visibility. The possibilities of variation are large indeed, and the cartographer would do well to exercise his ingenuity and experiment with various alternatives before settling upon any one design or kind of symbol.

Point Symbols. A point symbol is any kind of mark or device that by its graphic characteristics refers more to a "place" than it does to a line or area. Obviously, in order to be seen any symbol must occupy area on the map, and sometimes point symbols actually cover considerable cartographic space; nevertheless, when the characteristics being symbolized are located at a "place" or are aggregated or summarized for a "location," even if that be a state or nation (e.g., the population of France), the device usually employed is called a point symbol.

The vocabulary of point symbols is large and differentiation among them is accomplished by variations in (a) basic shape or visual character (e.g., a square as contrasted with a circle), (b) color, and (c) size. Point symbols may represent any phenomenon having territorial extent or simply location. A dot may represent a city; a triangle, a triangulation station; a circle, the population of a city or the production of an industrial plant. The variations and uses are legion. A point symbol may represent either simply the nominal quality of a place, as, for example, a city or an airport, but the symbolization may be extended to include ordinal and interval scaling. Figure 5.2 is an illustrative outline with some examples of the various combinations possible. Normally we combine only nominal and either interval or ordinal characteristics; it is possible, however, to combine all three (not shown in Fig. 5.2). For example, a point symbol representing the location of a city might be assigned to a nominal category (e.g., industrial) by its color, to an ordinal category (e.g., major) by its shape, and to an interval category (e.g., so many inhabitants) by its size. Such complexity in coding becomes difficult for the average map reader to understand, and in ordinary circumstances would probably not constitute effective communication. Although there is an almost unlimited number of ways point symbols may be used, this kind of device is not adaptable for showing directly a geographical ratio or relationship.

Line Symbols. In elementary terms the graphic characteristics of a line immediately suggest several possibilities: connectivity of some sort from one end to the other, similar characteristics along its course, and differences from one side to another, to name but a few. The line symbol is most versatile and is widely used in cartography. Like point symbols, lines can be used to represent nominal, ordinal, and interval characteristics of geographical data or combinations of them. These are graphically outlined in Figure 5.3. The basic vocabulary for lines also consists of differences in character or shape (e.g., smooth versus angular or dotted versus continuous), color, and size, i.e., width.

The most obvious use of lines as carto-

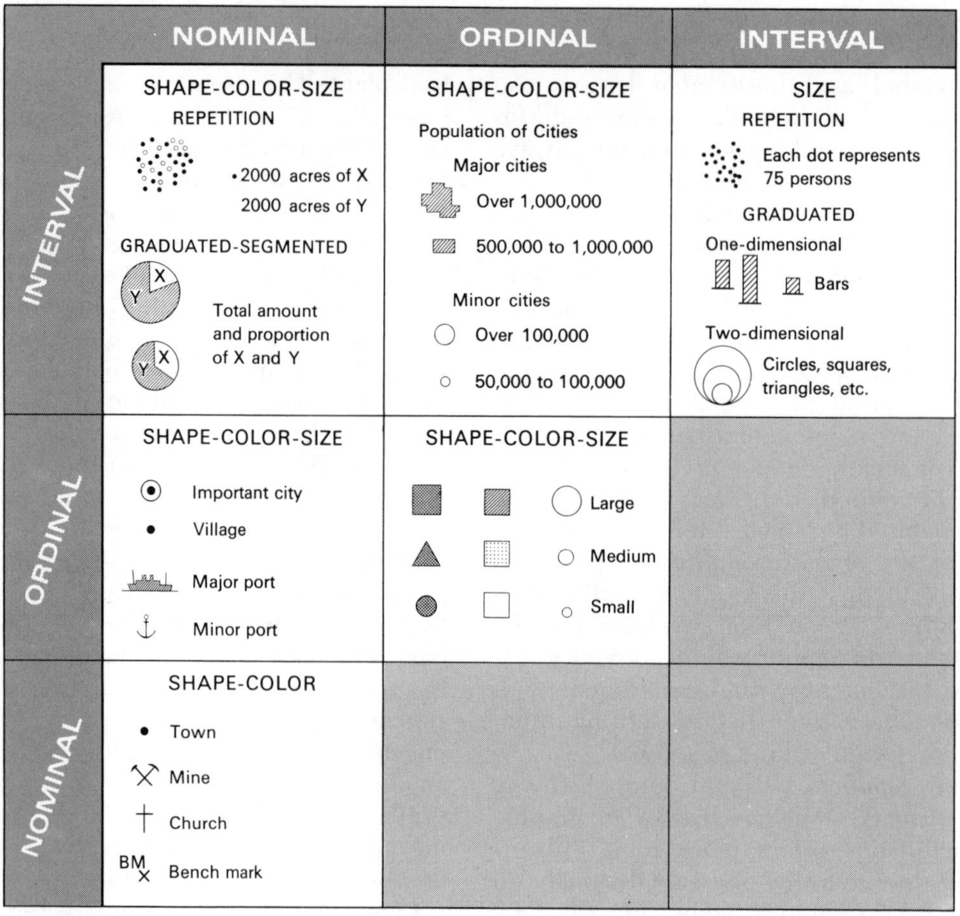

	NOMINAL	ORDINAL	INTERVAL
INTERVAL	SHAPE-COLOR-SIZE REPETITION • 2000 acres of X 2000 acres of Y GRADUATED-SEGMENTED Total amount and proportion of X and Y	SHAPE-COLOR-SIZE Population of Cities Major cities Over 1,000,000 500,000 to 1,000,000 Minor cities Over 100,000 50,000 to 100,000	SIZE REPETITION Each dot represents 75 persons GRADUATED One-dimensional Bars Two-dimensional Circles, squares, triangles, etc.
ORDINAL	SHAPE-COLOR-SIZE ⊙ Important city • Village Major port ⚓ Minor port	SHAPE-COLOR-SIZE Large Medium Small	
NOMINAL	SHAPE-COLOR • Town ✕ Mine † Church BM✕ Bench mark		

Figure 5.2 Illustrative outline of the various ways *point* symbols can be employed to portray nominal, ordinal, or interval differentiation or combinations of them. The bolder lettering in each cell suggests how the symbolism may be varied so that the desired differentiation may be communicated. Only a few of the many possibilities are shown as examples.

graphic symbols is to portray the nominal characteristics of linear geographical phenomena, such as rivers, roads, or boundaries. Such lines are fundamentally alike in that they commonly represent the nature of a feature without indicating any ranking or measurable characteristics. They may be used for many kinds of purposes. For example, a line may represent a road and show exactly where it lies on a topographic map, whereas at a smaller scale it may be drawn so as merely to show that two places are connected by a road. On the other hand, another kind of line (e.g., a boundary) may be employed to separate unlike areas. Thus, a line can be used to join things, a line along which there is some constant quality; or it may be used to separate things, a line along which there is no constant quality.

Ordinal and interval differentiation of linear or connective phenomena is accomplished in a variety of ways. Traffic movement, commodity flow, and migration are commonly represented by lines of varying width or graded visual importance and are generally called flow lines. An arrow alone may represent nothing but movement, but by varying width or some other design characteristic we may symbolize differences in actual movement or merely in some quantitative characteristic that does not involve movement, such as roadbed ca-

pacity. As a matter of fact, the tapering line of a river on a map shows both the nominal quality of the feature, a river as opposed to a boundary, but by being tapered it also implies the ordinal aspects of being "more" in one area as opposed to "less" in another.

A more complex use of the line is as a symbol along which the same value of some measurable phenomenon exists or is assumed to exist. Usually used in sets, this kind of line symbol actually portrays phe-

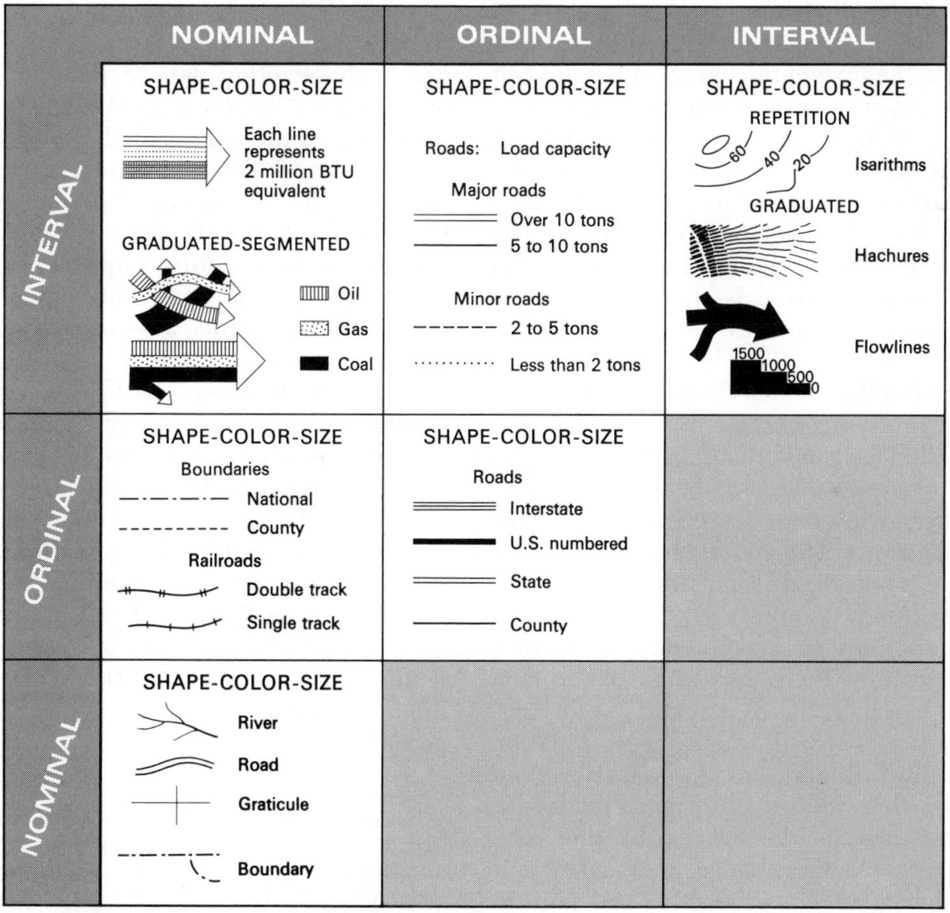

	NOMINAL	ORDINAL	INTERVAL
INTERVAL	SHAPE-COLOR-SIZE Each line represents 2 million BTU equivalent GRADUATED-SEGMENTED Oil Gas Coal	SHAPE-COLOR-SIZE Roads: Load capacity Major roads Over 10 tons 5 to 10 tons Minor roads 2 to 5 tons Less than 2 tons	SHAPE-COLOR-SIZE REPETITION 60 40 20 Isarithms GRADUATED Hachures Flowlines 1500 1000 500 0
ORDINAL	SHAPE-COLOR-SIZE Boundaries National County Railroads Double track Single track	SHAPE-COLOR-SIZE Roads Interstate U.S. numbered State County	
NOMINAL	SHAPE-COLOR-SIZE River Road Graticule Boundary		

Figure 5.3 Illustrative outline of the various ways *line* symbols can be employed to portray nominal, ordinal, and interval differentiation or combinations of them. The bolder lettering in each cell suggests how the symbolism may be varied so that the desired differentiation may be communicated. Only a few of the many possibilities are shown as examples.

nomena differentiated on an interval scale, but in many instances the primary communication is ordinal differentiation. For example, the best known of such linear symbols is the line of equal elevation above sea level, best known by its common name, the contour. The interval scaling of elevation (point data) throughout an area is shown by the positions of the individual contours, but their relative spacing also graphically indicates ordinal categories of slopes or gradients of "more" or "less" (By measurement and calculation, interval scaling of these data is also obtainable.) Although first used in the sixteenth century, they did not come into widespread use until the eighteenth. Edmund Halley's map (1701) of "curve lines" joining places with the same compass declination (Fig. 1.7) popularized this line symbol, and since then many other kinds of data, varying in amount from place to place, have been symbolized by these kinds of lines. To the prefix *iso* (Gr. *isos,* equal) is commonly added the term describing the phenomenon being mapped. Thus, lines of equal compass declination are called isogones (Gr. *gonia,* angle), and lines of equal temperature are called isotherms (Gr. *therme,* heat); there are scores of such terms. The generic term, or the collective name of all such lines, is isarithm (Gr, *arithmos,* number).

In addition to symbolizing at various map locations the point values of the data being mapped, the systematic use of isarithms in sets also allows the representation of a third dimension. As was observed above, this allows the map reader to deduce certain derivatives of the relation between the change of quantity and the change of place, such as the ratio of quantity change with distance, that it, gradient.

Area Symbols. Like point and line sym-bols, area symbols can be employed to portray nominal, ordinal, and interval scaling of geographical phenomena or combinations of these classes. Figure 5.4 illustrates in graphic outline form examples of the use of area symbols. The design of many area symbols, especially those used to communicate nominal categories, has become conventional through long use.

The graphic vocabulary of area symbols consists of variations in color and pattern. Color is a very complex phenomenon and is treated more fully in Chapter 11; suffice it here to observe that colors (including gray) can be either dark or light, subdued or brilliant. Variations from dark to light tones are commonly called *value* differences and variations in brilliance are commonly called *intensity* differences (see Chapter 11). When we are cartographically classifying geographical phenomena on interval or ordinal scales, the higher ranks and greater amounts are normally represented by darker values or greater intensities. Sometimes, in order clearly to differentiate among a large number of classes, it is necessary to employ some sort of repetitive marks, which collectively form what is called a visual pattern. Patterns may consist of lines or dots or various other combinations of marks.* Nominal differentiation is best symbolized by variation in pattern (e.g., lines versus dots) if the map is monochromatic or, if not, by differences of hue (e.g., brown versus green).

---

*Simple parallel lines tend to cause the viewer to have disturbing eye movements (see Chapter 11), and it is generally good practice to use parallel lines sparingly. Dot patterns are visually more stable and differentiate areas more clearly. Naturally, when a large number of patterns is required, parallel lines must occasionally be utilized. They should be surrounded by other kinds of patterns are frequently as possible.

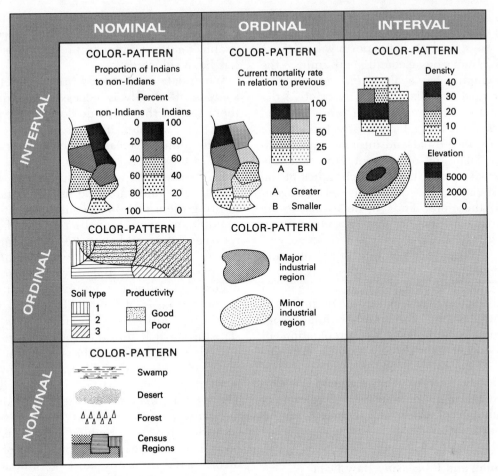

Figure 5.4 **Illustrative outline of the various ways *area* symbols can be employed to portray nominal, ordinal, and interval differentiation or combinations of them. The bolder lettering in each cell suggests how the symbolism may be varied so that the desired differentiation may be communicated. Only a few of the many possibilities are shown as examples.**

STATISTICS IN CARTOGRAPHY

The most significant thing that has happened to cartography, as it has to many fields of learning in recent years, is the increasing importance of quantitative descriptive and analytical methods. Distribution maps have long been largely statistical, but their sophistication has increased steadily. Cartographers are now called upon to map residuals from regression, correlation coefficients, and many other consequences of the application of statistical analytical methods to geographical problems.

An ever increasing amount of a cartographer's effort is devoted to mapping data that require statistical manipulation in order to obtain a variety of descriptive measures. Much of the data with which he works consist of samples obtained in various ways, and realistic portrayal of these data cannot be accomplished unless the cartographer has a clear understanding of their relative reliability as descriptive devices. Furthermore, in the manipulation of the data, in the selection of categories, in the use of ordinal and interval scales, and in the planning of the symbolization, the map maker must use the techniques of the statistician. The accomplished cartographer, therefore, finds it necessary to be familiar with some of the basic concepts of statistical method.

It is clearly not possible in an introductory book, where the main theme is cartography rather than statistics, to go into much detail concerning the variety of statistical techniques the cartographer might find useful. Nevertheless, the preparation of distribution maps usually requires familiarity with the general concepts involved. The student is urged to acquire a good book on statistical method to use as a reference. A number of titles are included in the bibliography.

Refining and Processing Raw Data

The basic approach to refining and processing data for distribution maps is by way of the kinds of symbols or techniques that may properly be used to convey the information. The vocabulary of symbols was discussed in the preceding sections of this chapter, and it is only necessary here to reiterate the fact that the cartographer has a wide choice indeed. Consequently, his first action, after selecting or preparing the map base, is to decide upon the mode of presentation. He must decide if it is to be a static (all the data as of one period or place) or a dynamic (change of time or place) map, and he must select the symbolism (point, line, or area, or combination).

The next step is to process the crude statistics in such a fashion that they become usable for the system of presentation that has been selected.

When data are obtained from a variety of sources, it usually is necessary to equate them so that they provide comparable values. For example, different countries use different units of measure such as long tons or short tons, U. S. gallons or Imperial gallons, hectares or acres, and so on. Frequently the units must be further equated to bring them into strict conformity. If, for example, we were preparing a map of fuel reserves, it would not be sufficient to change only the volume or weight units to comparable values, but it would also be necessary to bring the figures into conformity on the basis of their BTU ratings. It is also frequently necessary to process the statistics so that unwanted aspects are removed. A simple illustration is provided by the mechanics of preparing a density map of rural population based on county data. The total populations and areas of incorporated divisions as well as the totals for countries may be provided in census tables; if so, the areas and populations of the incorporated divisions must be subtracted from the county totals. Another illustration is provided by the well-known regional isothermal map. If the objective of the map is to reveal relationships among temperatures, latitudes, and air masses, for example, the local effects of elevation must be removed from the reported figures. This involves ascertaining the altitude of each station and the conversion of each temperature value to its sea-level equivalent.

After the statistical data have been made comparable, the next step is to convert them to mappable data. This may, of course, not be necessary for many maps, such as the isothermal map referred to above, because the data need only to be plotted and isarithms drawn. On the other hand, ratios, per acre yields, densities, percentages, and other sorts of indices must be calculated before plotting. The employment of variations in sizes of some kinds of symbols, such as graduated circles or squares, requires the determination of logarithms or roots.

A desk calculator that can add, subtract, multiply, and divide is of constant use to a cartographer working with distribution and statistical maps. Its operation is relatively simple. If a great number or if more difficult calculations are required, it is well worth the initial tabulating time to process the data with the aid of punch cards for subsequent machine processing. An increasing amount of data of all kinds is becoming available in tape and punch-card form, and the cartographer need only obtain copies of the "decks" of cards containing the necessary information. The entire processing can be done by programming in a computer, and the result can be printed tabulations that need only be plotted. Effort spent in initial search for data in the form easiest to process will often save considerable total expenditure of time. Furthermore, rapid progress is being made in introducing the cartographic method into the capabilities of the computer, and automatic plotting machines and other devices for producing graphic results are already widely used. Crude distribution maps now can be prepared automatically, including some kinds of isarithmic maps (see Chapter 7). If these facilities are available to the cartographer, he should become familiar with their capabilities.*

Other than electric desk calculators, probably the most frequently used aids are the slide rule and mathematical tables of roots, logarithms, and conversions. For many purposes, quite acceptable accuracy can be attained with the slide rule, and percentages, ratios, division, and multiplication can be computed almost as fast as

the figures can be read. The manipulation of the slide rule for these purposes can be learned in a short time. Squares, cubes, and roots can also be derived from the slide rule, or can be determined from mathematical tables such as those in Appendix C.

If a large number of measurements or calculations involves the same kinds of units on a variable scale, it is frequently a saving of time to prepare a nomograph such as is illustrated in Figure 5.5. Nomographs should be reserved for calculations involving two or more variables. Figure 6.6 is an example of a nomograph involving four related variables.

Absolute and Derived Quantities

One of the most important changes in cartography during the past two-hundred years has been the introduction of thematic cartography. Man became conscious of the variety of phenomena on the earth even before the flowering of the Classical Era, and it was only natural that he should also have been interested in variations in amount from place to place. His observations were generally so incomplete and unreliable, however, that it was not possible to map the results in any consistent fashion until comparatively recently. With the development of accurate enumeration and observation and the rise of statistical method, ways of mapping quantitative distributions quickly developed. Early thematic cartography during the first part of the nineteenth century was largely concerned with these kinds of maps. With but very few exceptions (one is the isarithm), most of the methods of portrayal by point, line, and area symbols were devised and

*See Waldo R. Tobler, "Automation in the Preparation of Thematic Maps," *The Cartographic Journal,* 2 32–38, (1965):, and J. C. Robertson, "The Symap Programme for Computer Mapping," *Ibid.,* 4 108–113, (1967).

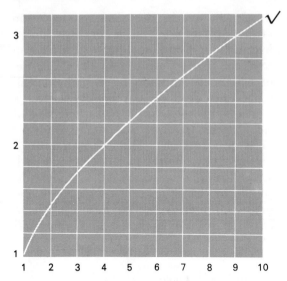

Figure 5.5 A simple nomograph to provide plotting values. The nomograph in this example is constructed to show the square roots of the plotting range on the X axis for constructing graduated squares. Of course, the range may extend to any limit. The vertical dimension is scaled to fit the actual drawing scale. Thus, all that is necessary is to find the appropriate point on the X axis, and the actual distance on the graph from the base to the curve is the proper map distance. When a large number of values need to be plotted, it usually saves time to make up such a nomograph. One can be made for any series that varies systematically.

employed since 1800 to portray ordinal and interval categories of geographical distributions. Each year there is a greater production of these kinds of thematic maps and it is to be expected that the rate of increase will continue to rise. Consequently, it is imperative that the cartographer become adept at selecting appropriate methods of communicating quantitative data.

All quantitative maps fall in one of two classes: they portray either absolute, observable quantities on the one hand, or calculated or relative, that is, derived, quantities on the other. Examples of the first class are maps showing variations in the numbers of people, the production or consumption of goods, or even the elevations of the land surface above sea level. The quantities are simply those observed concerning a single nominal class, and they are expressed on the map in relative or absolute terms according to ordinal or interval scaling. In this group many combinations are possible, and several kinds of values may be presented at once; but in no case are the data expressed as a relationship, except as the amount or kind of data may vary from place to place, thus providing the basis for a kind of visual correlation.

In the second group of quantitative maps are those which portray derived values (for example, averages, percentages, and densities); on these, the mapped values express either some kind of summarization or some sort of relationship between two or more kinds of data. This group, showing derived rather than absolute values, includes four general classes of relationships.

Averages. The first, and probably the most common, are the maps of averages or measures of central tendency, such as are obtained by the reduction of large amounts of statistical data, for example, weather observations, well records, or numbers and sizes of farms. In this group there are several common kinds of averages such as the mode, the median, and the mean. These will be considered in a bit more detail later in this chapter, but it is worth pointing out here that generally the most important in cartography is the mean. It is usually symbolized by $\overline{X}$ in the following equation

$$\overline{X} = \frac{\Sigma X}{N}$$

where ΣX is the summation of all the X values, and N is the number of occurences of X.

Ratios. The second class includes all those maps of *ratios,* such as proportions, percentages, or rates, in which some element of the data is singled out and compared to the whole. These are illustrated by maps such as those showing the percentage of rainy days, the proportion of all cattle that are beef cattle, mortality per 1000 persons, or the rate of growth or decline of some phenomenon. In this group the numerical value mapped will ordinarily be the result of one of the following basic kinds of operations:

$$\text{a ratio} = \frac{n_a}{n_b}$$

$$\text{a proportion} = \frac{n_a}{N}$$

$$\text{a percentage} = \frac{n_a}{N} \times 100$$

where n_a is the number in one category, n_b the number in another category, and N the total of all categories.

As used by the cartographer such ratios, proportions, and percentages commonly have the characteristics of a spatial average, since 12 persons per square mile is a ratio obtained by dividing the total number of people by the total number of square miles. (This kind of ratio is the basis of the familiar density concept treated below.) Rates and ratios, of course, need not be quantities related to the land. For example, assume the numerator were the total number of dairy cows in a county: we could use as the denominator such quantities as the number of farms, the number of farm operators, or the total number of cattle. These would result in the ratios of the number of dairy cattle per farm or per farm operator, or the proportion of dairy cattle to total cattle.

Maps of these kinds of relative quantities are made to show variations from place to place in the relationship mapped, and they are usually prepared from summations of statistical data either over area or through time. The appropriateness of the quantity obviously depends upon the specific use to which it is being put, but a few words of caution are in order. Percentages, rates, and ratios, when mapped on the basis of enumeration units, are usually assumed by the map reader to extend more or less uniformly throughout the enumeration unit. If the phenomenon does not in fact so occur, then the ratio mapped may be quite misleading, just as it may if too few of the items occur. Thus, a value of 100% of farms with tractors could be the result of only one farm with one tractor in a large area.

Quantities that are not comparable should never be made the basis for a ratio. For example, the number of tractors per county is next to meaningless, and we ought not even calculate the number of tractors per farm by dividing the total number of tractors by the number of farms in a county unless the farm sizes (or some such significant element) are relatively comparable. Common sense will usually dictate ways to insure comparability.

Densities. The third class of maps of related quantities consists of those commonly called *density* maps. These maps are employed when the major concern is focused on the relative geographical spacing of discrete phenomena. Examples are maps of the number of persons (or trees, cows, etc.) per square kilometer or mile, or the average spacing between phenomena such as service stations or mail collection points. The density (D) is computed by

$$D = \frac{N}{A}$$

where N is the total number of phenomena occurring in an enumeration unit, for example, a county, and A is the area of the unit. The average spacing between phenomena, another way of looking at density, is computed by

$$\bar{S} = 1.0746 \sqrt{\frac{A}{N}}$$

where $\bar{S}$ is the average spacing of the items or the mean distance between them in linear values of the same units used for A when an hexagonal spacing is assumed, this being the most economical of area. If we assume simply a square arrangement, the interneighbor interval, as it has been called, is simply the square root of the reciprocal of the population density expressed in suitable linear units.

The density class is more closely related to the land than the first two, and the significant element in the relationship is area. Thus, for example, 5000 persons in an area of 100 square miles is a density of 50 persons per square mile; if arranged hexagonally within the 100 square miles, each individual would be about 0.15 mile from his neighbors, and if arranged rectangularly each individual would be slightly closer. In many instances, a density value derived from the (1) total number within and (2) the total area of a statistical unit is not so significant as one that expresses the ratio between more closely related factors. For example, the relation of the number of people to productive area in predominately agricultural societies is frequently found to be more useful than is a simple population to total area ratio. If the data are available we can easily relate population to cultivated land, to productive area defined in some other way, or to that spatial segment that is important to the objective of the analysis.

When working with densities and average spacings, the cartographer is limited in the detail he can present by the sizes of the statistical units for which the enumeration of the numbers of items has been

made. As a general rule, the larger the units (such as townships, counties, or states) the less will be the differences among the values. In many cases the initial data must be supplemented by other sources in order to present a distribution as close to reality as possible.

Estimating Densities of Parts. When preparing density maps it is not uncommon for the raw data to have been gathered for enumeration districts within which there is considerable unevenness of distribution. This is not apparent from the raw data, of course. If supplementary data exist which allow us to estimate with reasonable accuracy the density in one part, the value to be assigned to the other can easily be calculated. The following description is essentially as prepared by John K. Wright.[*]

Assume, for example, an enumeration unit with a known average density of 100 per square mile. Assume, further, that examination of other evidence has shown that this unit may be divided into two parts, m, comprising 0.8 of the entire area having a relatively low density, and n, comprising the remaining 0.2 and having a relatively high density. If, then, we estimate that the density in m is 10 persons per square mile, a density of 460 to the square mile must be assigned to n in order that the estimated densities m and n may be consistent with 100, the average density for the enumeration unit as a whole.

The figure 460 for the density in n is obtained by solving the following equation:

$$\frac{D - (D_m a_m)}{1 - a_m} = D_n$$

or

$$\frac{100 - (10 \times 0.8)}{0.2} = 460$$

[*]John K. Wright, "A Method of Mapping Densities of Population with Cape Cod as an Example," *The Geographical Review,* **26**, 103–110, (1936).

where D is the average density of the unit as a whole, D_m the estimated density in m, a_m the fraction of the total area of the unit comprised in m, $1 - a_m$ the fraction comprised in n, and D_n the density that must accordingly be assigned to n.

D_m and a_m are estimated approximately. It is not necessary to measure a_m accurately, since the amount of error in a rough estimate is likely to be less than the amount of error in the best possible estimate of D_m.

Study of neighboring areas sometimes gives a clue to a value that may reasonably be assigned to D_m. For example, other maps may show what would appear to be similar types of distribution prevailing over D_m, a part of the unit, and over the whole of E, an adjacent unit. It would be reasonable, therefore, to assign to D_m a density comparable with the average density in E.

Having assigned estimated but consistent densities to two parts of an area, we may then divide each or one of these parts into two subdivisions and work out densities for the latter in the same manner; and the process may be repeated within each subdivision.

The method is merely an aid to consistency in apportioning established values within the limits of territorial units for whose subdivisions no statistical data are available. The method can easily be applied in the mapping of any phenomena for which statistics are available only by larger units. The table in Appendix E enables us to solve the fundamental equation without either multiplication or division. Figure 5.6 shows the refinement that can be made.

Potential. The fourth class of distribution

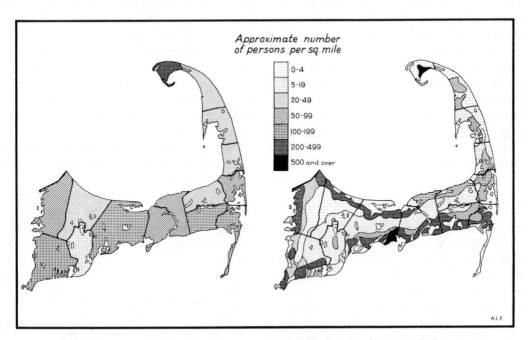

Figure 5.6   The left-hand map shows the density according to whole census enumeration units, and the right-hand map shows the refinement that can be developed by the system described in the text for estimating densities of parts. (Redrawn from *The Geographical Review*, published by the American Geographical Society of New York).

maps comprises those called *potential* maps. These maps assume that people and some kinds of human activity interact or influence one another, directly with the numbers or magnitudes involved and inversely with the distance between them. Because this assumption is derived from the physical laws governing the gravitational attraction of inanimate masses, it is called a gravity concept. It has been applied to such things as population, prices, and other economic and cultural elements.

The value of the potential at any point is the sum at that point of the influence of all other points upon it plus its influence on itself. The potential P of place i of phenomenon X will be

$$P_i = X_i + \sum \frac{x_j}{D_{ij}}$$

where x_j is the value of X at each of the places involved and D_{ij} is the distance between place i and j. The preparation of a potential map requires that this summation be repeated for each place. The resulting map values will be in numbers of units of X at particular points. It is apparent that the relationship can easily be amended by inserting a constant in the equation or by defining in special ways the distance between the places.

Averages and Indexes of Variation

Often in cartography we are called upon to map summaries of distributions. These summaries may refer to spatially distributed phenomena that occur throughout a "statistical" unit area, for example, a census enumeration district, or to a series of observations at one place taken through time. Always when dealing with summaries of observations (over area or through time), we should also be concerned with the amount of variation that exists in the set of observations being summarized.

The cartographer's concern with the na-ture of averages and their quality, as shown by their indexes of variation, extends beyond the simple facts. In communicating geographical characteristics by way of a map, the cartographer has numerous opportunities to vary the manner of presentation so as to portray properly the character of the distribution. For example, distribution maps may "look" very precise, or they may be made to look general, or titles and legends may be designed to draw the map reader's attention to the critical characteristics of the data so that he will not be misled. In order that he can properly communicate the geographical facts, the cartographer must understand the nature of the distributions with which he is working.

There are various kinds of averages, or measures of central tendency, and a larger number of indexes of variation. One is more appropriate than another depending primarily upon the manner in which the variable being mapped is scaled. The study of these measures and their calculation is more appropriate in a course in statistical method or from a book that has that as its primary aim.[*] Only the more important will be mentioned here, and any extended discussion will be limited to the mean and the standard deviation, generally the most used and significant in quantitative distribution mapping.

Table 5.1 shows the most appropriate average with which to summarize a distribution scaled in a certain way, together with a commonly used associated index of variation.

Table 5.1

Scaling Method	Average	Index of Variation
Nominal	Mode	Variation ratio
Ordinal	Median	Decile range
Interval	Mean	Standard deviation

[*]Reference is again made to Freeman, *op. cit.*, Section B, pp. 29–67.

In a nominal distribution the mode is the class that occurs most frequently. For example, if we were mapping the distribution of forests, grasslands, shrublands, etc. on the basis of data acquired by small areas, we would expect that each small mapping unit (area) would contain some of each class, such as grass on the uplands, forest in the river valleys, and so on. We would assign to each such small unit the quality of the modal class (the one that occurred with the greatest geographical frequency) that covers the greatest amount of area. Similarly, we might wish to map the average occurrence of some phenomenon that varied through time at a series of places such as the frequencies of rainy days, cloudy days, and sunny days. The modal class would be the class that occurred most often.

The variation ratio (v) is a statistic that indicates the proportion of nonmodal cases, so that the nearer v is to zero the better the quality of the mode as a summarizing statement. It is calculated by

$$v = 1 - \frac{f_{modal}}{N}$$

where f_{modal} is the frequency or number of occurrences in the modal class and N is the total of all occurrences. If we are concerned with geographical frequency, then the calculation is

$$v = 1 - \frac{a_{modal}}{A}$$

wherex a_{modal} is the area occupied by the modal class (the class occuring in the largest area) and A is the total area.

The median is that point in an ordinal scale that is in the middle, that is, which neither exceeds or is exceeded in rank by more than half the total observations. It is probably less widely used in cartography than either the mode or the mean. Suppose we were to rank in each county the quality of the farms on a five-category scale from top to bottom. The median quality would

be that rank in which half the farms were above and the other half below it in rank. For example, if the data showed that there were 42 of A class, 1 of B class, 2 of C, and 1 of D, the median class would be A; if the data showed 10 of A, 12 of B, 3 of C, and 21 of D, the median class would be C. The median rank could be mapped or symbolized, of course, in a variety of ways. Just as was pointed out in connection with the mode, we can also map the median rank of occurrences through time at a series of places, such as a scaling of days in terms of their conduciveness to growing some crop or their desirability for carrying on some activity.

The decile range (d) is simply a statement of the number of ranks included between the first and the ninth deciles.[*] It is obtained by

$$d = d_g - d_1$$

in which d_g is the rank below which 90% of the cases occur, and d_1 the rank below which 10% of the cases occur. In the illustration above, in the first ranking the decile range would be 0, indicating that the median (A) is an excellent statement of the average, while in the second illustration the decile range would be 3, showing that the median C is not a good statement of the average.

The Mean

The arithmetic mean is the most frequently used average in cartography. Most of the maps of temperature, pressure, precipitation, income, yield, and the other elements common in physical and human geography are based upon means derived in one way or another. The mean is obtained by summing the values of the items

[*]A decile is a particular kind of *quantile*. A *quantile* is obtained by dividing a distribution into segments; quartiles divide it into four categories, deciles into ten, and so on, to centiles (or percentiles) that divide it into one hundred.

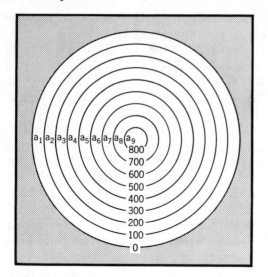

Figure 5.7 A conical island delineated by 100 foot contours.

we wished to obtain the average elevation above sea level from an ordinary contour map of elevation. Figure 5.7 may be used as an illustration, and it is assumed to be a perfectly conically shaped island, whose form and elevations are shown by evenly spaced contours at a 100 ft interval. It is clear from inspection that different elevations occur more or less frequently. In this instance, the geographic frequencies are the relative areas of the occurrences; consequently, it is necessary to obtain the areas of the spaces between the contours. This can be done most accurately by measuring them with a planimeter. The results obtained are expressed in any convenient square units, such as square inches. The measurements obtained from the original drawing of Figure 5.7 (before reduction for printing) are shown in the first two columns of Table 5.2. The mean elevation may be determined in two ways: by calculation or by graphic analysis.

As was observed earlier, the general expression for any mean is

$$\overline{X} = \frac{\Sigma X}{N}$$

In geographical distributions, whenever the

and dividing by the number of items. This procedure may be followed if the items are strictly comparable, but if the data are grouped in any way, then the items must be weighted. In the geographic analysis of problems, by means of maps, we are often required to obtain means from mapped data in which totals are difficult if not impossible to obtain. For example, suppose

Table 5.2   Calculation of the Mean Elevation of the Island Shown in Figure 5.7

Elevation Classes— Feet	Mid-Value of Elevation Classes—Feet	Map Areas of Elevation Classes—Square Inches	Cumulated Areas—Square Inches	
(Z)	(Z_m)	(a)	($a_1 + a_2 + a_3 \ldots$)	(aZ_m)
0–100	50	3.32	3.32	166
100–200	150	2.92	6.24	438
200–300	250	2.55	8.79	638
300–400	350	2.18	10.97	763
400–500	450	1.79	12.76	806
500–600	550	1.37	14.13	754
600–700	650	0.99	15.12	644
700–800	750	0.59	15.71	443
800–900	850	0.20	15.91	170
Total		15.91		4822

X values are in any way related to areal extent, they must, of course, be weighted for their frequency: this is most easily done by multiplying each X value by the area of its extent, summing these products, and then dividing the sum by the total area. The general expression for any geographic mean is, therefore,

$$\overline{X} = \frac{\Sigma aX}{A}$$

in which ΣaX represents the sum of the products of each X value multiplied by its area, and A is the total area, that is, Σa. From Table 5.2

$$Z = \frac{4822}{15.91} = 303.1 \text{ ft}$$

When we wish also to use a cumulative frequency graph for other purposes, it may be desirable to accomplish the result graphically instead of arithmetically. A cumulative frequency graph is constructed by plotting the values of the distribution on the Y axis of an arithmetic graph in the order of their value against the extent of their progressively cumulated areas on the X axis. The paired data from the first and fourth columns of Table 5.2 may be graphed as is shown in Figure 5.8 by plotting the value of the upper-class limit of the first elevation category, Z_1 (100), on the Y axis and the area a (3.32) on the X axis. The next pair, Z_2 (200) and $a_1 + a_2$ (6.24), is similarly plotted. The addition of each area value to the sum of the preceding areas (after arranging them in the ascending order of Z values) assures that the curve, which results from joining the plotted points with a smooth line, will rise to the right. The area under the curve (bounded by the curve, the base line, and an ordinate erected at the value of the total area) may then be measured and expressed in square units, such as square centimeters or inches. When this total is divided by the length of

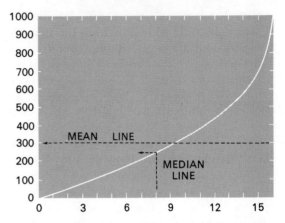

Figure 5.8   A cumulative frequency graph. The light horizontal dashed line is the top of a rectangle of the same area as that included between the curve and the base line. The height of the rectangle on the Y axis is the geographic mean. The Y value of the curve at the point of intersection with the light vertical dashed line erected at the midpoint of the cumulated areas plotted along the base is the geographic median.

the base line from the origin to the total area (measured on the graph in linear terms of the same units used to express the area under the curve), the quotient is the height of a rectangle with the same area as that under· the curve, one side of which is the length of the base line. The height of the rectangle when plotted on the Y axis of the graph is the mean height of the curve and is the geographic mean. In Figure 5.8 the area under the curve (on the original drawing, before reduction) is 9.63 sq. in. The length of the base as plotted was 6.36 in. Division of the area by the base 9.63/(6.36) results in a quotient of 1.51 in. This value, when measured on the Y axis from the base, gave a rectangle with a height having an elevational value of 302 ft, which is the geographic mean. In Figure 5.8 this height is shown by the light, dashed horizontal line. The difference of not much over 1 ft between the mean determined arithmetically (303.1 ft) and the mean determined graphically (302 ft) results from the graphic process and the consequent rounding of numbers, as well as the fact that mid-values of classes were used in one computation, whereas upper class limits were used in the other. This kind of graph is also occasionally called a hypsographic curve because it has been used to determine mean land heights (mean hypsometric values) exactly as illustrated here. It may, of course, be applied to any other kinds of data expressed with an interval scale that has actual or assumed continuous areal extent.

The cumulative frequency graph may also be employed to determine geographic percentiles, quartiles, and so on, as well as the geographic median, of a series scaled on an interval basis. The median is determined by erecting an ordinate at the midpoint of the base line and reading the Y value of its intersection with the curve. The median elevation of the island on Figure 5.7, as derived from the graph in Figure 5.8, is close to 280 ft, that is, half the map area is shown as above and half below that elevation. Percentile and quartile values are similarly obtained by subdivision of the base and determining the Y values of ordinates erected at such points. The curve can also be used to find the total area of all those sections lying above or below any given elevational value.

Many problems of cartographic analysis include data with open-end classes, that is, the top or bottom class may be only available on a "less than. . ." or a "more than. . ." basis. This is the case with the series of elevations shown by the contours on the island in Figure 5.7 wherein the contour of highest value is 800 ft. By definition, the area inside that contour is above (that is, more than) 800 ft, but by exactly how much is not shown. In other cases, the neat line or map edge may cut through a statistical unit area and produce a similar result. When the limits of all classes are not given, the mean value cannot be accurately ascertained, but judicious estimate and interpolation will usually make significant errors unlikely.

The Standard Deviation

Just as the mean is the most frequently used kind of average in cartography, the standard deviation is the most important index of variation. It is especially important because of (1) its descriptive utility, (2) its wide employment, (3) its use in the evaluation of the reliability of measures computed from samples, and (4) its utilization as an aid in the positioning of isarithms.

Many kinds of phenomena, when scaled on an interval basis, show a similarity in the frequency with which particular values occur in the series. The values that occur with the greatest frequency are usually those near the mean of the series. The greater the difference (or deviation) of a value from the mean of its series, the less frequently that value is likely to occur.* When an "ideal" series is graphed by plotting the frequencies against the values, the result is a *normal curve* of a *normal distribution*. It is illustrated in Figure 5.9 by an ordinary (not cumulative) frequency graph on which the frequency of occurrence of the values is plotted on the Y axis against the values on the X axis. Of particular significance in mapping data that have been obtained by taking samples is the fact that if a series of random samples is taken from a normal statistical "population," the means of the individual samples also tend to have a normal distribution. The standard deviation is a way of describing the dispersion of the values in a normal distribution.

*Although a great many phenomena show a "normal" distribution of values, some do not. A description of other kinds of distributions is more appropriately a topic for study in statistics.

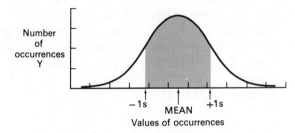

Figure 5.9   A frequency curve of a normal distribution. The number of occurrences are plotted on the Y axis and the values on the X axis. The shaded area shows the proportion (68.27%) of the occurrences within one standard deviation on either side (plus or minus) of the mean.

The standard deviation is the square root of the mean of the squared deviations from the mean. The difference or deviation of each item in the distribution is squared and the sum of these squares is known as the *variance*. The square root of the variance is the standard deviation. It is the range on either side of the mean value of a normal distribution, which includes approximately two-thirds (68.27%) of all the values in the series: and no matter how the original observations are distributed, the range of two standard deviations on either side of the mean will include at least three-quarters (75%) of the observations. For example, the standard deviation of the elevations on the island in Figure 5.7 is 211 ft. This means that about two-thirds of the island has an elevation within 211 ft of the mean; or, stating it another way, about two-thirds of the island lies above approximately 92 ft (303−211 ft) and below approximately 514 ft (303 + 211 ft), that is, lies between those two elevations.

The standard deviation may be easily computed by the following formula:

$$s = \sqrt{\frac{\Sigma(X - \overline{X})^2}{N}}$$

in which s is the standard deviation, and $(X - \overline{X})$ is the difference between each

value and the mean. When working with initial data, if we do not wish to determine the differences between each value and the mean, we may use the following formula:

$$s = \sqrt{\frac{\Sigma X^2}{N} - (\overline{X})^2}$$

As was pointed out in connection with the determination of the mean, when the data occur in varying frequencies, as they do in geographical distributions, then it is necessary to take this into account. The easiest way to do this is to use the areal extent as the frequency. As an illustration, the standard deviation of the elevations of the island shown in Figure 5.7 is calculated below. The necessary data are shown in Table 5.3, and it will be seen that mid-values are employed. The expression then becomes

$$s = \sqrt{\frac{\Sigma aZ^2}{A} - \left(\frac{\Sigma aZ}{A}\right)^2}$$

in which ΣaZ^2 is obtained by first squaring each Z value, multiplying it by the area it represents, and then summing the products. The term $(\Sigma aZ/A)^2$ will be recognized as the square of the geographic mean:

$$
\begin{aligned}
s &= \sqrt{\frac{2,171,975}{15.91} - \left(\frac{4822}{15.91}\right)^2} \\
&= \sqrt{136,516 - 91,857} = \sqrt{44,659} \\
&= 211 \text{ ft}
\end{aligned}
$$

The standard deviation has wide use in both descriptive and inference statistics. In cartography, when we are drawing isarithms to symbolize a distribution based on mean values (see Chapter 7), a high standard deviation indicates that the isarithms should be generalized.

The Standard Error of the Mean. Many maps are made of distributions wherein the data are obtained by sampling. For example, a map of average temperatures is based upon temperature records for a particular period and thus is but one sample of the time through which such temperatures have occurred. The particular sample available will ordinarily not provide the true mean because it contains only a portion of the total of all possible values. The likelihood of the mean being a different value can be inferred from the standard deviation of the values in the sample. This inference is made by calculating the standard deviation of the mean. It is usually called the standard error of the mean and is symbolized by $s_{\bar{x}}$. It is obtained by dividing the standard deviation of the sample values by the square root of the number of the sample values that were used in calculating the mean. Its expression is

$$s_{\bar{x}} = \frac{s}{\sqrt{N}}$$

where s is the standard deviation of the values of X, and N is the number of X values that entered into the calculation of the mean of X.[*] It is clear that the smaller the number of values used to arrive at a mean, the larger will be the standard error, that is, the greater the likelihood that the mean is incorrect. The cartographer must keep in mind the variations in the reliability of the means whenever he is employing isarithms to portray the distribution of a variable on an interval scale.

[*]When we are estimating the standard deviation of a total population from a sample, the standard deviation is calculated with $N-1$ instead of N.

Table 5.3   Calculation of the Standard Deviation of Elevations on the Island shown in Fig 5.7

Mid-Value of Elevation Class — Feet (Z_m)	Map (or actual) Area of Class — square inches (a)	(aZ_m)	(Z_m^2)	(aZ_m^2)
50	3.32	166	2 500	8 300
150	2.92	438	22 500	65 700
250	2.55	638	62 500	159 375
350	2.18	763	122 500	267 050
450	1.79	806	202 500	362 475
550	1.37	754	302 500	414 425
650	0.99	644	422 500	418 275
750	0.59	443	562 500	331 875
850	0.20	170	722 500	144 500
Total	15.91	4 822		2 171 975

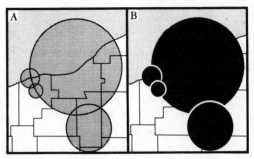

MAPPING POINT AND LINEAR DATA

This chapter and the two following deal with the fundamental elements involved in the symbolization of the vast array of geographical data that can be communicated cartographically. In order to introduce this portion of the book properly, a few general remarks are made here concerning distribution, statistical, and qualitative and quantitative maps. Following this brief introductory material, the remainder of this chapter will deal with the presentation of point and linear data.

All maps are "distribution maps" in the sense that it is impossible to represent relative geographical location without showing the distribution of something. On the other hand, the term "distribution map" is widely employed in a much more restricted sense. A large group of maps, mostly thematic, employ graphic symbols in a surprising variety of ways to show the general or detailed characteristics of a single class of geographical phenomena, in marked contrast to the reference class of map which is mainly prepared in order to portray the more general locational characteristics all at once, so to speak, of a variety of past or present geographical data. Many distribution maps deal with statistics derived from censuses, etc., and are called "statistical maps." Because many geographical phenomena can be differentiated without being obviously numerical and without being subjected to ordinary statistical manipulations, it is not meaningful to use the term "statistical map" except when it is restricted to the portrayal of data ordinarily included in the term "statistics."

Maps showing distributions in numerical form are one of the cartographer's stocks in trade. They are capable of surprising variety and can be used to present almost any kind of data. Few maps can be made that do not in some way present quantitative information, even if the amounts involved result from so simple an operation as the ordinal differentiation of symbols for cities of different size for an atlas map. Column after column of numbers in tabular form frequently are excessively forbidding, and the statistical map usually can present the important geographical characteristics of the material in a far more understandable, interesting, and efficient manner. Tabular quantitative materials of various kinds, ranging from governmental censuses, to the reports of industrial concerns, to the results of our own survey, exist in staggering variety. With such a wealth of material, it is to be expected that the cartographer would find that a large percentage of his effort is devoted to preparing this type of map. Many of the distributions mapped are derived parameters, that is, generalizations of numerical arrays of the sort suggested in the preceding chapter, and the cartographer must proceed warily lest he stumble intellectually. It is necessary for the student to appreciate that figures *can* easily lie cartographically if they are improperly presented to the map reader. As one author has observed, "One of the trickiest ways to misrepresent statistical data is by means of a map."[*]

Hasty evaluation of data, or the selection of data to support conclusions unwarranted in the first place, is thoroughly unscholarly and results in intellectually questionable maps. Their production is reprehensible, for they may be, and unfortunately sometimes are, used by their authors or others to

[*]Darrell Huff, *How to Lie with Statistics*, W. W. Norton and Company, Inc., New York, 1954.

support the very conclusions from which they were drawn in the first place. The exhortation, frequently implied or stated in this book, "to prepare the map so that it communicates what is intended," is not contradictory to the preceding; it is merely a matter of integrity.

Another source of dangerous error in cartography is the map that provides an impression of precision greater than is justified by the quality and quantity of the data used in its preparation. To keep from making this serious mistake, the cartographer is sometimes forced to invent special symbols such as dashed lines or patterns of question marks, which may detract from the aesthetic quality of the map but which will serve the more important purpose of preventing a map reader unfamiliar with the subject matter from falling into the common trap of "believing everything he sees."

Distribution maps employ a great variety of symbolism and to try to classify with precision all the variations in the geographical concepts that can be symbolized is all but impossible; the multiplicity of catego-

ries would reach astronomic numbers. Consequently, we cannot illustrate in a practical manner the variety of ways a given symbol may be used. Instead, this and the following chapter will illustrate the means by which the various fundamental classes of geographical data (point, line, area, and volume) can be and usually are represented.

The examples of distribution maps that appear in this book by no means completes the list, since the numbers of possibilities of combination and presentation are almost infinite. Rather than attempting to give directions for the preparation of this vast array of map types, attention is focused here on the basic and most frequently used classes and on the principles involved in their preparation. Other illustrations of their application in specific subject-matter situations may be found in great variety in books and periodicals.*

---

*A very useful reference, profusely illustrated and documented, is F. J. Monkhouse and H. R. Wilkinson, *Maps and Diagrams*, 2nd edition, University Paperbacks, Methuen and Co. Ltd., London, 1964.

DISTRIBUTION MAPS

All maps on which data are differentiated nominally (and many in which part of the objective is to show data scaled ordinally, or even according to interval categories) are clearly *qualitative* in that one of their main functions is to display the relative location of different kinds of geographical phenomena. These may range from classes of precipitation (rain, snow, hail, drizzle, for example) to types of rock, and from categories of employment to predominant races. Their fundamental objective is to present the geographical locations of such categories and, consequently, adequate base information is usually of vital importance to such maps. Opposed to qualitative

maps are those commonly called *quantitative*. These are the maps in which the data are arrayed according to interval and ordinal scales, and they range from amounts of precipitation to thicknesses of rock formations, and from levels of employment to ratios of racial composition. They naturally cannot escape being somewhat qualitative since they must display a kind of phenomenon, but their major objective is not the nominal differentiation of geographical data but the presentation of spatial variations in amounts. This class is much more complex, since, as was observed in the last chapter, the possibilities for manipulation of the data are much greater. Base data requirements are always fundamental but are usually not quite so great for this class of maps because it is not unusual that the major focus of interest is on the numerical variations within the distribution of the phenomenon being mapped rather than on the precise location of details.

Qualitative Distribution Maps

In principle, qualitative distribution maps, which have as their main objective the communication of nominal categories of geographical data, are relatively simple from the point of view of symbolism. Differences can be represented merely by changing the appearance of the symbol. Colored or uncolored lines can be dotted, dashed, or varied in other ways; point symbols can be dots, stars, anchors, triangles, and so on, almost indefinitely; area symbols can employ variations in color, such as hue, value, or intensity (see Chapter 11), or variations in pattern, such as texture, arrangement, and orientation.

The selection of symbols for nominal differentiation also poses the problem, common to all distribution maps, of symbolizing without much value contrast. Clarity and visibility are the result of contrast, and of the various kinds of contrast, value (degree of darkness) is visually the most important. Value changes are generally inappropriate, however, in nominal symbolization on maps because of the universal tendency to assign quantitative meaning to value differences. Generally a darker symbol looks more "important" to the map reader. Of course, such emphasis could be used to advantage if the cartographer were desirous of drawing attention to one or more of the various nominal or nominal-ordinal categories he is mapping. The use of hues complicates the problem even further, and is described at some length in Chapter 11.

Quantitative Distribution Maps

Quantitative distributions employ the variety of numerical measures considered in the preceding chapter. The majority are based on absolute numbers or on simple summaries such as averages. The numerical data can refer to point, linear, area, or volume geographical quantities and can be represented by marks that can be classed as point, line, and area symbols.

When approached from the point of view of the frequency of use, the cartographic symbolization of quantitative distributions becomes relatively simple because only a few classes of symbols are appropriately used on the great majority of maps. Generally speaking, to show variations from place to place of the amounts in a single class of data, it is usually desirable to employ only one basic form of graphic device and then vary it in terms of size, frequency, or visual value to portray the differences in quantity from place to place. To be sure, we can combine several forms of quantitative symbolism on a single map, but this should be approached warily, since it is easy to make the communication too complex for the map reader.

The remainder of this chapter will be concerned with the more common techniques used to symbolize variations in quantities of geographical point and linear data. The emphasis is placed on the fundamental categories of symbols and principles involved and in their use, rather than listing the possible variations. For example, we can employ in certain ways a variation in the size of a symbol to portray summary data such as the populations of cities, but the symbol form can run an almost unlimited gamut from a star to a circle, to a little drawing of a skyscraper, to a standardized human figure, and so on; nevertheless, there are basic principles involved in this class of symbolism that are applicable no matter what the symbol form. The symbolization of geographical areas and volumes, both actual and abstract, is dealt with in Chapter 7.

MAPPING QUANTITATIVE POINT DATA

Quantitative point data can be symbolized in a great variety of ways. Any real or conceptual quantity that can be thought of as existing in variable amounts from place to place can be symbolized simply by assigning a unit value to some point symbol and then putting the right number of these same-sized symbols in the right places on the map. This technique results in what is generally called a dot map because simple, round dots are the most frequently used symbol; but any point symbol can be used. The kinds of quantities that can be portrayed in this way can range from slope values, to people, and even to percentages. Differently colored or shaped dots (or other marks) can be used to show geographical mixtures; but fundamentally, all we are doing is repeating a chosen symbol to portray the geographical frequency.

A second system for representing point data is to summarize the quantities that apply to points or areas and then to symbolize the variations in the quantities by variations in the sizes of a chosen symbol, such as a square or a circle. This is generally called graduated circle (or star, or square, etc.) map. Details can be added by segmenting each symbol to show proportions, and a variety of classes of data can be included on one map by varying the symbol form.

A third method for representing a continuous array of point data, such as the elevations of all points above sea level, or the temperatures of all places, is by assuming that such continuous arrays form the surface of a volume and then showing the configuration of the "statistical surface" by area symbols or by linear symbols called isarithms. From the area symbols or the positions of the lines we can infer the values at points. Such a system has other useful characteristics, and this method will be treated in the next chapter.

The Dot Map

The simplest of all maps using point symbols is the one wherein the data are presented by varying numbers of uniform dots, each representing the same amount. It is possible, of course, to substitute little drawings of men, or sheep, or cows (or whatever is being represented) for the simple dot. This generally reduces the amount of detail that may be presented, but it is sometimes desirable for rough distributions for maps for children. This kind of map is capable of showing more clearly than any other type the details of locational character. It provides an easily understood visual impression of relative density, readily accepted by the reader, and easily interpreted on an ordinal scale; but it does not provide him with any absolute figures.[*]

A second advantage is the relative ease with which such maps may be made. No computation is ordinarily necessary beyond that of determining the number of dots required, which merely necessitates dividing the totals for each enumeration unit by the number decided upon as the unit value of each dot.

Although the usual map reader does not believe he experiences particular difficulty in interpreting the dot map, recent research indicates that it is not as straightforward a

[*]Theoretically, it would be possible to count the dots and then multiply the number by the unit value of each dot to arrive at a total, but in practice it would be done only under duress.

kind of symbolization as many have thought. It seems especially difficult for the untrained reader to estimate relative visual densities on an interval scale with much success, although ordinal decisions appear not to be difficult. The latter is probably the most important function to be performed by the dot map.

It may be an important part of the cartographer's objective to have the reader gain considerable information from the map regarding the various densities, relative as well as absolute. If so, the reader probably will need help. Most dot maps stop with simply a legend that tells the reader the magnitude of the unit value of the dot used. Of considerable assistance would be a more complete legend made up of several boxes containing appropriate numbers of dots and labelled as to the actual density of the phenomenon being mapped as represented by these "samples."

Dot maps ordinarily show only one kind of fact, for example, population or acres of cultivated land: but by using differently colored dots or differently shaped point symbols, it is sometimes possible to include several different distributions on the same map. Of course, if there is no mixing of two types of data they both may be shown on the same map.*

The Size and Value of the Dot. If the visual impression conveyed by a dot map is to be realistic, the size of the dot and the unit value assigned must be carefully chosen. The five dot maps shown here have been prepared from the same data; only the size or number of dots used has been changed. The maps show potato acreage in Wisconsin.

If the dots are too small, as in Figure 6.1, then the distribution will appear sparse

*A good analysis of the conceptual variations in dot mapping is given by Richard E. Dahlberg, "Towards the Improvement of the Dot Map," *International Yearbook of Cartography*, **7**, 157–167, (1967).

and insignificant, and patterns will not be visible. If the dots are too large, then they will coalesce too much in the darker areas, as in Figure 6.2, and give an overall impression of excessive density that is equally erroneous. It appears in Figure 6.2 that there is little room for anything else in the region. Furthermore, when dots are gross they dominate the base data and generally result in an ugly map.

Equally important is the selection of the unit value of the dot, and naturally the two problems (size and value of the dot) are inseparable. The total number of dots should neither be so large that the map gives a greater impression of accuracy than is warranted nor should the total be so small that the distribution lacks any pattern of character. These unfortunate possibilities are illustrated in Figures 6.3 and 6.4.

The selection of unit value and size of dots should be made so that in the denser areas (of a dense distribution) the dots will just coalesce to form a dark area. Figure 6.5 is constructed from the same data as the preceding examples in this section but with a dot size and unit value more wisely chosen. Of course, if the distribution is sparse everywhere, then even the relatively dense areas should not appear dark.

On relatively large-scale dot maps showing land use phenomena, it is possible to relate the dot size to the scale of the map by making the dot cover the scaled area value. For example, we might make a map showing the area planted in wheat with a unit value of 1000 acres and a dot size that covers exactly 1000 acres at the scale of the map. We should not follow such a procedure blindly, however, since the elusive quality, "relative importance," of areal phenomena is often dependent upon factors other than strict area relationship.

The data needed for a dot map consist of the enumeration of the number of items to be mapped in unit areas, usually civil divi-

Figure 6.1   A dot map in which the dots are too small so that an unrevealing map is produced. Each dot represents 40 acres in potatoes in 1947.

Figure 6.3   A dot map in which the unit value of the dot is too large so that too few dots result; a barren map revealing little pattern is produced. Each dot in this example represents 150 acres.

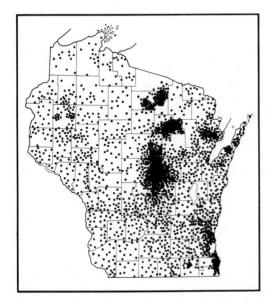

Figure 6.2   A dot map in which the dots are too large so that an excessively "heavy" map is produced. An erroneous impression of excessive potato production is given. The same data and number of dots are used as in Fig. 6.1.

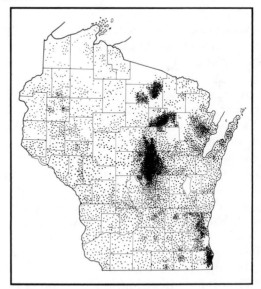

Figure 6.4   A dot map in which the unit value of the dot is too small so that too many dots result; an excessively detailed map is produced. The dots are the same size as those in Fig. 6.3. Each dot in this example represents 15 acres.

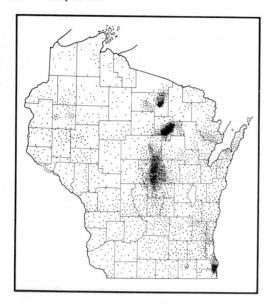

Figure 6.5 A dot map in which the dot size and dot value have been more wisely chosen than in the preceding examples. Each dot in this example represents 40 acres.

sions used as census statistical units. Rarely is it possible to employ so large a map scale that each single item may be shown by a dot, although theoretically the farmhouses on a topographic map might be termed a kind of dot map; rather, it is usually necessary to assign a number of the phenomena to each dot. This is called the unit value of the dot and is obtained merely by dividing the total number of items in each statistical division by the chosen unit value. For example, if a county had a total of 16,000 acres in corn, and a unit value of 50 acres per dot had been chosen, then 320 dots would be placed in the county to symbolize the corn acreage.

Professor J. Ross Mackay developed an ingenious nomograph to assist in determining the desirable dot size and unit value.*

*J. Ross Mackay, "Dotting the Dot Map," *Surveying and Mapping*, **9**, 3-10 (1949).

This graph, shown in Figure 6.6, requires a knowledge of the sizes of dots that can be made by various kinds of pens. This information is presented in Table 6.1. By varying the relation between dot diameter and unit value, the cartographer can settle upon a compromise between the two that will best present the characteristics of the distribution.

The Use of the Nomograph. The nomograph may be used in several ways, but perhaps the easiest is first to select three unit areas on the proposed dot map that are representative of (1) a dense area, (2) an area of average density, and (3) an area of sparse density. A tentative unit value can then be selected and divided into the totals for each of the three statistical divisions. The map area of one of the divisions can be estimated in terms of square inches, and the number of dots per square inch can be deduced.

As in the previous example, assume that a county contains 16,000 acres of corn; a unit value of 50 acres per dot is chosen, which results in 320 dots to be placed within the boundaries of the division; and assume further that the statistical division on the map covers one-half square inch. This would mean that the dots would be placed on the map with a density of 640 dots per square inch. An ordinate at 640 is erected from the X axis of the nomograph. A radial line from the origin of the nomograph to a given dot diameter on the upper right-circular scale will intersect the ordinate. The location of the intersection on the interior scale will show the average distance between the dots if they were evenly spaced. The height of the intersect on the Y axis will indicate what proportion of the area will be black if that dot diameter and number of dots per square inch were used. Also shown is the "zone of coalescing dots," at or beyond which dots will fall on one another.

If the initial trial seems unsatisfactory for each of the three type areas selected for experimentation, either the unit value or the dot size, or both, may be changed. We can enter the graph with any of several assumptions and determine the derivatives. It is good practice while using the nomograph to dot the areas (on a piece of tracing paper) in order to see the re-

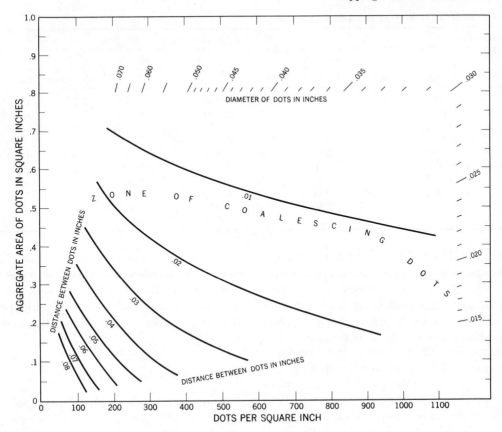

Figure 6.6   A nomograph showing the effects of varying the relation between dot size and dot density. Any combination of numbers of dots and diameters of dots that falls in or beyond the "zone of coalescing dots" will produce dots on top of one another. (Courtesy J. Ross Mackay and *Surveying and Mapping*.)

sults and to help visualize the consequence of other combinations.

The cartographer should remember that the visual relationships of black to white ratios, and the complications introduced by the pattern of dots, make it difficult, if not impossible, for a dot map to be visually perfect. The best approach is by experiment after narrowing the choices by use of the nomograph.

Locating the Dots. Theoretically, the ideal dot map would be one with a large enough scale so that each single unit of the point data could be precisely located. Ordinarily dot maps are small scale, but sometimes, if the data are sparse enough, a unit value of one (for example, paper mills) can still be used. It is usually necessary, however, to make the unit value of the dot greater than one, and the problem then arises of locating the one point symbol that represents several differently located units.

It is helpful to consider the group of several items as having a kind of center of gravity, and then to place the symbol at that point. Usually, all the cartographer knows from the original data is the total

Table 6.1   Dot Diameters in Inches of Various Pens

Barch-Payzant		Leroy		Pelican — Graphos type				Wrico	
				R		O			
Number	Diameter	Number	Diameter	mm	in.	mm	in.	Number	Diameter
8	0.012	00	0.013	0.4	0.016	0.2	0.008	7,7A	0.018
7	0.018	0	0.017	0.5	0.020	0.3	0.012	6,6A	0.025
6	0.025	1	0.021	0.6	0.024	0.4	0.016	5,5A	0.027
5	0.036	2	0.026	0.7	0.028	0.5	0.020	4,4A	0.036
4	0.046	3	0.035	0.8	0.031	0.8	0.031	3,3A	0.048
3	0.059	4	0.043	1.0	0.039	1.0	0.039	2,2A	0.062
2	0.073	5	0.055	1.25	0.049	1.25	0.049		
1	0.086	6	0.067	1.50	0.059	1.60	0.063		
		7	0.083	1.75	0.069	2.00	0.079		
				2.00	0.079	2.50	0.098		

number of dots to place within a unit area. Consequently, he must draw upon every available source of information to assist him in placing the dots as reasonably as possible, such as topographic maps and other distribution maps that he knows correlate well with the one being prepared. The quality of the finished map depends largely upon the ability of the cartographer to bring all the pertinent evidence to bear on the problem of where to put the dots.

Considerable detail can be introduced into the map if it is prepared on the basis of the smallest civil divisions and then greatly reduced. Placing the dots with reference to minor civil divisions can be done easily by using translucent material for the map and putting under it a map showing the minor civil divisions to serve as a guide. Only the larger administrative units need ordinarily be shown as base data on the finished map. Special care must be exercised not to leave the guiding boundary areas relatively free of dots, since if this is done they will show up markedly in the final map as white zones. We should also refrain from inadvertently producing worm-like lines of dots, unwanted clusters, or "tweed" patterns when placing the dots. Such regularity can easily occur and is quite noticeable by contrast with its amorphous surroundings.

A dot map in which the requisite numbers of dots are evenly spread over the unit areas, rather like a patterned area symbol, although numerically correct, could better use some symbol system other than repetitive dots.

The Graduated Symbol Map

Graduated symbols, that is, symbols that are differently sized, are widely employed, and by far the most common is the graduated circle. The variation of size is employed either to symbolize amounts at specific locations or when the totals referring to enumeration units are of more interest than details of location. Thus, they are useful (1) when point data exist in close proximity but are large in aggregate number, such as the population of a city, (2) for symbolizing totals of quantities such as tonnage, costs, and traffic counts, or (3) for representing the aggregate amounts that refer to relatively large territories. In the latter instance, the territory is considered merely as a location even though it obviously has areal extent. The area of a statistical unit is naturally a two-dimensional geographical quantity, but when data referring to it are aggregated and symbolized by a point symbol (in combination with other statistical units), the areal unit has, in a sense, been reduced to only a locational or point value for mapping purposes.*

The graduated circle is one of the oldest of the quantitative point symbols used for statistical representation. Near the beginning of the nineteenth century, it was used in graphs illustrating the then new census materials, and its first appearance on maps was in the third decade of that century. Since that time, it has been near the top of any list of quantitative point symbols in the frequency of its use; and its ease of construction and interpretation makes it likely that it will continue to be widely utilized.

The Ratio of Circle Sizes. When the graduated circle was first employed for interval scales, the variations in the actual areas of the circles were made uniformly comparable to the numbers they represented. For example, if two statistics had values to be represented that were in a 1:2 ratio, that is, the magnitude of the second was twice that of the first, the second circle was constructed so that its area was twice that of the first. Since the area of a circle is $\pi\,r^2$, and since π is constant, the method of construction was simply to extract the square roots of the data and then construct the circles with radii or diameters proportional to the square roots. The unit radius value, by which the square roots are made into appropriate plotting dimensions, may be any desirable unit, such as so many millimeters or hundredths of an inch; it is selected so that the largest circle will not be "too large," and the smallest, not "too small." As long as the square roots are all divided by the same unit radius value, the areas of all the circles will be in linear proportion to the sizes of the numbers they represent.

If the cartographer did not wish to show relative sizes of each individual statistical

quantity, he could proceed as in ordinal classification, or he could classify the range and then show all individuals in one class by a "standardized" circle constructed to the size of the midpoint of that class.

Extensive research in the psychophysical aspects of cartographic symbols has demonstrated that the perceptual response to circular areas is not a linear function; instead, the ordinary observer will *underestimate* the sizes of the larger circles in relation to the smaller ones. For example, to use the previous illustration, if the magnitudes of two quantities were in a 1:2 ratio and the actual areas of two circles representing them were in the same ratio, then a map reader would think the second was significantly less than twice the size of the first. When we make the areas of the circles strictly proportional to the numbers they represent, we have, therefore, in effect reduced the *visual* sizes of the larger circles in relation to the smaller, or, to look at it another way, we have increased the visual significance of the smaller circles relative to the larger. This is unfortunate, since the primary purpose of making such maps is to symbolize geographical quantities so that the map reader will obtain a realistic impression of the distribution being mapped.

We may compensate easily for underestimation. Instead of extracting the square roots of the data and making the radii of the circles relative to the square roots, the procedure is (1) to determine the logarithms of the data, (2) multiply the logarithms by 0.57, (3) determine the antilogarithms, and (4) divide the antilogarithms by the chosen unit value for the radii of the circles.* Figure 6.7 shows the difference in the sizes

---

*This is an example of the inconsistency that results when we try to match up point data and point symbols, and linear data and line symbols, for example.

*This procedure was devised by Professor James J. Flannery. Note that the square root of $\log n = \log n/2$ or $\log n \times 0.5$. Multiplying the logarithms by 0.57 instead of 0.5 serves proportionately to increase somewhat the relative sizes of the larger circles so that they appear in proper relation to the smaller ones.

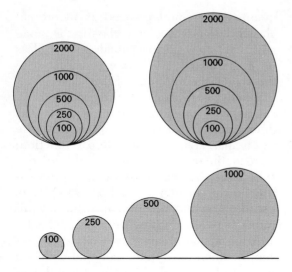

Figure 6.7 The two nested sets of graduated circles prepared from the same data show the difference between circles scaled, left, in linear relation to n (i.e., according to square roots), and, right, to compensate for underestimation. A key or legend designed as in the lower series (the same set of circles as the first four in right above) is probably easier than a nested set for the map reader to use.

of two sets of circles constructed in the two ways.

An example will demonstrate the procedure. The first four columns of Table 6.2 show the calculations needed to determine the numbers

which, when divided by the chosen unit radius value, will give the plotting radii of the circles. The antilogs provide directly the plotting values and need only be converted to map dimensions by dividing by some convenient unit value. The map shown in Figure 6.8 has been constructed from the data shown in Table 6.2, using a unit radius value chosen so that the circles do not overlap.

The table in Appendix F provides the antilogarithms of $n \times 0.57$. Consequently, we need only enter the table with the data and the tabular values provide the plotting values that need only to be divided by a suitable unit radius value.

Magnitude of the Unit Radius. The cartographer often finds that with a desirable unit radius value for the map as a whole, the circles in one part may fall largely on top of one another if their centers are placed approximately at the locations of the data they are to symbolize. Two of several ways to approach this problem are shown in Figure 6.9. The two small maps are taken from the same data as that for Figure 6.8, but with the unit radius value has been enlarged.

The selection of the unit radius value with which to scale the circles is important and should be done with the aid of some preliminary experimentation. The ideal value is one that will provide a map that is

Table 6.2   1950 Populations of the City of Cleveland and other Cities over 30,000 in Northeastern Ohio Outside the Cleveland Metropolitan District

City	(1) 1950 Population (n)	(2) log n	(3) log $n \times 0.57$	(4) Antilog of log $n \times 0.57$	(5) Drafting Radii of Circles— Unit Radius Value: $300 = 0.10$ in. Col. 4 ÷ 300
Akron	274 605	5.43872	3.10007	1 261	0.42
Canton	116 912	5.06785	2.88867	774	0.26
Cleveland	914 808	5.96133	3.39796	2 500	0.83
Elyria	30 307	4.48159	2.55451	359	0.12
Lorain	51 202	4.70927	2.68428	483	0.16
Mansfield	43 564	4.63909	2.64428	441	0.15
Warren	49 856	4.69775	2.67772	476	0.16
Youngstown	168 330	5.22616	2.97891	953	0.32

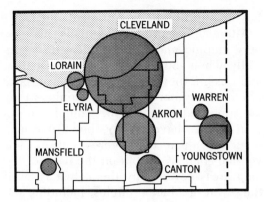

Figure 6.8 The relative sizes of some cities in northeastern Ohio prepared from the data in Table 6.2.

neither "too full" nor "too empty." For example, Figures 6.10 and 6.11 show the area of land available for crops in some counties. When a small unit radius is chosen, the circles are too small to show much, as is illustrated in Figure 6.10, and the impression is given that there is practically no cultivated land in those counties. When a large unit radius is chosen, the representation does not reveal much, and the impression is given that practically all the land is cultivated, as in Figure 6.11.

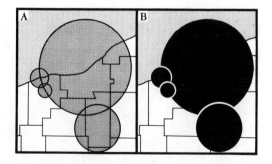

Figure 6.9 Two ways of solving the problem of overlapping circles. The data are the same as those used for Fig. 6.8, but the unit radius value for the circles is smaller than that chosen for Fig. 6.8, resulting in larger circles. In A, the circles are allowed to overlap, being "transparent." In B smaller circles are "above" larger circles.

Legend Design. There are many ways of arranging a series of circles in a legend to give the reader an "anchoring stimulus" to assist him in estimating the magnitudes represented by the various sizes of circles on the map. The most economical of space is the nested set of circles, such as those shown in Figure 6.7, but preliminary testing indicates that the side-by-side array is more efficient for the reader. The selection of the sizes of circles to be included in the key is dependent, of course, on the range of sizes on the map, but there is some evidence that quartile intervals are best in the nested set.

Squares, Cubes, Spheres, and other Point Symbols

Instead of graduating circles, we may employ almost any other figure, geometric or pictorial. The circle is the easiest to construct and scale, but the requirements of design may make some other figures desirable.

The desirability of presenting several nominal categories at once, indicating precise location and reducing overlap, has led to the use of triangular shaped segments of circles in various attitudes (Figs. 6.12 and 6.18). Such point symbols may be scaled ordinally into small, medium, and large categories, or they may be scaled on an interval basis. Either sort of scale can be represented by varying the length of the radii of the circle segment. Several such symbols can be "pointed at" one location, and they may be differentiated nominally by color, pattern, or both.*

The square is relatively simple to use as a symbol to represent magnitudes, since all that is necessary is to obtain the square

*A variety of examples of the use of this sort of point symbolization appears in the *United States and Canada*, Oxford Regional Economic Atlas, Clarendon Press, Oxford, 1967.

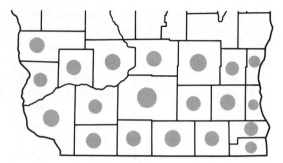

Figure 6.10 Area available for crops by counties. The unit radius is too small.

roots of the data and then scale the sides accordingly. This will grade the areas of the squares in linear proportion to the numbers they represent. Whereas there is clear evidence that the phenomenon of underestimation affects the map reader's impression of circles so scaled, the evidence is less impressive by far with respect to squares.

It is not uncommon that a cartographer is faced with a range of data so large that with interval scaling he cannot effectively show both ends of the range by graduated circles or squares. If he makes the symbols large enough to be differentiated clearly in the lower end of the scale, then those at the upper end will overshadow everything else and ruin the communication. Many attempts have been made to surmount this difficulty by employing a symbol form drawn to simulate a volume rather than an area, such as pictorial cubes or spheres, as shown in Figures 6.13 and 6.14. This has been done by scaling the symbols according to the cube roots of the data, that is, by making the sides of the front of a perspective cube or the circle bounding a sphere proportional to the roots. If the symbols were actually three dimensional, their volumes would be in strict linear proportion to the numbers they represented. The cube roots of a series cover a much smaller range than the square roots, and this, of course, is what makes it ostensibly possible to por-

tray graphically the larger range of data.

Although sphere-like symbols may be very graphic and visually pleasing, especially when enhanced by good execution and design, they do not serve the purpose of effectively portraying the numbers they are supposed to represent. A number of recent studies have clearly shown that map readers evaluate the spheres, not on their volume comparison but on the basis of the map areas covered by them, that is, as if they were graduated circles, including the characteristic underestimation. One team of investigators puts it clearly: "In so far as cartographic volume symbols are used to create an immediate impression of volume, [for example] population magnitude, their three-dimensionality has no effect. The trick may be technically attractive, but it is not applicable to the 'human factor.' The space saved on the map by constructing the symbols as projections of solids, is counterbalanced by an almost exactly corresponding loss of desired impression."[*] To what extent cube-like symbols suffer a similar perceptual fate is not yet known. When a cartographer is faced with an "excessive" range of data, he should not attempt the graphically impossible.

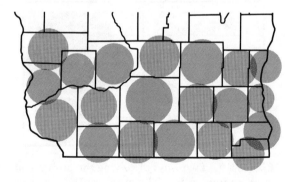

Figure 6.11 Same data as in Fig. 6.10 but the unit radius is too large.

[*]Gosta Ekman, Ralf Lindman, and W. William-Olson, "A Psychophysical Study of Cartographic Symbols," *Reports from the Psychological Laboratory of the University of Stockholm* (**91**), 12 (1961).

When other graduated symbols are employed, such as soldiers, stars, animals, and so on, care must be taken to scale them properly. Their heights or diameters should be made proportional (as outlined above) to the numbers by a system that ensures that the *areas* covered by the symbols are visually related to the magnitudes they represent.*

Segmented Graduated Symbols

More than one nominal class of data making up the proportions of a whole can be

*There has not been enough research yet to determine to what extent underestimation affects the perception of these more complex symbol types. Determination of relative sizes of some geometric symbols has been studied by Robert L. Williams. See *Statistical Symbols for Maps: Their Design and Relative Values,* Yale University Map Laboratory, New Haven, 1956.

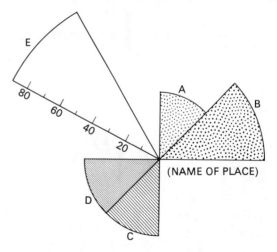

Figure 6.12   Examples of several ways to employ circle segments. *A* and *B* in the upper right quadrant can indicate related characteristics (e.g., passenger and freight traffic), *D* and *C* in the lower left different nominal categories, and *E*, with interval scaling, yet another category.

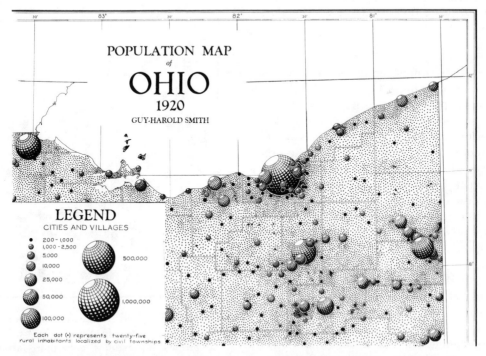

Figure 6.13   A portion of a population map of Ohio (1920) drawn by Guy-Harold Smith. Compare with Fig. 6.8. (Courtesy of the author and *The Geographical Review*, published by the American Geographical Society of New York.)

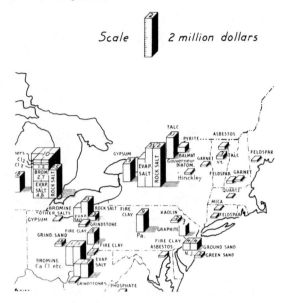

Scale — 2 million dollars

Figure 6.14 Cube-like symbols, called block piles, developed by Erwin Raisz. The graphic scaling on the front and top, together with the numbers, aids their perception. (From a map by E. Raisz, taken from *Mining and Metallurgy*, AIME.)

shown by segmenting the graduated symbol in some fashion. For example, the ethnic makeup of an urban population, or the value added by several categories of manufacturing, can be portrayed this way. The commonest way of doing this is to employ graduated circles and subdivide them as a pie is cut. There is no generic term for this kind of symbol, but the subdivided circles are often called "pie charts."

Any relation of one or more parts to the whole can be shown visually by the pie chart. For example, Figures 6.15 and 6.16 show the total amount of farmland in each county and, at the same time, show what percentage of that total is available for crops. The procedure merely requires that the percentage be determined and that, by using a "percentage protractor," the various values be marked off on each circle.

A percentage protractor may be easily constructed by drawing a circle and then subdividing the circumference according to percentages in the desired detail with dividers. A small hole at the center will make it possible to place the center of the protractor over the center of each graduated circle. The appropriate percent can be marked off at the periphery.

It is important that the subdivision of each circle begin at the same point; otherwise, the reader will have difficulty in comparing the values. Also important is the selection of the portion to be shaded or colored. As is illustrated in Figures 6.15 and 6.16, this can have considerable influence on the effect gained by the reader.

If only two classes of data are being represented by a pie chart, and if they are of equal importance to the purpose being served by the map, we may use a back-

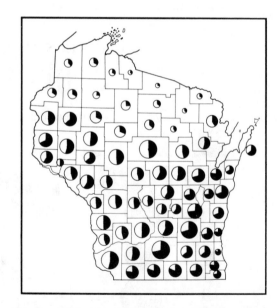

Figure 6.15 Land in farms and percentage available for crops in Wisconsin by counties The circles represent the land in farms. The percentage available for crops has been blacked in on each circle.

ground with a neutral tone so that the visual contrasts of the two segments of each circle with the background are approximately equal. Figure 6.17 has been constructed in this manner.

Many other, more complex kinds of segmentation can be employed. For example, a kind of map conceptually similar to the pie chart can be constructed by employing one or more concentric circles to represent portion of the total. The diameters of these interior circles are scaled according to the original data in the same manner as the circle representing the total. The visual impression is not very efficient, however, since percentages are not easily read. Naturally, graduated squares, rectangles, and other figures may be segmented in various ways to show parts of a whole. Figure 6.18 is an example of the detail that can be included in the legends of such maps.

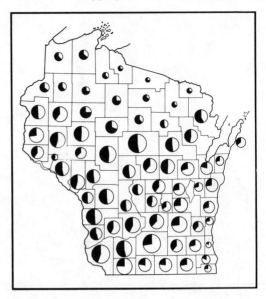

Figure 6.16 Same data as in Fig. 6.15, but in this map the percentage not available for crops has been blacked in on each circle.

Directional and Time Series Point Symbols

A variety of more complex cartographic presentations can be accomplished by combining graphs and maps so that variations in directional components or changes through time can be given the added quality of relative geographical location. For example, the upper diagram in Figure 6.19 shows, for each of the stations indicated, the amount of darkness, daytime cloud, and sunshine throughout the year. The lower diagram shows, among other things, the directional component and the percent frequency of winds for August in the same area.

There is almost no limit to the amount of information we can somehow encode in these kinds of point symbols. Of paramount importance, however, is the question of how much of the information is available to the reader. As long as the reader can obtain from the map a significant geographical element, that is, an appreciation of similarities and differences from place to place, then the map is serving its purpose. If that part of the information is not readily obtainable, then the map is a failure and the data would probably be more easily made available in tabular form.

MAPPING QUANTITATIVE LINEAR DATA

The term quantitative linear data includes a multitude of kinds of geographical phenomena ranging from the movement of commodities and people to the load-limit capacity of a roadbed. Naturally, most such maps employ some kinds of line symbols to portray the data, but commodity movement has sometimes been portrayed with point symbols.

It is occasionally difficult to decide

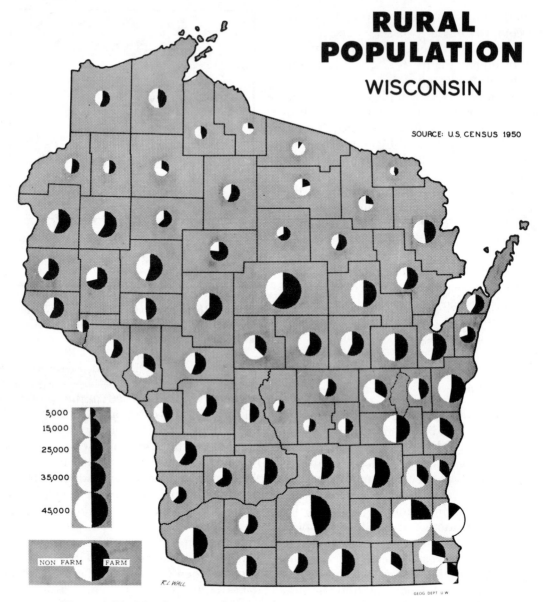

Figure 6.17   Pie chart map showing the proportion of rural farm (black) and nonfarm (white) populations in counties. The sizes of the circles are scaled in proportion to the total rural population. (Map by Robert L. Wall.)

whether some data are really point or linear. For example, with respect to continuous distributions, geographical gradient is one of the important geographical concepts, and there are numerous kinds: the slope of the land, variations from place to place of air temperature or atmospheric pressure, and many other more complex kinds, as derive from concepts such as population or economic potential. Such quantities may be thought of as spatial vector quantities in that at each point of the distribution on the map there is a direction, a magnitude, and a sense. To use the example of the slope of the land surface: there is a degree or percent of slope, a direction of maximum value in relation to the compass (such as NE-SW), and a sense in that the potential energy flow is in one of the directions.

These kinds of quantitative data can be mapped in numerous ways, depending largely upon whether the cartographer is more concerned with giving emphasis to the directional aspects at places or to the overall spatial interrelation of the magnitudes. For the former (for example, wind velocities), we will commonly employ lines of some kind that show by their orientation and character the direction and magnitude

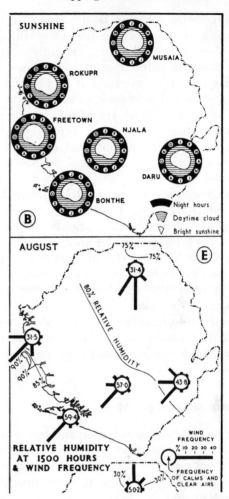

Figure 6.19 Cartographic representation of proportions of directional and sky condition data. (From J. I. Clarke, *Sierra Leone in Maps*, University of London Press, Ltd., London, 1966.)

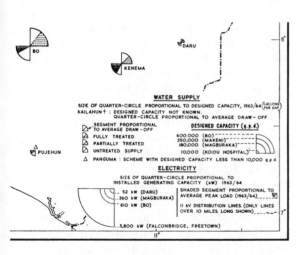

Figure 6.18 The legend and a small portion of a map illustrating the amount of information that can be coded in segmented, graduated point symbols. (From J. I. Clarke, *Sierra Leone in Maps*, University of London Press, Ltd., London, 1966.)

of the potential or actual flow; for the latter (for example, atmospheric pressure), we will usually conceive of the total mass of data as forming a volume distribution and then portray the continuous "statistical" surface of that volume by some means. The best known example of this general class of quantitative distribution, which can be represented in a variety of ways, is the surface configuration of the land. This can be thought of as an aggregation of an infinite number of related linear magnitudes and gradients, at points that can be shown directly by a type of gradient or vector line symbol, called a *hachure,* or indirectly by a set of lines called *isarithms* (contours). Representing geographical volumes by portraying their continuous real or assumed statistical surfaces is treated in the next chapter.

Another class of quantitative linear data is not continuous. In this group belong those data that refer to particular "connections" or "routes." For example, the flow of petroleum from the Middle East to Europe, of vacationers from the urban northeast of the United States to Florida, or the draft capacity of a canal system, all occur along specific routes either in the actual sense with the canals or in the more general sense with the commodity or population flows.

Vector or Gradient Maps

This class of maps has lagged far behind others except in relation to the land surface. The slope of the land can be shown in a variety of ways. (When it is summarized and averaged for areas, it ceases to be linear.) These are outlined in Chapter 8. When the vector quantity is considered as a linear value, it is portrayed by the hachure, a line symbol that is varied in some fashion (thickness or frequency per lineal unit) to represent the varying magnitudes of the slopes, and it is located in such a way as to indicate the direction. The sense is not directly shown graphically, but it is usually implied by the relationships among the symbols or by other symbols on the map. The hachure (see Figures 8.4 and 8.5) was widely employed during the nineteenth century to show the surface configuration, but its use has largely died out.

Naturally, the hachure-like symbol need not be restricted to showing the gradients on the land surface; it can be employed to show the directions and magnitudes of any sort of geographical gradients, from temperature to population potential. It has not been used very often in this way, however, since cartographers and geographers have generally thought that the gradients were adequately portrayed by the arrangements of sets of isarithms, which provide considerably more information than just gradients. It is reasonable to suppose that this attitude will continue, not only because the isarithm is extremely efficient as a symbol but because it can be more easily automated than can the hachure-like symbol. Nevertheless, if emphasis on the gradients of a distribution is paramount, the hachure-like symbol can be employed. As noted in Chapter 8, an isarithmic map must be constructed first.

Flow Maps

Maps showing linear movement are commonly called flow maps. There are two main types: those that simply symbolize the direction and frequency by means of a line or lines, usually with an arrowhead, and those that portray varying amounts of flow by varying the design of the line. These types of linear representation are relatively old. One of the major uses of the first kind of symbolism is the representation of mean movements of the atmosphere by means of "streamlines," which are curving lines, usually with arrowheads to show the directional sense. The first use of these was by Edmund Halley, in the late seventeenth century, to show the prevailing winds over ocean areas. Differentiation, if any, is usually nominal, and this can be accomplished by differing width of line, varying color, or employing variations in design, such as dashed, solid, or dotted lines. Ordinal and interval scaling is not common but can be accomplished. An example of nominal scaling in a streamline map is shown in Figure 6.20.

Another class of flow map, characterized by interval scaling, came into widespread use in European thematic cartography around the middle of the nineteenth century during the period when Europe was becoming "statistics conscious." The development of rail transportation during the early part of that century and the rapidly increasing movement of goods by both rail and canal, together with the rise of statistical symbolism, contributed to the use of this easy-to-understand symbol. Today, flow lines are used in the same fashion to represent the flow or movement of everything from mineral raw materials to automobile traffic.

The techniques used in the construction of flow maps are easily observed from well-made examples such as those illustrated here. Interval scaling is most often accomplished by merely graduating the thickness of the lines in proportion to the values by using a convenient unit width. The unit width is selected so that division of the data by the unit width results in a number representing a map dimension in inches or millimeters, in the same fashion that a unit value is used to determine the drafting sizes of graduated circles. For example, if we were representing the flow of freight traffic between two cities, we could use a unit value of 1 mm for each 1 million tons; therefore, if the traffic total were 4.5 million tons, a line 4.5 mm wide would be shown between the two cities. There appears to be no tendency toward underestimation in map readers viewing flow lines of variable width. Lines may be shown as smooth curves, as in Figure 6.21, or as angular lines, as in Figures 6.22 and 6.23.

Movement along an actual route may be represented or, as in "origin and destina-

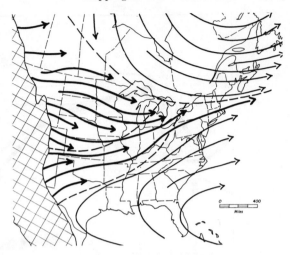

Figure 6.20 Streamlines showing mean resultant airflow over North America in January east of the Rocky Mountains. Nominal differentiation (directional source) is accomplished by variation in width. (Courtesy J. R. Borchert and the *Annals of the Association of American Geographers.*)

tion" maps, the terminal points may simply be connected by straight lines. Arrowheads at the ends of lines are often used to show the directions of movement, although the varying thickness and angle with which "tributaries" enter frequently show flow adequately. Arrows may be placed along the lines, as in Figure 6.24, to show the direction or even to show relative movement in opposite directions, as in Figure 6.22. Lines can increase or decrease in width as the values change, but between tributaries, a line should maintain a uniform width. Tributaries should, of course, enter smoothly in order to enhance the visual concept of movement.

In some instances, the range of the data is so large that a unit width value capable of allowing interval differentiation among the small lines would render the large ones much too large. It is, consequently, sometimes necessary to symbolize the smaller lines in some way, such as by dots or dashes (see Fig. 6.24).

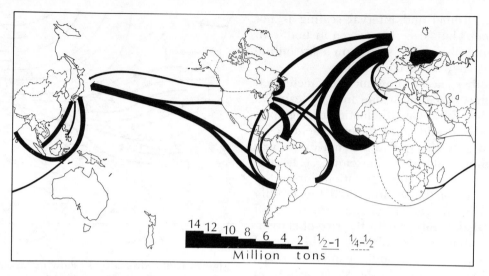

Figure 6.21   A portion of a flow-line map showing the movement of iron ore. Map by G. B. Lewis. (From G. Manners, "Transport Costs, Freight Rates, and the Changing Economic Geography of Iron Ore", *Geography*, 52 (1967), 260-279.)

Instead of using a system of symbolization for interval scaling that depends upon variable width, we can employ alternative methods. For example, we may replace the single line with a variable number of parallel lines, letting each of the lines represent a unit value. Accordingly, to use the previous example of a traffic total of 4.5 million tons, we could assign one line to each million tons; we would then draw 4 parallel lines between the two cities, and the half could be represented by a dashed line. Figure 6.25 is an example of this method; note that it employs an uneven class interval. Another alternative is to categorize the data and symbolize the various classes with symbols varying in pattern and value. The ultimate in direct quantitative symbolism is reached in the map shown in Figure 6.26. In this map each dot represents approximately 100 vehicles in one mile in a twenty-four hour period.

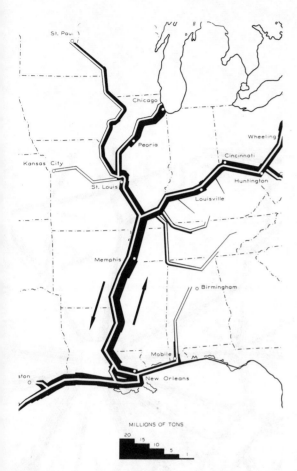

Figure 6.22   A portion of a flow map showing 1949 tonnage of barge and raft traffic in the United States. The legend has been moved. Note how direction of movement is indicated. (From Edward L. Ullman, *American Commodity Flow*, University of Washington Press, Seattle, 1957.)

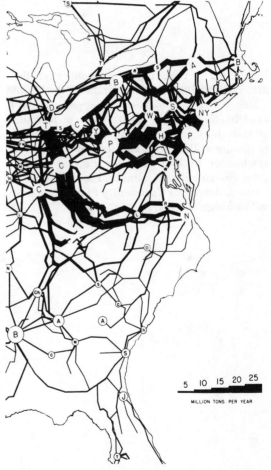

Figure 6.23   Part of a flow map showing annual tonnage on class one railways. (Data employed copyrighted by H. H. Copeland and Son; map prepared and copyrighted by Edward L. Ullman.)

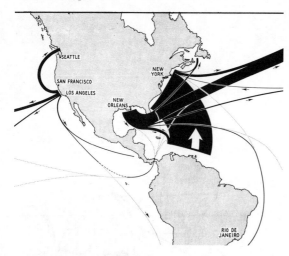

Figure 6.24 A portion of a quantitative flow map symbolizing 1947 tanker traffic of the United States. Since the thickest line represents well over 20,000,000 tons, small values of less than 200,000 tons cannot be adequately shown by proportional thickness. Consequently, they are symbolized by line character (dots or dashes). (Map drawn by R. P. Hinkle. From Edward L. Ullman, *American Commodity Flow*, University of Washington Press, Seattle, 1957.)

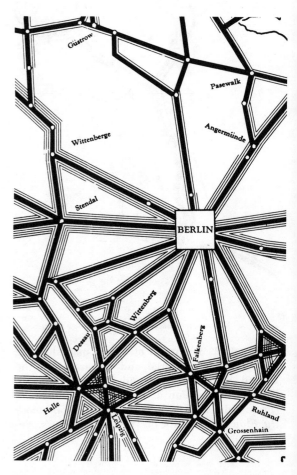

Figure 6.25 Part of a map showing daily freight movement on German railroads in the 1920's. The heavy line alone represents less than 1000 tons daily; one thin line on one side 1000 to 2,500 tons; two thin lines, 2500 to 5000; three, 5000 to 10,000; four, 10,000 to 25,000; and five, over 25,000. The thin lines above and to the left of the central heavy line represent west and southbound movement; lines below and to the right represent north and eastbound movement.

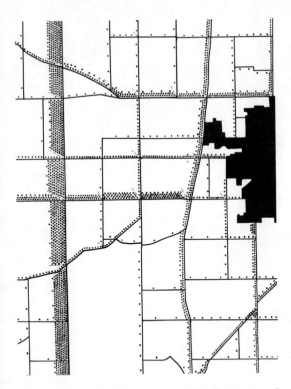

Figure 6.26 Part of a large-scale flow map of highway traffic in a segment of southeastern Wisconsin west of Kenosha. Each dot represents 100 vehicles in a unit distance of one mile (at map scale) in an average 24-hour period. (Map by Gwen M. Schultz.)

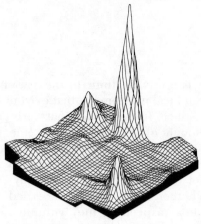

MAPPING AREA AND VOLUME DATA

There is no limit to the variety of geographical phenomena that exist over area as distinct from those that are generally thought of as being point or linear in character. This bewildering array is made more difficult for the cartographer because it is not easy to decide in many instances whether he is more concerned with the "area" aspects or the "volume" aspects of a distribution. As in any method of classification, the system of divisions adopted here regarding geographical phenomena (point, line, area, and volume) is arbitrary, since the class limits are positions in a continuum. Consequently, it is difficult to fit some phenomena into the groups. This is particularly true of geographical area and volume data.

The allocation of a to-be-mapped phenomenon to a class often depends upon your point of view or the characteristic that is singled out for emphasis. For example, a road is usually thought of as a linear phenomenon, but at a very large scale and for some purposes, the fact that its surface character is different from its surroundings may be the important matter, as, for example, in the study of variations in albedo. Similarly, a city is frequently thought of as a geographical point phenomenon but a given array of cities in a region can be considered as distinguishing that area from another with a different frequency of city incidence. The fundamental point to be kept in mind is that we must first determine the attribute that is to be given the emphasis before the mapping system can be chosen. When approached from this point of view, the cartographic representation of areas and volumes is not difficult.

Mapping Qualitative Area and Volume Data

As was observed in the discussions having to do with mapping point and linear data, the mapping of geographical data that are scaled or ordered nominally is relatively straightforward. Forests versus grasslands, national areas, culture regions, all the myriad ways we can distinguish the character of one geographical region from another are accomplished by utilizing area symbols that graphically distinguish the one from the other. These symbols may be colors, patterns, or combinations. Colors can vary in hue, value, or intensity, and patterns can vary in texture, arrangement, and orientation. The use of color and pattern symbols involves many complications; these matters are treated in Chapter 11.

Nominal differentiation of area data often presents the cartographer with a problem for which there is no easy solution. Frequently, nominal categories are not geographically exclusive, that is, two or more categories often occur in the same area. Examples are botanical species, races, and land use characteristics. Accordingly, if the cartographer does not want to generalize sufficiently to remove the mixture, he is required to employ symbolization that somehow represents geographical mixture or overlap. This may be done in a number of ways, as suggested in Figure 7.1, none of which is suited to all circumstances. If the cartographer is not careful, this practice can lead to incredibly complicated maps.

If color is being employed, it is possible to choose colors that give the impression of mixture. For example, a red and a blue when superposed appear purple, and that

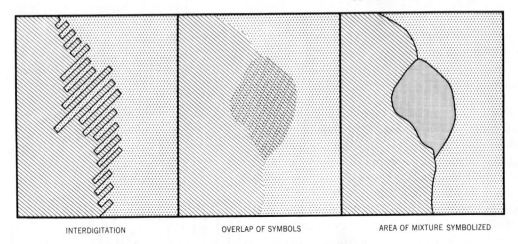

INTERDIGITATION   OVERLAP OF SYMBOLS   AREA OF MIXTURE SYMBOLIZED

Figure 7.1 Several methods of showing geographical mixture or overlap with area symbols.

color looks like a mixture. On the other hand, a color produced by mixing blue and yellow produces green and that color does not look like a mixture of its components.

Strictly nominal characterization of geographical volumes in cartography is uncommon. There are many such, ranging from air masses and kinds of precipitation (e.g., snow versus rain) to glaciers and other kinds of rock, but when represented on maps these phenomena are usually treated as simply occurring in given areas (area symbols), or their quantitative characteristics are emphasized.

Kinds of Quantitative Area and Volume Distributions

As was pointed out above, a single geographical distribution can often be thought of in several ways, and it is fundamentally important that the cartographer initially make a decision regarding the conceptual class of representation he will employ. In mapping quantitative area and volume data, there are two basically different methods of representation. The distinction be-

tween the two can be clarified by an example involving population.

We may derive the ratio or density between numbers of persons and the areas in which they live by utilizing statistical unit areas (census enumeration districts, for example), as previously described, as a basis for density values. The dominant interest in the ratios may be merely their occurrence and magnitudes as individual values related to the areas they represent. On the other hand, in spite of the fact that the factor of population is obviously discrete, that is, not continuous, the paramount concern may be with the spatial assemblage of the ratios as they constitute a continuously distributed phenomenon similar to a temperature distribution. Consequently, the representation of a concept of density may be treated cartographically in either of two ways, depending upon whether the interest is (1) as an area distribution where the focus of interest is on the specific ratio values at particular places, or (2) as a volume distribution where the focus of interest is on the geographical organization of the magnitudes and directions of the gradients, that is, in its surface configuration. Needless to say, the cartographer may sometimes try to gain both objectives.

The Statistical Surface

The distinction between quantitative area distributions and volume distributions, although logical and very useful for pedagogical purposes, is generally ignored in the literature. Instead, both are usually lumped together in what has been called the concept of the statistical surface.* If the magnitudes of ratios assigned to unit areas or the sample values of continuous distri-

*See Arthur H. Robinson, "The Cartographic Representation of the Statistical Surface," *International Yearbook of Cartography,* **1**, 53–62 (1961); and George F. Jenks, "Generalization in Statistical Mapping," *Annals Association of American Geographers,* **53**, 15–26 (1963).

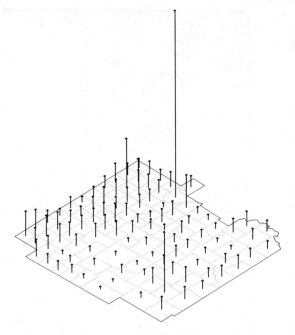

Figure 7.3   Elevated points, the relative height of each of which is proportional to the number in that unit area in Fig. 7.2. (Courtesy George F. Jenks and *Annals of the Association of American Geographers*.)

butions (e.g., temperatures at weather stations) are conceived to have a relative vertical dimension, the sample values or the unit area values can both be visualized as forming a three-dimensional surface.

For example, Figure 7.2 shows an array of numerical data that can either be a set of ratios, such as population densities, or a set of observations, such as temperatures at weather stations, each arbitrarily located at the center of a unit area. Figure 7.3 shows the erection of a column in each unit area the length of which is proportional to the number in that unit area in Figure 7.2. In Figure 7.4 the relative magnitudes of the values in the unit areas are emphasized by erecting over each a prism that has a base shaped like the unit area and a height that is scaled as the length of the line in Figure 7.3. In a very real sense, this is a three-

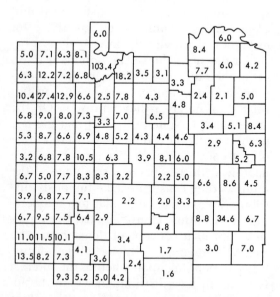

Figure 7.2   An array of statistical data in unit areas. The numbers are actually rural population densities in some minor civil divisions in Kansas, but the numbers could be sample values of some other continuous distribution. (Courtesy George F. Jenks and *Annals of the Association of American Geographers*.)

dimensional form of the familiar two-dimensional histogram. Figure 7.5, on the other hand, illustrates another way of showing the same distribution, but here the emphasis is put on the magnitudes and directions of the gradients. Both Figures 7.4 and 7.5 can be thought of as the representation of a statistical surface consisting of a series of numbers each of which has an *x, y,* and *z* characteristic. The *x* and *y* values refer to the horizontal, or planimetric, locations and the *z* values to the assumed relative heights above some horizontal datum, such as the plane of the map.

Mapping the Statistical Surface

When the emphasis in the mapping of the statistical surface is on the locational

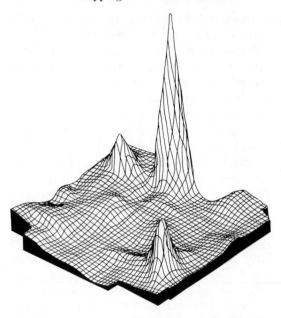

Figure 7.5   A perspective view of the smoothed statistical surface produced by assuming gradients among all the numbers shown in Fig. 7.2. The elevated points in Fig. 7.3 would all just touch the surface shown here. (Courtesy George F. Jenks and *Annals of the Association of American Geographers.*)

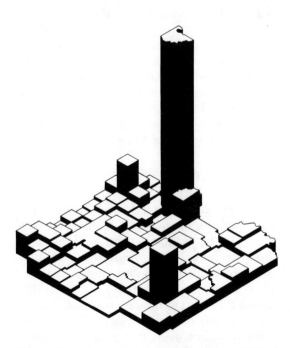

Figure 7.4   A perspective view of the statistical surface produced by erecting prisms over each unit area proportional in height to the numbers shown in Fig. 7.2. (Courtesy George F. Jenks and *Annals of the Association of American Geographers.*)

representation of the *z* values, that is, when the array of numbers is conceived as a quantitative area distribution, the system of representation is called *choropleth* mapping (from Gk. *chōros,* place, and *plēthos,* magnitude). There are two kinds of choropleth mapping. In one, called *simple choropleth* mapping, the primary objective is to symbolize the magnitudes of the statistics as they occur within the boundaries of the unit areas, such as counties, states, or other kinds of enumeration districts.

In the other subclass, called *dasymetric* mapping (from Gk. *dasys,* thick or dense, and *metron,* measure) the primary objectives are to focus interest on (1) the location and *z* magnitudes of areas having relative *z* uniformity, regardless of the unit area boundaries, and (2) on the zones between which there occur more or less abrupt changes in these magnitudes. Both kinds of maps employ area symbols, and since they are quantitative, the darker visual value (gray tone) or the more intense color is assigned to the greater magnitudes.

When the emphasis in the mapping is on the gradients among the z values, their magnitudes and directions, a basically different kind of symbolization system is employed, called *isarithmic* mapping (from Gk. *isos*, equal, and *arithmos*, number). There are also two kinds of isarithmic mapping, isometric and isopleth, but the distinction between them is based on the characteristics of the data (as is explained later in the chapter) rather than the nature of the distribution or objective of the mapping system. All isarithmic mapping has the same appearance and is primarily concerned with portraying gradients. There are many examples of isarithmic maps and probably the best known is the familiar contour map showing the relative elevations and gradients of the form of the land, that is, its surface configuration. Since the isarithmic technique can be applied to the kinds of z values that may also be mapped by choropleth or dasymetric methods, before he can settle on the mapping method,

the cartographer must first decide which characteristics of the data are paramount in his set of objectives.

Whichever of the three methods is employed, it is always necessary to group, that is, to generalize, the data in order to symbolize them. This calls for the selection of a system of class intervals and the designation of class limits. These important topics are treated separately in the latter part of this chapter.

The simple choropleth, the dasymetric, and the isarithmic methods are shown comparatively in Figure 7.6. The diagnostic characteristics, that is, the basic differences in appearance, among them are clearly evident. In the simple choropleth map the geographical locations of the class limits employed coincide with the boundaries of the unit areas. In the dasymetric map the map positions of the class limits are independent of the boundaries of the enumeration districts, but the lines separating one class from another have no numerical value. In the isarithmic map the geographical positions of the lines showing the class interval employed are independent of the unit areas; they are lines along which the value is assumed to be constant.

CHOROPLETH MAPPING

The Simple Choropleth Map

The simple choropleth map is, in effect, a spatially arranged presentation of statistics that are tied to enumeration districts on the ground. Tabular statistics are very convenient for many purposes, but frequently we want to refer to the array geographically rather than alphabetically and to be able easily to compare magnitudes in various places. This kind of map is made simply by symbolizing in some manner the quantitative magnitude that applies to the enumeration district or some other sort of unit area.

The simple choropleth map requires the least analysis on the part of the cartographer. This technique may be used for many kinds of ratios, and when the statistical units are small it provides sufficient variability in the map (if there is variability among the data) to take on some of the visual character of a continuous distribution. Ordinarily, however, it is employed when the problems of doing any other sort of map make it out of the question. For example, when mapping at a large scale, the cartographer may not know enough about the details of the distribution, or at a small

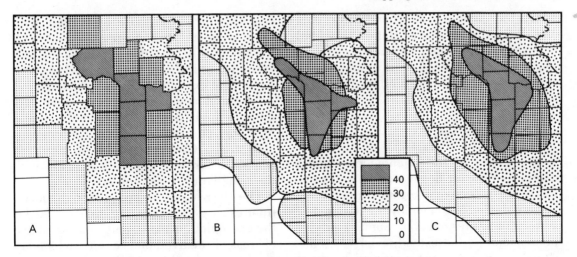

Figure 7.6 Examples of the three ways one can represent the set of _z_ values that refer to enumeration districts, or unit areas: _A_, a simple choropleth map; _B_, a dasymetric map; and _C_, an isarithmic map. The addition of area symbols to map _C_ as shown here is not absolutely necessary. Many isarithmic maps show only the numbered lines, which, by their relative positions, show the gradients and their directions.

scale, the regional variations may be too inconsistent to attempt an isarithmic or dasymetric map. In effect, the choropleth map presents only the spatial organization of the statistical data, with no effort being made to insert any inferences into the presentation.

Absolute numbers alone are not ordinarily presented in a simple choropleth map. For a great many classes of data, the absolute quantities of a phenomena (such as persons, automobiles, farms, and vacationers) are naturally functions of the areal size of the enumeration district. When we are concerned with the geographical distribution, we are normally concerned with comparisons of ratios involving area (density) or ratios that are independent of area (percentages or proportions), that is, when the effect of variations in the size of the enumeration districts has been removed.

The quantity mapped is always some kind of geographical average that is assumed to refer to the whole of a unit area.

Where class limits call for a change of symbolization, it occurs only at the unit area boundaries in a simple choropleth map. Since the boundaries are usually quite unrelated to the variations in the phenomenon being mapped, this adds a further measure of geographical obscurity to that supplied by the practice of dealing entirely with averages. Generally, the smaller the unit areas the more revealing will be the simple choropleth map. On the other hand, if the size of the unit areas is very small and the interest of the reader is perforce focused on the configuration of the statistical surface and the gradients thereof, then it would be better to map the data by isarithmic methods.

Choropleth mapping is "safe," in that we simply symbolize the quantities where they are, but it is the least informative of the methods of mapping these kinds of distributions. The production of simple choropleth maps is easily automated. Because of the speed and relatively low cost with which the computer can handle the data and direct the automatic plotter, this kind of map will no doubt increase greatly in importance. Figure 7.7 is an example of a simple choropleth map produced entirely automatically.

The Dasymetric Map

The nature of the phenomenon being mapped, or the inefficiency of the unit areas employed to derive the ratio values, may make it desirable to employ the dasymetric system of representation. Some kinds of phenomena do not extend, even when abstractly conceived, across certain kinds of areas. Neither do they have distributional

characteristics like those associated with the sloping-surface-concept associated with a continuously changing set of values. For example, agricultural population density in a region of extreme variation in character of agricultural land is likely to show relatively uniform high density in favorable areas together with sparse occupance in unfavorable areas. Changes would be rapid in zones that are not likely to coincide with the boundaries of the unit areas.

The dasymetric map, although made from exactly the same initial data used for the choropleth map, assumes the existence of areas of relative homogeneity, which are also assumed to be separated from one another by rapid changes; to use the analogy

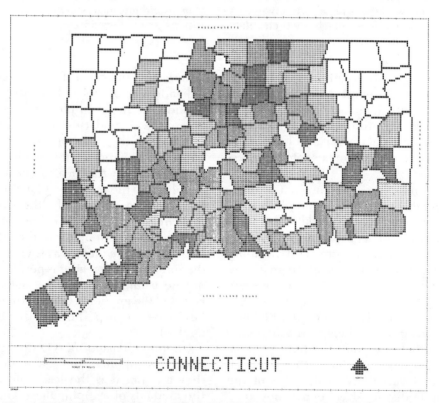

Figure 7.7 A machine-produced, simple choropleth map of Connecticut showing the average sale price per house by towns. (Map prepared by the Laboratory for Computer Graphics, Harvard University.)

of showing the surface configuration of a volume, these zones of rapid change would be escarpments on such a surface. By subdivision of the original statistical unit areas, additional detail may be added to the presentation. These additions are made on the basis of whatever knowledge the cartographer has respecting the data. This may be founded on field knowledge or any other kinds of data the cartographer may obtain.

The arithmetic involved in reconciling the numerical assumptions with the original unit area data has been explained and illustrated in Chapter 5. It should be emphasized that there is nothing in the data itself that will indicate to the cartographer the kinds of subdivision he should make or where the zones of rapid change occur. Instead, this kind of information must come from other knowledge concerning the spatial relationships among the data and other geographical distributions. It may be observed in this connection that it is very much more difficult to program the dasymetric process for automation than it is the simple choropleth, since it is not only complicated but each application is unique.

The phenomena found in association with the distribution being mapped and upon which the cartographer must base his rearrangement of the basic unit area data are of two kinds which may be called (a) limiting variables and (b) related variables. Without a knowledge of these two classes of phenomena, the cartographer cannot modify the bare statistics supplied by the enumeration and he must be content with the simple choropleth technique. It is not necessary that there be a cause and effect relationship between the distribution being mapped and either of the two classes of variables; it is only necessary that there be a geographical association that enables us to insert a greater degree of geographical reality into the generally barren statistics of the simple choropleth map.

Limiting Variables. These are variables that set an absolute upper limit on the quantity of the mapped phenomenon that can occur in an area. For example, suppose we were mapping percent of cropland and the data showed that county Y had 60% of its total area as cropland. If 15% of the area of the county were devoted to urban land uses, it is obvious that that area could not be cropland. Exclusion of the urban area allocates all the cropland to the remaining sections of the county.

Suppose, furthermore, that a second limiting variable in county Y were woodland, the areal percentage of which was known for minor civil divisions such as townships. If a township had 60% of its area in woodland, then the maximum cropland it could have would be 40%. Other areas of the county would, therefore, need to have a much higher percent of cropland to make possible the original statistic of 60% for the county as a whole.

The functioning of limiting variables in the mapping process can be seen more easily when placed in a diagrammatic cross-classification, as in Figure 7.8. If cropland classes are: less than 10%, 10 to 30%, 30 to 50%, 50 to 70%, and over 70%, then it is evident that the existence of urban land uses limits the cropland class to less than

		CROPLAND CLASSES—%				
		<10	10-30	30-50	50-70	>70
URBAN LAND USE	Yes					
	No					

Figure 7.8 Diagram showing the possibilities for the occurrence of cropland classes in the presence of urban land use. The shading shows those classes that cannot occur.

10%; where no urban land uses occur, all five cropland classes are possible. The classification system can be expanded to include a second limiting variable as shown in Figure 7.9. Assuming the same kinds of quantitative classes for woodland, it can be seen that the possibilities for the occurrence of cropland classes are severely limited.

The diagrams in Figure 7.10 show how a knowledge of the geographical occurrence of limiting variables enables the cartographer to add considerable geographic detail in dasymetric mapping to the simple choropleth representation. In *A* all that is known is that the county in question consists of 55% cropland. In *B* and *C* the limiting variables have been mapped: these consist of urban land use, consisting of 15% of the area of the county and an area of 50 to 70% woodland (occupying 42% of the area of the county). Using the formula for estimating densities of fractional parts and starting with urban land use, if we assume there is no cropland in the urban area, we may determine that the county outside the urban area consists of 65% cropland. That section, as shown in *B*, is almost half occupied by an area of 50 to 70% woodland. Assum-

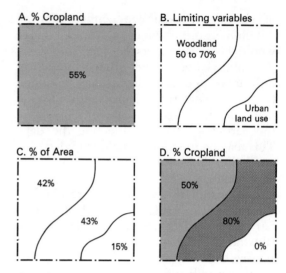

Figure 7.10   How a knowledge of limiting variables can assist in dasymetric mapping. See text for explanation.

ing that the most cropland that could occur in the woodland area would be 50%, we may next determine that the area between the woodland zone and the urban area is approximately 80% cropland. A choropleth map would have found the county in only one class; approached dasymetrically, there are three classes represented.

Related Variables. The functioning of related variables in the dasymetric mapping process is much more complex than is that of limiting variables. Related variables are those geographical phenomena that show predictable variations in spatial association with the phenomenon being mapped, but that cannot be employed in a limiting sense. To continue with the example of percent of cropland, we may have available a map of surface configuration that shows that the land form varies from level on the one hand to very hilly, steep country on the other. There would normally be a high positive correlation between level land and percent of cropland. Similarly, we could think of several other varia-

		CROPLAND CLASSES—%					
		<10	10-30	30-50	50-70	>70	
URBAN LAND USE	Yes						<10
							10-30
							30-50
							50-70
							>70
	No						<10
							10-30
							30-50
							50-70
							>70

WOODLAND CLASSES—%

Figure 7.9   Diagram showing the relation of two limiting variables to the possible occurrence of cropland classes. The shading shows those classes that cannot occur.

bles that would assist in predicting variations in the geographical occurrence of percent of cropland, such as types of farming regions and soil characteristics.

Related variables are more difficult to use properly because they cannot be employed in the strict manner possible with limiting variables. They are, however, very useful in helping the cartographer put some geographical "sense" into the bare statistics that are commonly gathered on the basis of enumeration districts, the boundaries of which often have little or no relation to the precise distribution of the phenomenon being mapped.

ISARITHMIC MAPPING

When the interest in a geographical distribution is focused (1) primarily on the *form* of the distribution, that is, the organization of the orientations, magnitudes, and directions (sense) of the myriad gradients which together constitute it, or (2) on the values at points of a truly continuous distribution, such as the land (elevation), air temperature, or pressure, the isarithmic method in one form or another is usually used. We can map these kinds of data by simple choropleth or dasymetric (if appropriate) methods, but it is not common.[*]

In isarithmic mapping, the distribution is clearly conceived of as a volume and, in order visually to comprehend a volume, it is obviously necessary to see the shape of the outside surface enclosing it. A geographical volume is consciously or unconsciously thought of as resting upon an underneath surface such as mean sea level. The z values of the distribution, which define the volume, are therefore departures from the assumed or actual base.[*] Together the z values suggest a three-dimensional surface (see Figs. 7.2 to 7.5). From the geometric form of this surface we can infer the extent of the volume resting on the base. In many instances, the actual *volume* itself is less interesting and of less concern than is the form of its surface. This is not always the case; for example, a road engineer is greatly concerned with actual volumes as in cut-and-fill problems. Nevertheless, the character of a three-dimensional geographical distribution is most clearly mapped by delineating its surface.

The symbolization of real or abstract three-dimensional surfaces is difficult, and more time and effort probably have been devoted to it than to all other problems of symbolization put together. The principles involved in delineating such a surface are best illustrated by beginning with the familiar example of the surface of the land. This continuous, sometimes steep, sometimes almost flat, and usually not level surface is so intimately connected to man's life that its character and delineation are of outstanding significance; consequently, numerous ways of depicting its configuration have been devised for various purposes (see Chapter 8). In the course of time it has developed that the most useful way is the isarithmic technique, which enables us (with training) not only to visualize the three-dimensional form of the land, but which also makes possible the derivation of a variety of useful data from the method used, such as slope or gradient, elevation and the recognition of ridges or troughs, and flat plains.

[*]For example, average elevation above sea level, and average air temperature and pressure, can be mapped by unit areas, but it is rarely done.

[*]In most cases, these values are thought of as positive magnitudes lying above the assumed base, but sometimes the reverse is true. For example, when we think of the volume of the sea, we visualize magnitudes below the level surface.

Isarithms

If the irregular land surface has been mapped in terms of *planimetry,* that is, the relative horizontal position of all points on the land, it is evident that there exists an infinity of points, each of which has, by reason of its location, an *x, y,* and *z* coordinate position with respect to the *datum* surface (spheroid) to which the earth's surface has been orthogonally transferred (see Fig. 8.7). By definition, the land surface is at all points either above or below the smooth assumed reference elevation, called a *datum.* If an imaginary surface, parallel to the horizontal datum and a given *z* distance from it, is assumed to intersect the irregular land surface, it must do so at all points having that *z* value. The *trace,* or the line of intersection, of these two surfaces will be a closed line. When this line, an isarithm, is orthogonally viewed, that is, perpendicularly projected to the map, it shows by its position the *x* and *y* locations of all the points that have the particular *z* value it represents.

Figure 7.11 shows, in perspective, a hypothetical island, evenly spaced *z* levels, and below, an isarithmic map of the distribution of the *z* values on the surface of the island. In this case, *z* is elevation above the average level of the sea, which is defined as 0; it is therefore a contour map. The lowest or outer isarithm represents the average position of the shoreline. The next isarithm in Figure 7.11 is the trace of the plane spaced 20 *z* units above 0; it is in the same location as the average shoreline would be were the sea 20 *z* units higher.

The configuration of a three-dimensional surface is symbolized by the characteristic shapes and patterns of spacings of a set of isarithms, especially when the intersecting *z* planes are equidistant from one another. Smooth, steep, gentle, concave, convex, and other simple kinds of gradients and combining forms may be readily visualized from isarithmic maps, as indicated in Figure 7.12. For example, the bends of contours always point upstream when they lie athwart a valley; they always point down slope when crossing a spur. The angles of slope, that is, the gradients or rates of change of elevation, of the land are shown by the relative spacings of the set of contours provided by equally spaced *z* planes. Profiles of the land surface along a traverse, or along a road or railroad, can easily be constructed from a contour map by working backward from the map to the profile. The recognition of detailed topographic forms and often even genetic structural details are readily revealed by the patterns of the contours on topographic maps.

It is not necessary that the *z* surface to be represented by isarithms be an actual visible, tangible surface, such as the land; the configuration of any three-dimensional surface may be mapped in the same way. For example, the form of a defined pressure surface in the atmosphere, such as the 500 millibar surface, may be mapped by passing *z* levels through the atmosphere. The shapes in nature are not visible, but the patterns of the isarithms show the gradients, the troughs, and the ridges of that surface in the same way that contours show the ups and downs of the land. The three-dimensional surface may even be an abstraction. For example, the *z* values may consist of some sort of ratio or proportion, such as persons per square mile. Anything that varies in magnitude and either actually exists or can be assumed to exist in continuous fashion over area constitutes a statistical surface. Its configuration can be mapped isarithmically.

Inferring the Statistical Surface. In order to map the traces of the intersections of horizontal levels with a statistical surface, it

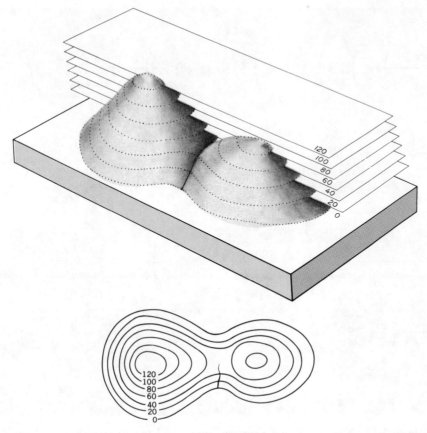

Figure 7.11   In the upper diagram, horizontal levels of given z values are seen passing part way through a hypothetical island. The traces of the intersections of the planes with the island surface are indicated by dotted lines. In the lower drawing, the traces have been mapped orthogonally on the map plane and constitute the representation of the island by means of isarithms (contours).

is necessary, as in following the proverbial recipe for rabbit stew, first to "catch" the statistical surface. This is easier said than done.

Only since the development of stereo-viewing equipment for air photographs has it been possible to specify in its entirety the infinity of points on the land, each of which has its z value of relative elevation. In *all* other cases the statistical surface must be obtained from a limited number of z values, and the totality of the surface

must be *inferred* from these. In a very real sense, the infinity of z values constitutes a statistical population or universe but, because of practical limitations, only a sample of these points is usually available. Since we cannot know precisely the characteristics of the universe from which the sample has been taken, any extension of the characteristics of the sample to the universe or to a particular part of it constitutes but an inference, the validity of which can only be fixed provided we have certain basic

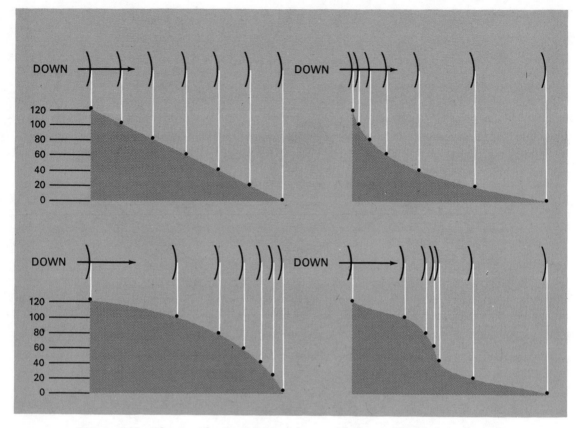

Figure 7.12 The isarithmic spacings above and the profiles below show, in each diagram, the manner in which the spacings of the isarithms show the nature of the configurations from which they result. All the forms may be described as variations of the general rule that if the isarithms are the traces of equally spaced z levels, the closer the isarithms are, the greater the gradient will be.

knowledge about the characteristics of the universe.

An example will clarify the foregoing. If air temperature is simultaneously observed at a series of weather stations, then the temperatures at the given station positions constitute the sample of z values. If it is desired to make a synoptic (instantaneous view) map of the total distribution of temperatures (by means of *isotherms*, that is, "contours" of temperature), it can be done inductively only by making assumptions and inferences as to the nature of the temperatures that existed at the infinity of points that lie between the stations, for which, of course, actual z values are not available. It is apparent then that the accuracy and representativeness of the given sample values, which the cartographer uses to locate the traces, are of considerable significance to the inference he draws of the total statistical surface. Various aspects of the probable validity of sample values are considered in subsequent sections.

Kinds of Isarithmic Mapping

It has been common practice in the past to give names to the isarithms employed for a particular kind of phenomenon. Thus we have isotherms (temperature), isobars (barometric pressure), isohyets (precipitation), and so on, and by now the number of such terms is overwhelming, to say the least. Such proliferation of the technical terminology serves no very useful purpose (except when necessary to distinguish among two sets of isarithms on one map), especially when it tends to direct attention from the more important fact that there do exist some fundamental differences between the two major uses of isarithms.

There are two classes of z values that differ in terms of the precision with which they can specify a statistical surface: (1) a. actual values that can occur at points; b. derived values that can occur at points; and (2) derived values that cannot occur at points.

Actual values that can exist at points are exemplified by data such as elevation above or depth below sea level, a given actual temperature, the actual depth of precipitation, and thickness of a rock stratum. These kinds of values do exist at points. Only actual errors in observation or in the specification of the xy positions of the observation points can affect the validity of the sample values. If accurate quantities were known for *all* points, then each isarithm portraying the statistical surface could have only one position on the map.

Derived values that can exist at points are of two kinds. One kind consists of averages or measures of dispersion, such as means, medians, standard deviations, and other sorts of statistics derived from a time-series of observations made at a point. We can calculate a mean monthly temperature, an average retail sales figure, or an average land value for some particular place; the resulting numbers, although representative of magnitudes at the point in question, cannot, by their very nature, actually exist at any moment. They are subject to various kinds of error, particularly with regard to the extent of the sampling from which the derivation was made. A second kind of derived value that can occur at points consists of ratios and percentages of point values. Examples of these are the ratio of dry to rainy days that occurred at a particular place, or the percentage of total precipitation that fell as snow. Such ratios are also incapable of existing at any instant, but they do represent quantities that apply to the point for which they are derived. Like measures of central tendency or dispersion, they are generally subject to more error than simple actual values. If they are rigorously defined and uniformly derived, they approach the validity of actual point values.

Quite different in concept is the derived value that cannot occur at a point. Representative of this class are percentages and other kinds of ratios that include area in their definition directly or by implication, such as persons per square mile, the ratio of beef cattle to total cattle, or the ratio of cropland to total land in farms. With such a quantity, only an average value for a unit area can be derived. Consequently, although it is perfectly legitimate to assume a statistical surface specified by these kinds of quantities, such a surface is dependent only upon a series of average values for unit areas. Since each unit area represents a larger or smaller aggregate of xy points, no single point can have such a value. Nevertheless, in order to symbolize the undulations of the statistical surface by isarithms, it is necessary to assume the existence of such z values at singular points.

Because of the fundamental differences in the concepts of these two kinds of statistical surfaces, it is conventional to make a distinction among the isarithms employed to display their form. Different names are applied to them, and considerable confusion in definition and spelling exists in the literature. The following terminology seems to be consistent with present-day usage.

An *isarithm* is any trace of the intersection of a horizontal plane with a statistical surface. It is thus the generic term; it may also be called an *isoline* or *isogram*. Isarithms showing the distribution of actual or derived quantities that can occur at points (in which there is relatively little error) are called *isometric lines*. In contradistinction those isarithms that display the configurations of statistical surfaces that are based upon quantities that cannot exist at points, and that are likely to be subject to a somewhat larger inherent error of position, are commonly called *isopleths*.

An observation is in order here. Much less confusion will result if the student will transfer the distinctions just made from the symbols to the surfaces they delineate. An isarithm is defined simply as the trace made by the intersection of a horizontal *z* level with *any* three-dimensional surface. Whatever the nature of the surface may be, the function of the trace as a cartographic symbol is the same. Consequently, the "difference" between an isopleth and an isometric line is in reality an attempt to distinguish between the precision of two kinds of surfaces, rather than a distinction between kinds of cartographic symbols.

Isarithmic maps serve two purposes: (1) they may provide a total view of the configuration of the statistical surface of a volume distribution, for example, the form of the land; and (2) they may serve to portray the location of a series of quantities, an area distribution, for example, elevations at points. The ability with which they perform these functions is dependent upon

the validity and reality of the surface. For example, we can safely obtain values of elevations by interpolation from a contour map, within a certain margin of error. On many isarithmic maps this cannot be so readily done; this is especially true of large-scale isoplethic maps. In such maps the isopleths serve more as "form lines," to delineate the character of the surface and less as a method of portraying specific values at points.

Elements of Isarithmic Mapping

When we are called upon to prepare an isarithmic map, whether we are working with an isarithmic or isopleth surface, we follow much the same procedure. In each case the numbers on our map, constituting the series of *z* values, are spaced some distance apart, and whether they can or cannot actually occur at points, we must, by assumption, inference, and estimate, produce isarithms representing a continuous statistical surface.

Location of the Control Points. The location of each *z* value of the assumed statistical surface is called the *control point*. The *xy* positions and *z* values of these points constitute the statistical evidence, called in cartography the control, from which the locations of the isarithms are inferred. If the *z* value of a control point is correct but the *xy* position is incorrect, then a displacement of the isarithm will result. If the *xy* position is correct but the *z* value is in error, the same thing will happen if a sloping surface with the same gradient is assumed. This is illustrated in Figure 7.13. Clearly, both the positions of the control points and the validity of the value at the control point have considerable effect on the location of the statistical surface and hence on the positions of the isarithms.

The problem of choosing a location for

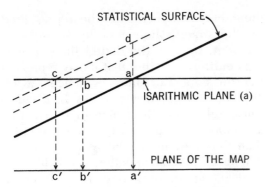

Figure 7.13   If *a'* is the *xy* location of a control point with *z* value *a*, and if the isarithmic plane has the value *a*, then *a'* will be the orthogonal position on the map of the isarithm. If the *z* value *a* is incorrect and really should be *d*, then the *a* isarithmic plane would intersect the surface at *c* and the isarithm on the map would be located at *c'*. If the *a* value were correct but the *xy* position of *a* should be at *b*, then *b'* would be the map position of the isarithm.

be chosen as the center. If the distribution within the unit area is known to be uneven, the control point is shifted toward the concentration. Center of area may be considered as the balance point of an area having an even distribution of values without any unevenness of the distribution taken into account. The center of gravity takes into account any variation of the distribution. Figure 7.14 illustrates the concept. The four diagrams of statistical divisions show possible locations of the center of gravity and center of area for uniform and variable distributions, with a dot map of rural population used as an example. They serve also to illustrate the problem of locating the control point in regularly and irregularly shaped divisions. In A, which is rectangular, the center of gravity and center of area would, of course, coincide at the intersection of the diagonals. Because the distribu-

the control point is not difficult when the statistical surface being mapped is based upon actual or derived quantities that occur at points. The location of each observation is then the location of the control point. This is not so, however, when the mapping distributions are derived from ratios, percentages, and the like that involve area in their definitions, such as, for example, the density of population (persons per unit of area). These kinds of quantities are derived from two sets of data based on unit areas. The resulting numbers, upon which the ultimate locations of the isopleths depend, refer to the whole areal unit employed, and each is "spread," so to speak, over the entire area of the unit. Therefore, there can be no *points* at which the values used in plotting the isopleths exist. Nevertheless, the lines must be located somewhere and upon some basis. In order to do this, positions of control points are assumed.

When the distribution is uniform over an area of regular shape, the control point may

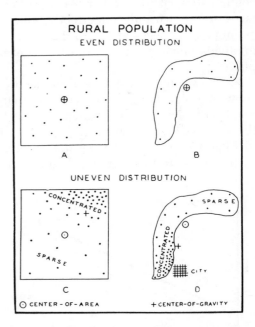

Figure 7.14   Placing control points. In each, a dot map is used to represent a distribution. (From Mackay, courtesy of *Economic Geography*.)

tion in *B* is even, the centers also coincide, but they lie outside the irregularly shaped area at some point that is more representative of the whole division than any point within it. The distributions are uneven in *C* and *D*, so the center of gravity is displaced away from the center of area. In each case, however, the center of gravity is probably the more reasonable location for the control point.

Gradient and Interpolation on Isarithmic Maps. The selection of the isarithmic method follows upon the assumption that the distribution being mapped is primarily composed of slopes. The surface of the land is visible either by stereophotogrammetric methods or in the field. The infinity of *z* values can, therefore, be directly or indirectly employed as a guide for the delineation of the form by the isarithms. The basic characteristics of the configuration of most other statistical surfaces are known only imperfectly, and an assumption is necessary as to the kinds of gradients between the *z* values at control points.

A cross section of a statistical surface that shows a profile along the top is an ordinary graph in which *z* is the ordinate and *xy* is the abscissa. Consequently, as is illustrated in Figure 7.15, if two *z* values at different *xy* positions are represented, then the gradient between them may be represented by the straight line *a* or by some other gradient, such as the dotted lines *b* or *c*. It is entirely possible that *c* may represent the true slope; but unless evidence is available to indicate that *z* varies in a curvilinear relation with change in *xy*, such an assumption is obviously more complex than the gradient shown by *a*. In science, whenever several hypotheses can fit a set of data, the simplest is chosen. As previously pointed out, the slopes of the actual land surface are known, in most cases, to bear certain kinds of curvilinear relationships to *xy* positions of the land. On the other hand, for most other statistical data distributed over the earth, this is not known. There are theories in some cases, such as that population density tends to be curvilinear (for example, *b* in Figure 7.15), but their validity has not yet been generally demonstrated. Consequently, in the majority of cases a linear gradient is assumed in the construction of isarithmic maps.

Interpolation is the name for the process of estimating the magnitude of intermediate values in a series, such as the *z* values along the lines shown in Figure 7.15. The control points of a statistical surface to be mapped by isarithms constitute the series, and when no evidence exists to indicate a nonlinear gradient between control points, interpolation becomes merely a matter of estimating linear distances on the map in proportion to the difference between the control point values. For example, Figure 7.16 is a map on which are located *z* values at *xy* positions *a*, *b*, *c*, and *d*. If the position of the isarithm with a value of 20 is desired, it will lie 3/10 of the distance from *a* to *b*, 3/7 of the distance from *a* to *c*, and 3/4 of the distance from *a* to *d*. Lacking any other data, the dotted line representing the 20-isarithm would be drawn as a smooth line through these three interpolated positions.

Professor J. Ross Mackay has illustrated

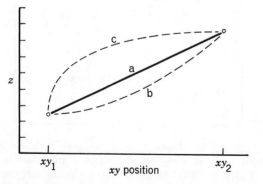

Figure 7.15   Three kinds of gradient between two control points.

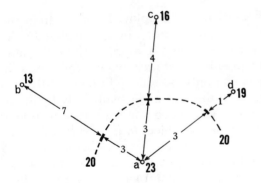

Figure 7.16 Interpolation between control points.

one of the common problems that arises when interpolating "by hand" among control points located in a rectangular pattern.* Much census data are based upon the rectangular minor civil divisions of the United States and Canada, and many other surveys use more or less rectangular subdivisions. When control points are arranged rectangularly, it is usual that alternative choices will arise concerning the location of an isarithm when one pair of diagonally opposite z values forming the corners of the rectangle is above and the other pair below the value of the isarithm to be drawn. Figure 7.17 illustrates the problem.

A careful examination of other relevant information may help to indicate which choice of the two alternatives is the better. If this is not possible, averaging of the interpolated values at the intersection of diagonals will usually provide a value that will remove the element of choice (see Fig. 7.18). We are forced, in the absence of other data, to assume the validity of the average.

If the cartographer has control over the shapes of the unit areas, he can prevent the problem of alternative choice by designing the pattern of unit areas in several ways so that the control points have a triangular pattern, such as are shown in Figure 7.19.

*J. Ross Mackay, "The Alternative Choice in Isopleth Interpolation," *The Professional Geographer,* V (New Series), 2–4 (1953).

Another related kind of problem, which is just as significant when interpolating from isarithmic maps, results from the fact that the steepest gradients on an irregular form twist and turn over the surface. Thus, the line representing the linear gradient on a curved surface will be a line normal (at right angles) to the isarithms and will be curved when orthogonally projected to the map plane (cf. hachures). In Figure 7.20, diagram *A* shows two straight parallel isarithms. Because the surface between those two isarithms must be a plane (with the assumption of linear gradient), the value of point *a* is interpolated merely by the relative distances to the adjacent isarithms. In diagram *B* point *a'* would be interpolated by the relative distances along the smooth curve that passes through *a'* and lies normal to the isarithms. Therefore, when the cartographer is plotting isarithms on a map by hand, he should "look ahead" and determine by the arrangement of z values the

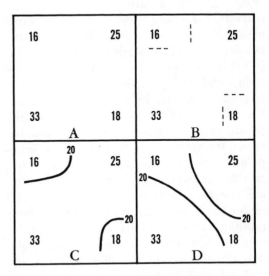

Figure 7.17 In *A* the z values are arranged rectangularly. In *B* the positions of the 20-isopleths between adjacent pairs have been interpolated. Diagrams *C* and *D* show the two ways the isopleth of 20 could be drawn through these points. (Redrawn, courtesy of Mackay and *The Professional Geographer.*)

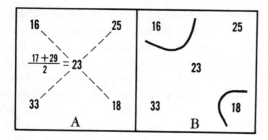

Figure 7.18 If the average of the interpolated values at the center is assumed to be correct, the isopleth of 20 would be drawn as in B. (Redrawn, courtesy of Mackay and *The Professional Geographer*.)

nature of the curvature of the surface and plot the positions of the isarithms accordingly.

In recent years several programs for computer interpolation of z at xy positions in isarithmic mapping have been developed. Machine printouts can either produce the isarithms as lines or as the edges of automatically produced area symbols (Fig. 7.21). As in all isarithmic mapping (except of the land surface), a fundamental assumption is required concerning the configuration of the statistical surface. This is made necessary by the use of sample points and, irrespective of whether the interpolation is done by machine or by hand, the character of the surface away from the points must be inferred. The relation between the spatial sampling plans, the interpolation models, the sizes of the samples, and the fundamental character of the "total" statistical surface is complex and is still under study in order to evaluate the error produced by the various alternative combinations. Nevertheless, machine interpolation is currently in use and will become increasingly common.

Error in Isarithmic Mapping

Error in maps is an elusive thing and depends rather a lot on the various

definitions we employ. It ranges from the generalized "error" in a coastline to the "error" involved in ascribing a mean value to an entire unit area. Such a complex subject can only be touched upon lightly in a book of this sort and usually only by implication. The process of isarithmic mapping, however, is subject to a considerable variety of errors that arise from factors involved in the mapping process. Two classes of these (control point location and interpolation) have already been touched on above because they are steps in the mapping process. Three others will be considered briefly here. Two of these are of concern in both isometric and isopleth mapping (the

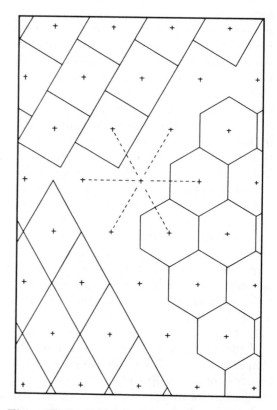

Figure 7.19 Control points arranged in triangular fashion remove the problem of alternative choice. (From Mackay, courtesy of *The Professional Geographer*.)

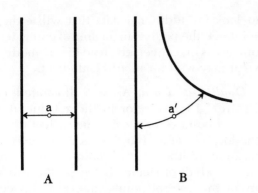

Figure 7.20 Interpolation of point a' in B is done by estimating distances along a smooth curve normal to the isarithms.

the surface, it is apparent that the larger the number of points the closer we can approach the actual surface in our isarithmic delineation. On the other hand, a large number of control points will, of course, provide a greater degree of detail and may create a problem in generalization. As observed earlier, however, it is better to generalize from a knowledge of detail than to be forced to do so by a lack of information.

When the control points are unevenly arranged so that there is a greater density of points in one area than in another, this will produce an unevenness in the consistency of the treatment. This may provide more detail than we would want in the area of dense control or it may lead to the converse, less detail in the area of sparse control than is wanted. We can reduce the detail, but we cannot add it unless there are more data. It is always good practice to show the control points on an isarithmic map to help the reader judge the quality of

number of control points and the quality of the data), while one other (the shapes of the unit areas) is of concern only in isopleth mapping.

Number of Control Points. When the configuration of a statistical surface is to be inferred from a sample of the z values from

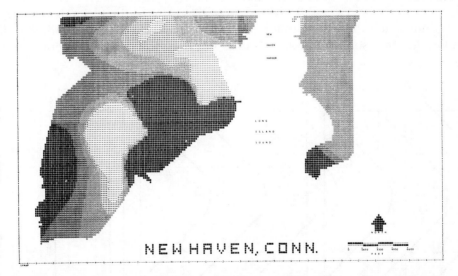

Figure 7.21 Part of a machine-produced map of New Haven, Conn., showing isopleths of the percent change in population, 1960–1967 based upon census tracts. The isopleths are located at the edges of the areas of pattern produced by the plotter. (Map prepared by the Laboratory for Computer Graphics, Harvard University.)

the map. This may be done by either showing the data points for isometric line maps or the unit areas (or control points) for isopleth maps.

Shapes of Unit Areas. The shapes of the unit areas used in isopleth mapping can also have a marked effect upon the positioning and orientation of the isopleths. Elongated unit areas can create, in some instances, strong gradients transverse to the orientation of the elongation, which in turn will induce the isopleths to lie in the direction of elongation. Figure 7.22 illustrates this phenomenon. In *A* a hypothetical series of unit areas is shown together with isopleths located with an interval of two. In *B* a smaller series of elongated minor civil divisions has been formed by aggregation from exactly the same distribution of data as shown in *A*. The plotted isarithms show the same pattern. But when this process is followed with a different orientation of the elongated unit areas in *C*, the isopleths are straightened and "pulled" into a different pattern.

There is not much the cartographer can do to allow for this sort of effect, except

to look for additional data that will either reinforce the pattern in an area of elongated unit areas or, conversely, lead him to modify what may well be an artificial effect.

Quality of Data. As was illustrated in Figure 7.13, any error in the z value at a control point can have as much effect upon the location of an isarithm as changing the location of the control point. There are several kinds of factors that affect the validity of the control point value and, hence, the certainty with which the cartographer can locate the isarithm. Whenever there is a question concerning the accuracy of the z value, this doubt is automatically projected to the map; it there becomes transformed to a question as to the xy position of the isarithm. There is, therefore, always a zone on the map within which any isarithm may be located, depending upon the certainty of the z values, and the width of this zone depends upon the amount of error in the z value, assuming that the control point (the xy position) is correct.

There are three kinds of errors that commonly affect the reliability of the z values from which an isarithmic map is made:

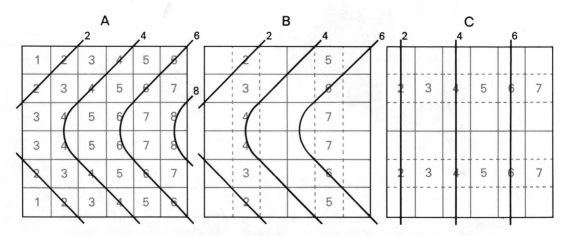

Figure 7.22 Diagram illustrating how elongated unit areas can sometimes affect the positions of isopleths. See text for explanation.

(1) *observational error,* (2) *sampling error,* and (3) *bias or persistent error.*[*] A fourth is related not so much to error as it is to the validity of the concept being presented by the map. This may be called, for the lack of a better term, *conceptual error.*

Observational error refers to the method used to obtain the *z* values. If they are derived by means of instruments operated or read by human beings, there is usually some inaccuracy, both in the instruments and in their reading by the observers. Observational error is not limited to instruments, however. A considerable amount of statistical data is based upon various kinds of estimates, such as, for example, a farmer's estimate of the extent of his cropland, the yield per acre, and the extent of soil erosion. Most statistical data are subject to observational error of one sort or another.

Sampling error is of several kinds. The most obvious is that associated with any map that purports to represent the distribution of an entire class of data, only a sample from which is actually known. Any map of mean climatological values, which is not specifically for a particular period, is, in effect, attempting to describe a total average situation (or statistical population) from only one relatively small time sample. Many kinds of data in censuses are collected by a sampling procedure. In this case, the statistical population exists at one time but the cost of ascertaining it in its entirety is too great; thus, a sample is taken.

Another kind of sampling error is involved in the spacing of control points and in the sizes of the statistical units for which mean values are computed. As pointed out above, the farther apart the control points

are, the less may be inferred about the nature of the distribution between the points, that is, the less valid is each *z* value as a representative of all the possible *z* values within the area to which it refers. On a statistical surface that is subdivided into unit areas for which means are computed, it is apparent that the larger the statistical unit the more likely it is to contain extreme values that will be merged into one average. The actual variations of an irregular statistical surface will therefore be progressively diminished as the number of sample means is diminished.

Bias or persistent error may be of many kinds. Instruments may consistently record too low or too high; the majority of weather stations may be in valleys or on hilltops; people show preferences for certain numbers when estimating or counting; and so on. Bias is difficult to ascertain but may have considerable effect upon *z* values.

Conceptual error in the *z* value may be illustrated by the use of mean values computed from a series of observations with which to make a map of mean monthly temperatures. The mean may not describe very well the actual fact, since the dispersion of the values around the mean may be large. For example, the standard deviation of mean December temperatures in central United States is around 4°F. There is therefore a high probability that approximately one-third of the time an individual December mean will be more than 4° above or 4° below the average of the December means. A cartographer constructing a map that purports to show by isarithms the distributions of mean December temperatures will, in effect, be forced to locate each isarithm near the center of a map zone or band represented horizontally by several degrees of temperature gradient. Furthermore, it is necessary to realize that if individual year-to-year maps had been made,

[*]A detailed treatment of the cartographic aspects of three kinds of statistical error is contained in David J. Blumenstock, "The Reliability Factor in the Drawing of Isarithms," *Annals of the Association of American Geographers,* **43**, 289–304 (1953).

approximately one-third of the time the isarithm would lie outside the zone delineated by an average of 8° of temperature gradient. On even a small-scale map, such a zone would be of noticeable dimensions. It follows that minute isarithmic wiggles and sharp curves would be an affectation of accuracy not supported by the nature of the data.

It is important to take into account the various kinds of possible error in the validity of the z values. In many cases, some of the kinds of sampling error have already been ascertained and need only be obtained from the sources, for example, the United States Census of Agriculture. Standard deviations and standard errors of the mean are often available or are not difficult to compute. Simple logic and common sense will often provide enough of an answer for highly generalized maps. The total effect of all the possible sources of error and inconsistency from one part of an isarithmic map to another can function toward only one end, that is, to *smooth out the isarithms.*

CLASS INTERVALS

Whether the cartographer maps a statistical surface by choropleth or isarithmic methods, he must select and apply a system of generalizing the array of data. Class intervals are the numerical categories of such a system, usually thought of as being bounded by class limits such as 0–2, 2–4, 4–8, and 8–16. On a choropleth map each class is symbolized differently, with pattern or color, and on an isarithmic map the limits are the z values of the isarithms.

It is virtually impossible to symbolize separately the unique value of each unit area, and it is impossible, of course, to employ an infinite number of isarithms. We can employ shading techniques or hachures to show the configuration of a surface, but the ready commensurability of that form of representation is slight. Most statistical surfaces are mapped with one of the objectives being to make it possible for the reader to appreciate the quantitative aspects. Class intervals on the choropleth map allow the reader to find for any chosen area its range between the given class limits, and on an isarithmic map they allow the reader to estimate with high probability the values at points to within one-half the interval between the bounding isarithms.

There are many different kinds of class intervals and the cartographer needs to analyse his communication problem carefully when choosing one.

Factors in Choosing Class Intervals

As Mackay has pointed out, class intervals are the mesh sizes of cartography with the chosen limits forming the screen wires.[*] We must choose the mesh size wisely so that the "size sorting" of the distributional data will be most effectively accomplished. Among the most important of the cartographer's concerns in this connection is his analysis of the problem as related to generalization. There is a direct relation between the number of classes and the amount of information that will be portrayed: generally, the greater the number of classes the more information will be revealed by the system. We must keep in mind, however, that much detail is not necessarily good, since excessive detail can easily draw the reader's attention away from perhaps more important aspects.

A second factor related to generalization has to do with the quality of the data. If the

[*]J. Ross Mackay, "An Analysis of Isopleth and Choropleth Class Intervals," *Economic Geography,* **31,** 71–81 (1955).

data are of good quality and, in the case of isarithmic mapping, if there are many control points, then there can be more detail included than if the data are of less quality. A third factor has to do with the significance of various parts of the range being portrayed. For example, we may be very concerned with the relative changes from place to place and this would lead us to use a smaller interval in the lower section of the range, since a change from 2 to 4 is the same relative change (100%) as is a change from 50 to 100. On the other hand, the interest may be in the absolute values of the higher part of the range and then an opposite choice would be made.

An important element in choosing class intervals as related to perceptual abilities is the practice of symbolizing them on choropleth maps (also sometimes on isarithmic maps), with area symbols that are graded in terms of visual value (darkness). The ordinary human eye is capable of distinguishing only a limited number of such steps, perhaps eight, between black and white (and usually less between any other color and white), and this places severe limitations on the number of classes that can be effectively portrayed. To be sure, we can add pattern (repetitive structure of marks) to the value representation, but even this does not make it possible to increase the number very greatly.

Class Intervals in Isarithmic Mapping. The choice of an interval in isarithmic mapping is affected by an important factor that does not arise in choropleth mapping. Relative gradients on isarithmic maps are easily judged only when there is a constant interval; the general rule that the closer the isarithms, the steeper the gradient, holds only under such circumstances. Therefore, if primary interest is in the configuration of the statistical surface, the interval should be constant.

The recognition of form on the statistical surface from the patterns made by the isarithms, such as concave or convex slopes, hills or valleys, escarpments, and the like, is difficult, if not impossible, when irregular intervals are used. Figure 7.23 illustrates how the nature of slopes is easily revealed by the spacings of lines from an equal steps interval in the one case and almost completely hidden by an unequal interval in the other. In some cases, when interest is concentrated in one portion of the range of values, an increasing or decreasing interval can be used. This requires much more mental effort on the part of the map reader if he tries to infer the form of the surface, but it does allow the cartographer to concentrate detail in one portion of the range, as illustrated in Figure 7.24.

An equal steps or constant interval is ordinarily employed on contour maps of the land and isometric maps of other phenomena. On the other hand, Mackay has found that an equal steps interval is the exception in isopleth mapping.[*]

Kinds of Class Interval Series

There is an infinite number of kinds of series that can be employed by the cartographer in his setting of class limits. It is almost impossible logically to classify the various series according to their mathematical characteristics or the procedures used to derive them, but we can separate them into three major groups according to their major utility.

One is the equal steps or constant series, which employs some kind of equal division of the data or the geographical area. As was observed above, this kind of series is basic to showing the surface configuration of a

[*]J. Ross Mackay, "Isopleth Class Intervals: A Consideration in their Selection," *Canadian Geographer,* 7, 42–45 (1963).

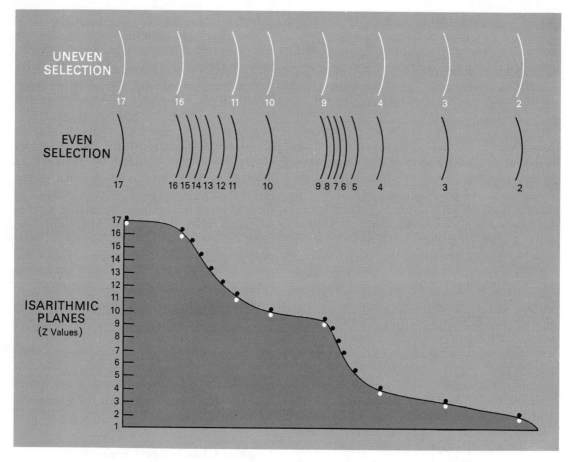

Figure 7.23   Only when an even selection of *z* value isarithmic planes, that is, a uniform interval, is chosen does the pattern of isarithms clearly reveal the shape of the statistical surface, as is shown by comparing the profile with the isarithmic map sections above.

volume distribution by isarithms. There are several possibilities in this class.

A second class of series includes those in which the interval becomes systematically smaller toward either the upper or lower end of the scale. Generally, as has been pointed out, the tendency is to put the greater detail at the lower end because in that section of the range a small absolute difference is a large relative difference. There seems to be a considerably greater fascination for relative differences than for absolute; many distributions are highly skewed, and fur-

thermore, in cartography, there is usually much more room for such detail in the regions of sparse occurrences! There is a great variety of series that increase or decrease toward one end of the range.

A third class is the irregular or variable series. This kind of series is employed when the cartographer wishes to call attention to various internal characteristics of the distribution (such as some values that may be significant in relation to other analyses), when he wishes to minimize certain error aspects, or when he wishes to high-

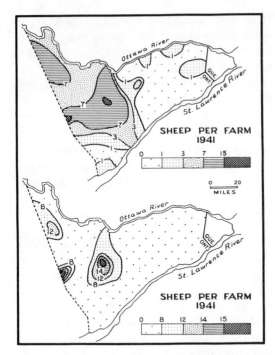

Figure 7.24 These two maps have been prepared from the same data but employ different class intervals. In the top map the interval increases in steps of 1, 2, 4, and 8, whereas those in the bottom map increase in the reverse order, 8, 4, 2, 1. The result is to provide detail and accentuate differences in the lower end of the scale in the top map and to reverse this relationship in the bottom map. (Redrawn from Mackay, courtesy of *Economic Geography*.).

light certain elements of the surface configuration that would not be properly dealt with were a constant or a regular ascending or descending series being employed. These kinds of series are often chosen with the aid of graphic devices, such as the clinographic curve, the frequency curve, or the cumulative curve.*

*Jenks and Coulson have devised a method of testing various class intervals series (as applied to a given distribution) against one another in a fashion analogous to the determination of the coefficient of variability. See George F. Jenks and Michael R. Coulson, "Class Intervals for Statistical Maps," *International Yearbook of Cartography*, 3, 119–134 (1963).

Before the general availability of computers, the analysis of the consequences of employing various kinds of class interval series for simple choropleth maps was inordinately time-consuming. Now, however, we can quickly obtain measures that state the degree of total error (deviation from class means), its distribution among classes, and the ratio of between-class variation to the total variation, for any number of classes desired.‡

Derivation of Class Interval Series

There are four types of constant series. The best known, as was mentioned previously, is the series of equal steps commonly used in contour and other isometric line maps as well as in many choropleth maps. When the isarithmic levels are uniformly spaced, the mapped isarithms show by their spacing the relative gradients. This kind of series is basic to showing the surface configuration of a volume distribution by isarithms. We obtain this kind of constant series merely by dividing the range between the high and low values by the desired number of classes to obtain the common difference. For convenience, the common difference is usually a round number. (Note that this is a special case of the arithmetic series discussed below.)

A second type of constant series is based upon the characteristics of the distribution itself. If we derive the mean and standard deviation, these may be utilized to set class limits such as the mean plus and minus one standard deviation, from one standard deviation to two standard deviations (above or below the mean), and so on. The more nearly normal the distribution the more

‡George F. Jenks, Fred C. Caspall, and Donald L. Williams, "The Error Factor in Statistical Mapping," *Abstracts* of Papers Presented at 64th Annual Meeting of the Association of American Geographers, Washington, D.C., 1968, p. 45.

useful this type is for graphically portraying the areas and magnitudes of departures from the average.

A third kind of constant series involves the employment of *quantiles,* a division of the array of data into equal parts; there may be any division, commonly used ones being quartiles (four), quintiles (five), sextiles (six), and so on, up to deciles (ten) or even centiles (one hundred). Quantiles are determined by arraying the values in the order of their magnitude, from the lowest to the highest. If we wished to obtain quartile values, for example, we count one-fourth of the way up from the bottom to the first quartile value, and so on. Quantiles are useful when the mapping is based on unit areas, especially choropleth mapping, but if there are major differences in the sizes of the unit areas, a large part of the value of the quantile series is lost.

A fourth type of constant series is what might be called equal area steps or, in a sense, "geographical quantiles." In this kind of series the area of the map is divided into equal sized regions, the number depending on the cartographer's choice. The determination of the class limits for this kind of series is most easily accomplished by employing a cumulative frequency graph, as described in a later section of this chapter.

Two groups of series with unequal steps are (1) arithmetic series in which each class is separated from the next by a stated numerical difference (not constant), and (2) geometric series in which each class is separated by a stated numerical ratio. The general form of the equation to produce class limits of either type of series is

$$L + B_1X + B_2X \ldots B_nX = H$$

where L = the lowest value
H = the highest value
B_n = the value of the nth term in the progression

It is only necessary then to obtain B_n by some method and then solve the equation for X for any given values of L and H to establish the class limits.

In arithmetic progressions the quantity Bn is obtained by

$$B_n = a + [\,(n - 1)\,d\,]$$

where a = the value of the first term
n = the number of the term being determined, that is, the first, second, etc.
d = the stated difference

The stated difference, d, may be any value; if it is a simple positive number, e.g., 2, the interval between the classes will increase upward at a constant rate of increase, that is, the smallest class will be at the bottom of the scale and the largest at the top. If we employ a simple negative number, e.g., -2, the opposite will be the case. If d is made a positive variable quantity, e.g., $2\,(n-1)$, then the class limits will increase at an increasing rate toward the upper end of the range; if d is made a negative variable quantity, e.g., $-2\,(n-1)$, the opposite will hold true. More involved variable quantities will produce more complex kinds of progressions such as those that increase upward but at a decreasing rate.

In geometric progressions the quantity B_n is obtained by

$$B_n = gr^{n-1}$$

where g = the value of the first non-zero term
n = the number of the term being determined, that is, the first, second, etc.
r = the stated ratio.

Needless to say, r may be any quantity. For example, if $L = 0$, $H = 64$, and $r = 2$, the progression will be the familiar 1, 2, 4, 8,

16, 32, 64; if $r = \frac{1}{2}$, the progression will be the reverse: 32, 48, 56, 60, 62, 63. The first series increases upward, thereby concentrating the detail in the lower end, and the second decreases upward, thereby concentrating the detail at the opposite end.

A third, less commonly employed progression is called the reciprocal or harmonic. This is a series in which the reciprocals of the terms in the series are in arithmetic progression. For example, if there were 5 classes, $L = 2$, and $H = 40$, the stated difference of the reciprocals, d, would be $0.5 - 0.025/5 = 0.095$. Subtracting 0.095, successively, beginning with 0.5, would provide a series of numbers 0.5, 0.405, 0.310, 0.215, 0.120, and 0.025. The reciprocals of these provide the class limits, namely 2, 2.5, 3.2, 4.6, 8.3, and 40. It will be seen that the detail is heavily concentrated in the very low range, and on account of this extreme concentration, this type of progression is not widely used.

Graphic Techniques for Selecting Class Intervals

When the unit areas from which isopleth and choropleth maps are made are approximately equal size, we can employ mathematical methods to arrive at class limits. This is especially so when the interest in the distribution is about evenly divided between presenting the statistics in place and presenting the surface configuration of the volume distribution. On the other hand, when the sizes of the unit areas vary considerably or when the major interest is focused on the character of the surface, then graphic methods are often helpful. Three of these involve the preparation of frequency diagrams from which we can extract critical values that may serve as suitable class limits. These are the ordinary frequency curve, the clinographic curve, and the cumulative frequency curve.

It must be emphasized that merely graphic approaches to the selection of class intervals may not be suitable for two reasons. On the one hand, the limits may be either ill defined or so oddly arranged in the series that it would be difficult for a reader to use them; and on the other, the arrangement of the values in numerical order (as is necessary in the preparation of the graphs) clearly destroys their geographical order or association. It is quite possible that some startling characteristic of a curve on a graph, such as a marked peak, depression, or flexure, may have no geographical significance whatever, since the data from which it derives may be widely and unsystematically dispersed.

The Frequency Curve. A frequency graph is prepared by arraying the z values of the data on the x axis of a graph plotted against the frequency of their occurrence on the y axis. Frequency in this sense means the number of times they occur, and if the areas of the unit areas are not widely different, the unit areas may simply be counted. If they are greatly different we should employ a cumulative frequency graph. As an illustration, Figure 7.25 has been prepared from the data shown in Figure 7.2.

Generally, the low points on a frequency diagram are thought to be the most useful as class limits because they tend to enclose larger groups of similar values. Often, however, there may be few such low points. Frequency graphs are ordinarily quite useful to illustrate the numerical characteristics of the distribution, but they do not often provide clearly desirable class limits.

The Clinographic Curve. A clinographic curve is prepared by arraying the z values of the data in numerical order on the y axis plotted against the cumulative area on the x axis. The y axis is scaled arithmetically

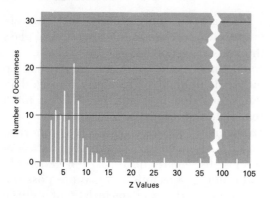

Figure 7.25 A frequency diagram prepared from the data geographically arrayed in Fig. 7.2, classed by whole numbers. If the data are continuous, as near the lower end, the plotted values may be joined by a curve instead of plotting each individual as shown here. (After Jenks and Coulson, *International Yearbook of Cartography.*).

(evenly spaced steps) while the x axis is scaled, in percent of total area, with a square root scale from 0% to 100%. We may use numbers of unit areas if they are of about equal sizes, otherwise we should employ the areas of the civil divisions. Figure 7.26 is a clinographic curve prepared from the data shown in Figure 7.2.

The critical points in a clinographic curve are those points where the slopes of the curve change. Ordinates at these flexures indicate the z values separating regions of different gradient, provided, of course, that the data adjacently plotted on the graph are in fact reasonably adjacent geographically.

The Cumulative Frequency Curve. A cumulative frequency curve is prepared by arraying the z values of the data in numerical order on the y axis plotted against the cumulative area on the x axis, as described in Chapter 5. Both axes are scaled arithmetically. Properly, the z value of each successive unit area value is plotted against the sum of its area and all preceding unit areas with lower z values. This means that the

graph will end on the right-hand side with an x value equal to the total area included in the geographical area being mapped. A curve so plotted must rise continuously until it reaches the last plotted point. Figure 5.8 is a cumulative frequency graph.

For class limits that are likely to be significant in terms of the surface configuration, the critical points are the tops and bottoms of "escarpments" on the curve, since these tend to outline areas of different gradient. Probably more important is the fact that the cumulative frequency graph may be used to determine other intervals. As described in Chapter 5, we may determine from it the geographical mean, the geographical median, or other functions of z values over areas such as "geographical quantiles" or equal area steps.

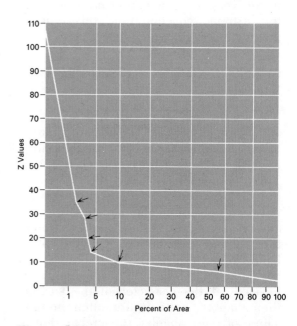

Figure 7.26 A clinographic curve prepared from the data geographically arrayed in Fig. 7.2. The y axis is scaled arithmetically, the x axis in square roots from 0 to 100% of area. The arrows point to places in the curve that could be used for class limits. (After Jenks and Coulson, *International Yearbook of Cartography.*)

MAPPING **8** THE LAND FORM

171

In the foregoing chapters, the representation of geographical data has been considered according to the principles of symbolism involved in the treatment of point, line, area, and volume data. Another approach would be to treat separately the various substantive categories of geographical information, such as population, agriculture, transportation, manufacturing, and so on, and then to discuss the various methods by which each can be symbolized. The difficulty with such a method is that there is considerable similarity in the means of representing diverse phenomena, which would necessarily result in a great deal of repetition. There is, however, one category, the land surface, that is so different from all the others as to make it almost imperative that it be treated separately. As may be expected, various aspects of the land surface form may be represented by point, line, or area symbols, or by combinations of them.

The representations of the x, y, and z dimensions of the earth's surface and their derivatives has, no doubt, always been of special concern to cartographers, but the earliest maps, and even those of the Middle Ages, showed little of it. This was probably because of the paucity of knowledge about landforms. To be sure, mountains may have been shown as piles of crags, and ranges appeared as "so many sugar loaves"; but until precise elevational and positional data became available upon which the cartographer could base his representation, the landform could not be well represented.

It should be noted here that, although this chapter is entitled Mapping the Landform, the principles of landform representation apply equally to the solid surface beneath the waters of the earth. There are special problems of data gathering in connection with mapping underwater surfaces, of course, but otherwise there is no basic difference. The discussion here will be focussed on the landform since its appearance is relatively familiar and the graphic potentialities of the various methods can be easily grasped.

There is something about the three-dimensional surface that intrigues cartographers and sets it a little apart from other cartographic symbolization. First, it requires more understanding and skill. Moreover, the land surface is a continuous phenomenon; that is, all portions of the solid earth necessarily have a three-dimensional form, and as soon as the land is represented, *all* of it must be represented, at least by implication. Also, the landform is the one major phenomenon with which the cartographer works that exists as a graphic impression in the minds of most map readers, and the map reader is therefore likely to be consciously or unconsciously critical in his approach to its graphic representation on the map.

Because of the relative importance to man of the minor landforms, the representation of landform together with other data has always been a great problem to the cartographer. If he shows the surface in sufficient detail to satisfy their local significance, then the problem arises of how to present the other map data. On the other hand, if the cartographer shows with relative thoroughness the nonlandform data, which may be more important to the specific objectives of the map, he may be reduced merely to suggesting the land surface, an expedient not likely to please either the map maker or the map reader. Furthermore, the development of aviation during this century has made increasingly important the effective and precise representation of the terrain. The pilot needs to be able to recognize the area beneath him, and the passenger has naturally become more interested than he formerly was in the general nature of that surface, for he may now see it. His height above it provides a reduction mindful of a map.

Perhaps an even more difficult task has been the depicting of landforms on smaller-scale maps. Small-scale landform representation is a major problem for atlas maps, wall maps and other general-purpose reference maps, as well as for those special-purpose maps in which regional terrain is an important element of base data. The smaller scale requires considerable generalization of the land form, which is no simple task, as well as the balancing of the surface representation with the other map data, so that neither overshadows the other. No less a creative task is the representation, in bolder strokes, of the landforms for wall maps, so that such important elements as major regional slopes, elevations, or degrees of dissection are clearly visible from a distance (see Fig. 8.1). To the geographer and geomorphologist are reserved the specialized techniques for scientific terrain appreciation and analysis.

For many years to come the representation of the land form on maps will be an interesting and challenging problem, since it is unlikely that convention, tradition, or the paralysis of standardization will take any great hold on this aspect of cartographic symbolization. This will probably be particularly true of terrain representation on special-purpose and thematic maps; each such attempt will be a new challenge, since in each case it must be fitted to the special, overall objective of the map.

The Historical Background

The story of the development of land-form representation is a recital of the

Figure 8.1   A much reduced monochrome print of a portion of a modern color wall map emphasizing surface. The detailed terrain is derived from photographing a carefully made, three-dimensional model. The map is reproduced by complex color printing analogous to process color. (Map by Wenschow, courtesy Denoyer-Geppert Company.)

search for methods suitable to a variety of purposes and scales. On large-scale maps the desirable symbolization is one that both appears natural and is capable of exact measurement of such elements of the land surface as slope, altitude, volume, and shape. The major problem arises from the fact that, generally speaking, the most effective visual technique is the least commensurable, whereas the most commensurable is the least effective visually. One of the major decisions of every survey organization has been in what manner to balance these opposing conditions. Although it is somewhat early as yet to judge, there is some indication that advances in color printing have enabled the cartographer to reach a relatively effective combination of techniques, without undue sacrifice of either desirable end. The newer, shaded relief, contour maps of the United States Geological Survey are a case in point (see Fig. 8.2).

The earlier representations of terrain in smaller-scale maps were concerned mainly with undulations of some magnitude and consisted of crude stylized drawings of hills and mountains as they might be seen from the side, such as those depicted in Figure 8.3. The perspective-like, oblique, or birdseye views became more sophisticated in the period from the fifteenth to the eighteenth centuries when the delineation of terrain developed along with the landscape painting of the period. The eight-

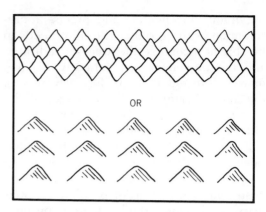

Figure 8.3   Crude, early symbolization of hills and mountains.

eenth century was the period when the great topographic surveys of Europe were initiated and, for the first time, map makers had factual data with which to work, albeit the first attempts were crude indeed. Delineation of the form of the land shifted from an employment of the oblique view to the plan view, perhaps because of the availability for the first time of extensive planimetric data. Symbols with which to present the form of the land without planimetric displacement were developed.

Large-Scale Representation. In 1799 the systematic use of a linear symbol called the *hachure* was developed by Lehmann, an Austrian army officer. As advocated by Lehmann, each individual hachure, rather like a vector symbol, is a line of varying width that follows the direction of greatest slope: that is, its planimetric position on the map would be normal to the positions of contours. By varying the widths of the hachures according to the angle of the slopes on which they lie, the magnitudes of the slopes (or steepness) may be indicated. The sense, or the direction of down (or up), is, however, not shown. When many hachures are drawn closely together, they collectively show the forms of the surface configuration. This turned out to be partic-

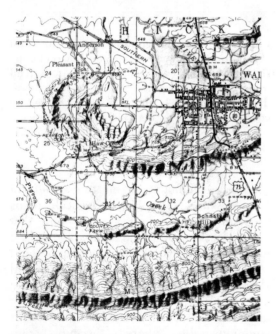

Figure 8.2   A small part of a modern contoured topographic map. The terrain is emphasized by colored shading. This monochrome reproduction, reduced from the original scale, cannot do justice to the excellence of the original. (Waldron Quadrangle, Arkansas, United States Geological Survey.)

ularly useful on the then recently initiated, large-scale, topographic military maps; and, for nearly a century, the hachure was widely employed.

Many varieties of hachuring were employed. Instead of varying the width of the line according to the slope, some used uniform-width strokes and then increased their density with increased slope. Hachuring provides an illusion of shading and the light source can be assumed to be orthogonal to the map (Figure 8.4) or from one side (Fig. 8.5). On small-scale maps or maps of poorly known areas, hachuring easily degenerates into "hairy caterpillars," an example of which is shown in Figure 8.6. These worms are not yet extinct.

From the earliest use in the sixteenth century of lines of equal depth of water (isobaths), Dutch, and later French, engineers and cartographers introduced the use of these kinds of lines to portray the under-

Figure 8.5   A section of the Dufour map of Switzerland in which obliquely lighted hachure shading is employed. Sheet 19 (1858), Switzerland, 1:100,000.

water and dryland configuration. By the early nineteenth century the contour was widely employed, but it was not until relatively late in the nineteenth century that contours became a common method of depicting the terrain on survey maps. One development, which grew out of the use of contours on large-scale maps, was their extreme generalization on small-scale maps, resulting in the familiar "relief map," together with "layer coloring" according to altitude. The colors, of course, are area symbols between isarithms.

After the development of lithography in the early part of the nineteenth century, it became possible to produce easily continuous tonal variation or shading. It was not until late in the last century that shading as a type of area symbol was widely utilized for the representation of the terrain, the shading applied being a function of the slope.

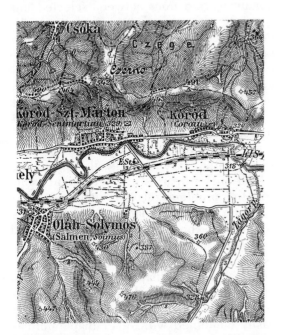

Figure 8.4   A section of a hachured topographic map with vertical illumination. Austria-Hungary, 1:75,000.

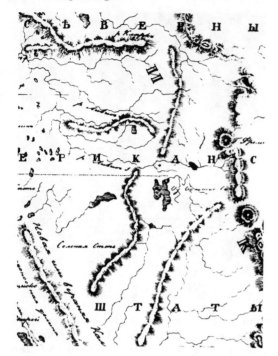

Figure 8.6 Genus *hachure*, species *woolly worm*. (From an old Russian atlas of western North America.)

By the beginning of the present century, the basic methods of presenting terrain on large-scale topographic maps (contouring hachuring, and shading) had been discovered, and the essential incompatibility between commensurability on the one hand and visual effectiveness on the other was readily apparent. For the last several decades the problem has been one of how to combine the techniques to achieve both ends. The newer topographic maps are the most effective yet produced.

Small-scale Representation. The representation of the land surface at large scales is concerned essentially with the three major elements of configuration: the slope, the height, and the shape of the surface formed by the combinations of elevations and gradients. The various methods outlined above, and their combinations and

derivatives, seem to provide the answer, more or less, for the problem at large scales; but the representation of land surface at smaller scales is quite another matter. Here the generalization required is so great that only the higher orders of form, elevation, and slope may be presented. As knowledge of the land surface of the earth has grown, so also has the need for a variety of methods of presenting effectively that surface at smaller scales.

No one method can satisfy all the small-scale requirements for land surface delineation. Consequently, with the growth of reproduction techniques and drawing media, the variety of the ways of depicting the land surface is steadily increasing, and it may be expected to continue to do so for some time to come. Some of the older methods are beginning to be discarded because of their rather obvious deficiencies.

Layer coloring, with or without spectrally organized gradations between selected and generalized contours, is one of the earliest devised techniques. It came into wide use during the latter part of the nineteenth century, and it has been more commonly employed than any other technique, mainly because of its relative simplicity. It leaves much to be desired. The character of the surface is presented only by implications of elevation; the generalized contours show little except regional elevations, which are ordinarily not very significant; and the problems of color gradation and multiple printing plates are sometimes difficult. Hachures were quickly employed at small scales after their introduction but, as was observed, they decline easily to the woolly worm. Shading can be very effective at any scale but it requires unusual skill and, in monochrome, cannot be easily combined with other map information. Without expensive multiple-color plates, this technique, when employed on a map with much other data, often becomes little more than an uneven background tone, which serves largely to reduce the visibility of the other map data.

In lieu of simple layer tinting, hachuring, and shading at small scales, many other techniques have been tried in more recent years to gain realistic effects. One of the more effective devices is that of drawing the terrain pictorially, as a kind of bird's eye view with a slight perspective, as in Figures 8.19 and 8.20. This sort of delineation requires a knowledge of landforms, considerable practice, and at least some manual skills. One who draws terrain must know what he is drawing, but he must also have the skill to interpret graphically the three-dimensional relationships he wishes to convey. The first requires training in such fields as landform analysis, geomorphology, and structural geology; the second requires more training in the visual arts than most cartographers have the good fortune to receive. There is practically no end to the combinations of perspective viewpoint, coloring, highlighting, shadowing, and line drawing that can be employed.

Instead of attempting to obtain realistic visual communication of the configuration of the land surface, the essential characteristics of the forms may be differentiated according to some nominal-interval scaling system. This may take the form of classes such as flat plains, tablelands, and low mountains, which have precise definition according to such factors as local relief (difference in elevation in a limited area), localization of gentle slopes, and so on. The areas so outlined may then be shown with area symbols, usually colors. The widespread use of this system is relatively new.

Many other systems of landform representation for special purposes have been tried or suggested. Most of them are relatively complex and intellectually involved. Their use is limited to the professional geographer and geomorphologist, whose knowledge of landforms is sufficient to interpret them. In this category would fall those methods that present certain aspects of the geometry of the land, such as the average-slope technique or the relative-relief technique.

A thorough understanding of the small-scale possibilities for delineating terrain requires familiarity with the fundamentals of the basic methods used to show terrain on topographic maps, namely, contours, shading, and hachures.

Contours

Representation of the form of the land by means of contours is the most commensurable system yet devised. As described in Chapter 7, contours are the traces obtained by passing parallel "planes" through the three-dimensional land surface and projecting these traces orthogonally to the plane of the map. A contour is, therefore, an isarithm of equal elevation above some sea level or other assumed starting elevation. The assumed horizontal surface of zero elevation is called the *datum*, which is the surface of a particular earth spheroid (a regular geometric figure) projected beneath the land. This surface is essentially that which would be assumed by a worldwide ocean that was not modified by localized variations in gravity.* It is apparent that the spheroid surface is not a flat surface but is curved in every direction. It is the problem of the mapper to establish the horizontal position on, and the vertical elevation above, this surface of a large number of

---

*This concept is an expedient, since the surface that would be assumed by a continuous ocean, with the present density and gravity differences retained, would not be quite that of a spheroid. It would have local undulations and would instead be the *geoid* surface. The establishment of this surface is one of the problems of geodesy.

points on the land. When enough positions are known and the curved datum surface has been transformed into a plane surface by means of a projection system (such as the polyconic), the map may then be made. The map reader *sees* the represented land surface orthogonally. Figure 8.7 illustrates these important relationships.

The representation of form by contours is an artificial system in the sense that it has few counterparts in nature and, therefore, it is not in the normal experience of the average person. Figure 8.8 shows the usual way we see form—by the interplay of light and dark—and, for comparison, a contour representation of the identical form.

Contours on a topographic map are remarkably expressive symbols if they have been correctly located and if the interval between them is relatively small. The most obvious expression is that of elevation, of course, but the spacing and orientation of the contours provide some visual clues to form.* To be sure, the efficiency with which a contour map portrays the landform depends to some degree on accident. It is quite possible that on a contour map with a 20 ft interval, a 19 ft feature such as a road cut or fill might not appear but one of 1 ft

could. Small hills, escarpments, and depressions can all be "lost" in between the contours. The larger the interval the more serious this possibility becomes.

We must remember that not all contour maps are of the same order of accuracy. Before the acceptance of the air photograph as a device from which to derive contours, the lines were drawn in the field with the aid of a scattering of "spot heights" or elevations. Consequently, they were often by no means precisely located.

It is apparent that much of the utility of contours depends upon their spacing, and the choice of a contour interval is not an easy task. Because the portrayal of the relative slopes (gradients) of the landform is one of the major objectives of contour maps, contour intervals are almost always equal-step progressions. Maps that portray two classes of landform areas differing markedly in relief, for example, a flood

*The elevation of all points on a map may be determined within one-half the contour interval. For example, if the interval between contours is 20 ft, then any point not on a contour must necessarily be above the lower and below the higher of the two contours between which it is located; accordingly, a point can be reckoned to the nearest 10 ft.

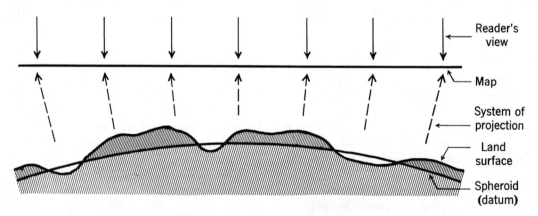

Figure 8.7   The building of a contour map. The positions of elevations are first projected orthogonally to the spheroid. The "map" on the spheroid is then reduced and transformed to a plane, which the reader views orthogonally. The curvatures are, of course, greatly exaggerated.

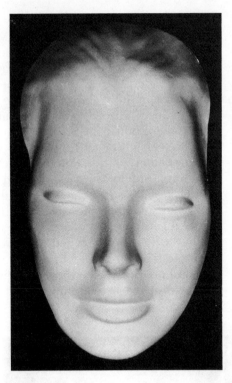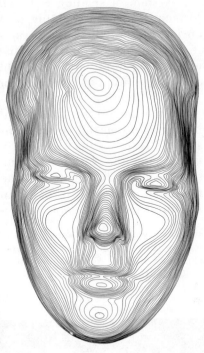

Figure 8.8   A plaster model and contours with a 1 mm interval precisely derived from it. The contours were obtained by photogrammetric methods (Courtesy G. Fremlin.)

plain and a hill land, may employ one interval for one area and a different one for the other. In areas of high relief the interval must necessarily be large because of the high range and the inability to accommodate many lines in a small space. As the interval is increased, the amount of surface detail lost between the contours becomes correspondingly greater. If, owing to lack of data or scale, the contour interval must be excessive, other methods of presentation, such as shading, are likely to be preferable. In some cases, lack of data requires the use of *form lines* in place of contours. These are discontinuous lines, which, by their arrangement, suggest slopes and the general configuration, but from which precise information regarding slopes and elevations are not to be read.

Although contours do not present quite so clear a visual picture of the surface as does shading, the immense amount of information that may be obtained by careful and experienced interpretation makes the contour by far the most useful device for presenting the land on topographic maps.

Shading

Shading, or, as it is sometimes called, plastic shading, is the gaining of a three-dimensional impression by employing variations in light and dark.* These variations, as in chiaroscuro drawing, are applied according to a number of different systems.

*The adjective "plastic" as it is used here derives from the German *plastik* (from Gk. *plastikos*) which refers to the modelled, three-dimensional effect; it has nothing to do with the modern synthetic materials for which the same term is used.

In theory the system of shading is based upon the principle that the lighting of a three-dimensional form will result in varying amounts of illumination on the varying slopes. The direction from which the light comes makes a significant difference. For example, vertical lighting would cause maximum illumination on the horizontal surfaces and minimum on the vertical surfaces. This would result in the logical progression, the steeper the darker. Oblique lighting produces more illumination on one side than the other of a form (Fig. 8.9).* In its simplest form, shading attempts to cre-

*In the actual practice of manual shading, it is usually not done systematically. John S. Keates, "Techniques of Relief Presentation," *Surveying and Mapping*, **21**, 459–463 (1961).

ate the impression, appropriately exaggerated, which we might gain from viewing a carefully lighted model of the land, as is illustrated in Figure 8.10. Of course, the usual shaded map is not quite the same as an area seen from above, since the observer is, in theory, directly above all parts of the map, there being no perspective.

The versatility of lithography gave the cartographer of the nineteenth century a medium previously lacking, and he was quick to take advantage of it. Lithography allowed smoother and easier application of shading with a crayon, and it was considerably faster than the tedious drawing of hachures, even if the hachures were not graded precisely according to slope categories. After the use of stone for lithography

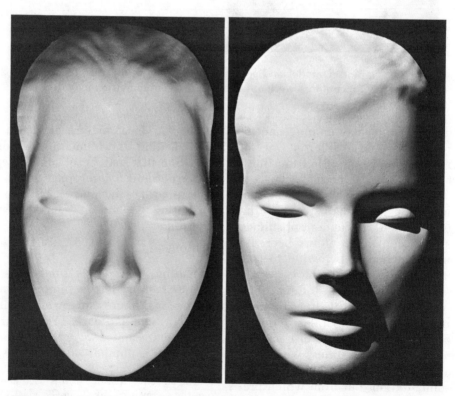

Figure 8.9   Two views of the same plaster model, one lighted vertically (left) and the other from the upper left ("northwest"). (Courtesy G. Fremlin.)

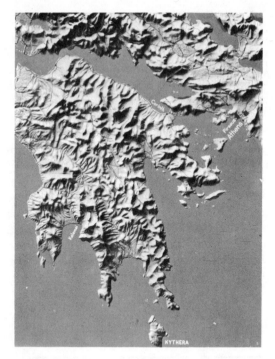

Figure 8.10   A much reduced portion of a topographic model.

declined, the gradation of light and dark (continuous tone) was accomplished by halftoning. The "terrain plate" is usually prepared separately and is halftoned (see Chapter 13); the line work of the map is then either combined with it and printed as a combination line and halftone, with one press plate, or it may be printed separately, as two impressions with different colors.

A continuous-tone drawing is generally decidedly preferable to photographing a model when the reproduction is to be a monochrome because much more contrast and sharper detail can be attained in the drawing, and because the cartographic artist can suitably generalize and put "life" into a rendering, whereas the camera cannot. The continuous-tone monochrome drawing may be done in wash (painted with a brush), by airbrush, carbon pencil, charcoal, or crayon; in any case it must be

reproduced by halftone, since it employs continuous tone (Fig. 8.11).

The drawing or the photograph of the model may also be reproduced in color, and various other effects are thus possible. For example, several identical halftones may be separated during the retouching stage; portions of them may then be printed in different colors, so as to achieve an effect of elevational layer tints, in addition to the realistic terrain. The coloring need not conform strictly to elevational lines but may be employed simply to distinguish lowlands from uplands in any particular region, as is often done on topographic maps.

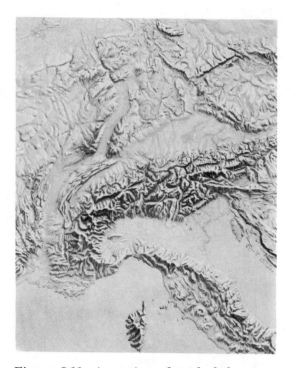

Figure 8.11   A portion of a shaded, monochrome, small-scale terrain drawing of Europe by Richard Edes Harrison. The drawing was prepared considerably larger than is shown here. (From *Eurasia*, of *Lands and Peoples of the World*, copyright 1958 by Ginn and Company. Reproduced with permission.)

It is, of course, not necessary that the vertically viewed terrain be drawn in such a way that its reproduction requires the halftone process. The obliquely illuminated hachure map, exemplified by the Dufour map (Fig. 8.5), is one example of the kind of copy that could be reproduced by a line cut. Ross board and Coquille board may also be employed to the same end (see Fig. 8.12).

One of the problems facing the cartographer who plans to employ shading is the direction from which the light is to come. A curious and not completely understood phenomenon is that with light from different directions, depressions and rises will appear reversed. Consequently, the direction must be chosen so that the proper effect will be obtained. When the light comes from an upper direction, elevations generally appear "up" and depressions, "down." In addition to providing the correct impression of relief, the direction of lighting is important in portraying effectively the terrain being presented. Many areas have a "grain" or pattern of terrain alignment that would not show effectively if the light came from a direction parallel to it. For example, a smooth ridge with a northwest-southeast trend would result in about the same illumination on both sides if the light were to come from the northwest. The cartographer must also select the direction of lighting and apply the darkening and highlighting so that the items of significance will be effectively portrayed. It is in this respect that shading is generally preferable to photographing a model (Fig. 8.13).

The utility of shading as a visual vehicle

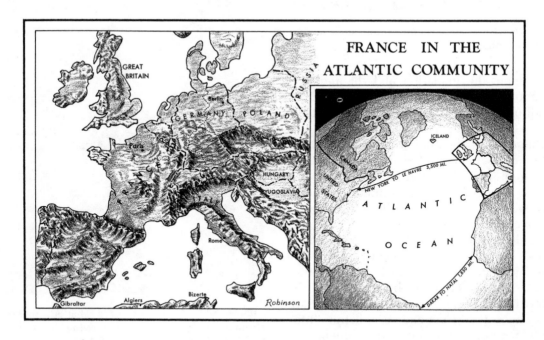

Figure 8.12   A pictorial sketch-map drawn on Coquille board. The reproduction is by ordinary line cut. (Reprinted by permission of the publishers from Donald C. McKay, *The United States and France*, Harvard University Press, Cambridge Mass.; copyright, 1951, by the President and Fellows of Harvard College.)

Figure 8.13   Some drawings for the purpose of illustrating how to shade in situations where drainage lines must cut through rock strata. (Courtesy Aeronautical Chart and Information Center.)

for presenting landforms has long been appreciated. It was to be expected that there would be attempts to make it in some way commensurable as well as visually effective. The first such attempt was the hachure. A number of other possibilities have been suggested, but none of them has been tried beyond the experimental stage. For example, Max Eckert attempted to use point symbols in the hachure-slope manner.[*] By using more carefully controlled dot sizes graded according to slope, he hoped to produce a map in which the amount of light that would be reflected from a vertical source would be accurately represented.

Two other proposals having to do with illuminated contours and a vertical lighting technique have been made by Professor Tanaka of Japan.[*] His latest suggestion is to illuminate contours systematically so

[*]Max Eckert, *Die Kartenwissenschaft,* Berlin and Leipzig, 1921, Bd. 1, pp. 585-590; see also Arthur H. Robinson, "A Method for Producing Shaded Relief from Areal Slope Data," *Annals of the Association of American Geographers,* **36,** 248–252 (1946).

[*]Illuminated contours are lines on a neutral gray background. On the light side of a hill they are lighter than the ground; on the dark side they are darker than the ground. K. Tanaka, "The Orthographical Relief Method of Representing Topography on Maps," *The Geographical Review,* **40,** 444–456 (1950).

that they provide an impression of shading with oblique lighting as well as being commensurable. Figure 8.14 is an example.

Hachuring. The hachure is a line symbol drawn down the slope, and it is varied in width or spacing in relation to the slope of the land on which it lies. The steeper the surface, the darker the representation. The original form of hachures was based on the assumption of a vertical source of illumination. Accordingly, some of the light would be reflected in the direction of origin (to the reader) in some proportion to the angle of the slope. A number of different slope-darkness relationships have been used. For example, the originator, Lehmann, established a system wherein the range between solid color and clear paper (white) corresponded to the range of slope angles from 45° to 0°. Consequently, a horizontal surface would be white, a 22 1/2° slope would be half white and half color, and a 45° slope would be solid color.

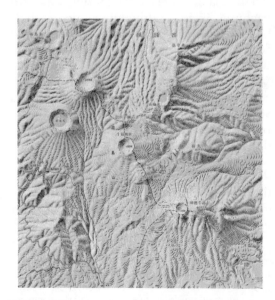

Figure 8.14   An example of Professor Tanaka's illuminated contour method. (Courtesy of *The Geographical Review,* published by the American Geographical Society of New York.)

The varying amounts of light and dark may be produced either by varying the thicknesses of lines, as in the Lehmann system, or by varying the spaces between lines of the same thickness. Many different combinations have been tried, but they all are based on the same general idea—a change of slope changes the amount of light reflected.

Not long after hachuring with vertical lighting was employed on topographic maps, it was discovered that much more realistic modelled effects would be attained by varying the line widths to give the effect of oblique lighting. The Dufour Map of Switzerland, started in 1833 and completed in 1866, is an outstanding example (Fig. 8.5). This method was followed in other surveys as well. In such maps the graphic quality took precedence over the desire to indicate precise slope, although attempts to combine them were extensively investigated.

The major difficulty experienced with hachures is that, although slope is their basis, it cannot practically be measured from the map, regardless of the precision underlying the representation.* Flat areas, whatever their location, appear the same, and only rivers and streams or spot heights strategically placed make it possible for the reader to tell valleys from uplands. Another difficulty of hachuring is that its effectiveness, when printed in one color, is dependent in large degree upon the darkness of the ink. Thus, a considerable problem is created, since, as darker inks are used to make the terrain more effective, the other may detail becomes correspondingly ob-

scured. Hachures are no longer used on topographic maps. They are still employed in small scales in atlases and on occasional special maps.

Analytical Shading. One of the interesting examples of the role the computer can play in cartography is provided by Professor P. Yoeli's experimentation with what he calls analytical hill shading.* Although not yet put to practical use, the method is nicely illustrative of the scientific elements in the attempt to portray the configuration of the land surface by light and shade.

Nearly one-hundred years ago, the equations having to do with the amount of light intensities at various points on an irregular surface were worked out. Stated simply, the intensity of light reaching an observer from a point on a three-dimensional surface under illumination by parallel rays is a function of the inclination and orientation of that part of the surface in relation to the directions of the observer's line of sight and the source of light. This remained in the realm of theory, however, because of the immense amount of calculation that would be needed to make use of the equations, as well as the practical problem of producing any graphic result. The computer now makes possible extensive computation and the basing of the degree of light and shade on a strict analysis of the surface configuration as delineated by contours.

In Professor Yoeli's experiment the surface to be shaded is divided into small rectangular segments or facets. From the contours may be calculated the average inclination and orientation of each facet. By then assuming a direction and an inclination of parallel light rays, the resulting rela-

---

*It is interesting to note that precise, effective hachuring depended upon a considerable knowledge of the terrain. In actual practice, contours were often drawn on the field sheets of a survey, and the hachures were drawn in the office from the contours. Thus, the original French survey for the nineteenth century 1:80,000 map had contoured field sheets but was published only in the hachured form!

*See the series of explanatory papers in *Kartographische Nachrichten,* **15** (1965), Heft 4; **16** (1966), Heft 1: **16** (1966), Heft 3; and **17** (1967), Heft 2. The first two appeared in English translation in *Surveying and Mapping,* **25,** 573–579 (1965), and **26,** 253–259 (1966).

tive brightness of each facet may be calculated. All that remains is to darken each
facet accordingly, and, if the facets are
small enough, a strictly analytical shading
results.

The elements in Figure 8.15 show the
original contour map, the appropriately
darkened facets, and the combination of
the two. As Professor Yoeli points out, the
visually ideal shading results when the
direction of the light can be shifted so as to
take advantage of the orientation of the
configuration. Even this can be programmed for the computer to a certain degree. So far the automatic darkening of the
facets as directed by the computer has
been attended by only limited success, but
no doubt, that problem will soon be solved.

Pictorial Representation

Even on ancient maps the major terrain
features were represented pictorially in
crude fashion. As artistic capabilities have
increased, and as knowledge of the character of the earth's surface has expanded, pictorial representation has become increasingly effective. Within the past fifty years,
this method of presenting the land features
has made great strides.

There are in common use today many
varieties of perspective delineation of the
land. Most of them stem from the attempts
made during the nineteenth century to portray the concepts that were being rapidly
developed by the growing science of geology. The earlier geologists illustrated their
studies and reports with cross sections in
order to show the structural relationships of
the rocks. The top line of a cross section is
a profile of the land, and it was only natural
that some pictorial sketching was occasionally added to make the appearance more
realistic. The next step was to cut out a
"block" of the earth's crust and view it as if
the observer were looking at it obliquely
from above so that two sides as well as the
top became visible. The block diagram has
today become a standard form of graphic
expression and its utility has been extended
to illustrate land use, land types, and
other kinds of earth distributions. Almost
any degree of elaborateness may be incorporated in a block diagram, ranging from
the successive geologic stages in the development of an area to multiple cross sections of the structure. Uncomplicated block
diagrams are not difficult to draw and even
the simplest is remarkably graphic (see
Fig. 8.16). For many examples of great vari-

Figure 8.15   A contour map, the shaded facets, and the union of the two in experimental analytical shading. The direction of light is from the northwest at an
inclination of 45°. (Courtesy P. Yoeli.)

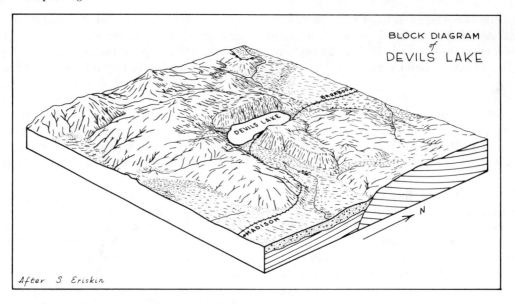

BLOCK DIAGRAM
of
DEVILS LAKE

DEVILS LAKE

BARABOO

MADISON

N

After S. Eriskin

Figure 8.16 A simple block diagram prepared for student field-trip use. The natural appearance of the surface forms on a perspective block makes the concepts easily understandable to anyone.

ety, the reader is referred to the many works of W. M. Davis, D. Johnson, and A. K. Lobeck. The latter produced a textbook that is required reading for anyone interested in developing his skill along this line.* The block diagram being obliquely oriented to the reader, and thus in perspective, is not a map in the sense that a map is viewed orthogonally and its scale relationships are systematically arranged on a plane projection. From the block diagram, however, have come map types that combine the perspective view of undulations of the land and the planimetric (two-dimensional) precision of the map. These are the landform or physiographic diagram developed primarily in the United States.

The categorization of most phenomena tends to submerge gradations, and the category of pictorial terrain maps is no exception. Within this general class there is a

*A. K. Lobeck, *Block Diagrams and Other Graphic Methods Used in Geology and Geography*, 2nd ed., Emerson-Trussell Book Company, Amherst, Mass., 1958.

large variety ranging from the perspective view utilizing a realistically appearing curved earth surface to the planimetrically correct pictorial map. There are many possible combinations of viewpoint and execution that can be employed to fit the purposes to which such maps are put. Nevertheless, we can recognize three main categories: the perspective drawing, the landform diagram, and the planimetrically correct pictorial map.

Perspective Drawing. The perspective drawing differs from the landform diagram in that the former attempts to show a portion of the earth as if seen in perspective from some distant point, while the landform diagram puts pictorial symbols on an otherwise conventional map base (Fig. 8.17). On the perspective drawing, the appearance of the whole is relative (as in a block diagram of a large section of the earth), and it has become very popular in recent decades. Drawings of this type are usually done on an orthographic projection

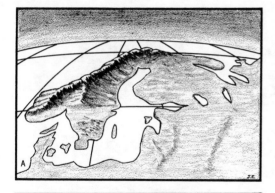

Figure 8.17 **Sketches showing the fundamental difference between the perspective drawing (at top) and the perspective map (at bottom).**

(and exaggerated) to a tremendous degree. Departures of the earth's surface up or down from the spheroid are actually very small, relative to horizontal distances, and they can hardly be shown at all at most medium and small scales. For example, the highest mountain on the earth, Everest, is a bit over 29,000 ft above sea level, or only about 5-1/2 miles. If the terrain of the continent of Asia were represented accurately on a three-dimensional model at a scale of 1:10,000,000 (a map about 4 ft square), Mount Everest would be only about 1/30 inch high! Consequently, almost all pictorial-terrain representation must greatly exaggerate and simplify the terrain. The problems of generalization, therefore, make it absolutely necessary that the cartographer be relatively competent in the field of the geography of landform.

Landform Diagram. One of the more distinctive contributions of American cartography is the pictorial map in which the terrain is represented schematically on an ordinary map base. Although this type of map is not limited to cartographers in the United States, it has reached its highest development through the efforts of A. K. Lobeck, Erwin Raisz, Guy-Harold Smith, and a few others.

Many combinations of media and scales are used for the perspective map. Simple line drawings with or without shading and highlighting make effective *sketch maps* to illustrate the broader concepts of land form relationships (see Figures 8.19 and 8.20). More schematic maps in which the pictorial treatment of the landform is treated more systematically are usually called physiographic diagrams or landform maps. The former attempts to relate the forms to their origin. The physiographic diagrams of A. K. Lobeck are of this type. In these are suggested, by varying darknesses and textures, the major structural and rock-type differences having expression in the surface

(for very small scales), or on an "oblique" photograph of a portion of the globe. The terrain is then modelled so that the earth's curvature is simulated and so that the entire drawing provides the impression of a view of the earth as seen from a point far above. Remarkably graphic effects can be created with this method of pictorial representation. They are particularly useful as illustrations of national viewpoints and of strategic concepts in a world that is growing smaller each year (Fig. 8.18).

The student should have clearly in mind why pictorial terrain must be generalized

Figure 8.18   A much reduced preliminary worksheet of a perspective drawing, by Richard Edes Harrison, of southern Europe as seen from the southwest. The map, for which this was the initial study, appeared in final form in *Fortune*, 1942.

forms. They do not have a particularly realistic appearance, and their common name "physiographic diagram" is appropriate (see Fig. 8.21).

Landform or land-type maps are those in which more emphasis is placed on the character of the surface forms with less attention to their genesis. This type of map is exemplified by those of Erwin Raisz, who has developed a set of schematic symbols with which to represent various classes of the varieties of landforms and land types (see Fig. 8.22).* There is, of course, no sharp distinction between the physiographic diagram and the landform map. All possible combinations of attention may be paid to the underlying structures, rock types, and geomorphic processes.*

Whatever the emphasis may be on such maps, the landforms are positioned without

perspective, but the symbols themselves are derived from their oblique appearance. All physiographic or landform maps have one major defect in common: the side view of a landform having a vertical dimension requires horizontal space, and on a map horizontal space is reserved for planimetric position. For example, if a single mountain is drawn on a map as seen from the side or in perspective, the peak or base and most or all of the profile will be in the wrong place planimetrically. This is well illustrated in Figure 8.20. This fundamental defect in physiographic diagrams and landform maps has, of course, been recognized by the cartographers who draw them, but has, properly, been justified on the ground that the realistic appearance obtained more than outweighs the disadvantages of plani-

*See E. Raisz, "The Physiographic Method of Representing Scenery on Maps," *The Geographical Review*, **21**, 297–304 (1931); E. Raisz, *General Cartography*, 2nd ed., McGraw-Hill, New York, 1948, pp. 120–121.

*Such landform maps can be prepared in such a way that the viewer can obtain from the map the relative heights of the several landforms. See Merrill K. Ridd, "The Proportional Relief Landform Map," *Annals Association of American Geographers*, **53**, 569–576 (1963), and *Map Supplement No. 3, op. cit.*

Figure 8.19 A small perspective map made with a combination of line and shading on Coquille board. Reproduced by line cut. (From Armin K. Lobeck, *Things Maps Don't Tell Us*, The Macmillan Company, New York, 1956).

position to express vertical dimension without producing planimetric displacement. In order to overcome this fundamental difficulty, it is sometimes desirable to substitute for the usual vertical or perspective profile across the land a line that gives a similar appearance but is not out of place planimetrically. This kind of cartographic legerdemain may be accomplished by mapping the traces of the intersections of the land surface with a series of parallel inclined planes (Fig. 8.23).*

*Adapted from Arthur H. Robinson and Norman J. W. Thrower, "A New Method of Terrain Representation," *The Geographical Review*, 47, 507–520, (1957), with the permission of the American Geographical Society. The original study was sponsored by the Department of the Army, Quartermaster Research and Development Center, Environmental Protection Division.

metric displacement. This is especially true on small-scale drawings in which the planimetric displacement is not bothersome to the reader and only occasionally causes the cartographer concern, for example, when some feature of significance is "behind" a higher area. At larger scales, however, the conflict of the perspective view and the consequent error of planimetric position make it desirable to adopt other techniques.

Planimetrically Correct Pictorial Map. A vertical profile across the surface of the ground, when viewed from the side and placed on a map, utilizes planimetric position in two dimensions, yet the top and base of the profile are actually in the same planimetric position on the earth. It is impossible, therefore, to utilize planimetric

Figure 8.20 A portion of a small perspective map made entirely with line work. Compare with Fig. 8.19 in which little systematic symbolism is employed. In this map classes of land form are given prominence by different treatments. (From Armin K. Lobeck, *Things Maps Don't Tell Us*, The Macmillan Company, New York, 1956.)

Figure 8.21   A portion of the small-scale *Physiographic Diagram of the United States*, by A. K. Lobeck, shown on the left. Note the schematic treatment of the surface. (Courtesy of the Geographical Press, formerly at Columbia University Press, now at C. S. Hammond and Company, Inc.) On the right is a vertically viewed shaded landform representation of the same area by R. E. Harrison. This is a portion of the drawing done at a scale of 1:5,000,000 for the *National Atlas of the United States* produced by the U. S. Geological Survey. (Courtesy U. S. Geological Survey.)

When a single horizontal plane is passed through the land surface, the mapped trace of the intersection is a contour. On the usual map the contour line is viewed orthogonally, that is, from directly above. If the plane is rotated on a horizontal axis, ab (see Fig. 8.24) and the successive traces plotted, and still viewed from above, the traces produced run the gamut from the contour trace, x, to a straight line, z. The trace produced by the inclined plane, y, when viewed from directly above, has much the same appearance as would a conventional vertical profile if it were being viewed in perspective at an oblique angle as in a landform diagram. If the angle of inclination from horizontal is designated as θ, then when θ is zero, the trace of the ground with the plane orthogonally represented

on a map is the conventional contour. When θ is 90°, the trace is a vertical profile, and its orthogonal representation is a straight line. When θ is greater than zero but less than 90°, (y in Fig. 8.24), the trace will be that of the intersection of the ground with an inclined plane. A series of such traces on parallel inclined planes produces an *appearance* of the third dimension while still retaining correct planimetry (Fig. 8.25).

The construction of an inclined trace is not difficult. A conventional contour map is required; the contours are needed to define the land surface. The contour map is first ruled with horizontal lines spaced at equal intervals. (These are actually contours of the inclined plane.) Then the intersection of the lowest land contour and lowest hori-

Figure 8.22 A portion of a small-scale landform map. Compare this with left-hand part of Fig. 8.21. Note the inclusion of descriptive terms. (From *Landforms of Arabia* by Erwin Raisz.)

zontal line is connected by a smooth line with the intersection of the next higher contour and the next horizontal line, and so on, as illustrated in Figure 8.26. The resulting line is the orthogonal projection to the map of the trace of the intersection of the land surface (as defined by the land contours), with a plane passed through the land inclined away from the viewer upward toward the top of the map. Fortunately, the successive horizontal lines represent successive elevations on a series of parallel inclined planes so only one set of horizontal lines is required to construct a series of "inclined contours," as illustrated in Figure 8.27.

Many people are familiar with the "flattening out" of rolling and hilly country when viewed from an airplane. Whatever the psychological reason may be, from man's relatively diminutive stature to the kinds of difficulties he experiences in traversing terrain, it is almost always necessary

to exaggerate the vertical dimension when representing landforms perspectively or visually at most map scales. As is shown in Figure 8.28, the traces of the intersection of the ground with planes inclined at different angles, when orthogonally projected to the map, may range from a contour to a conventional profile. If the angle between the inclined plane and the horizontal is designated as θ, then it is evident from Figure 8.28 that the smaller θ is, the more elonggated will the "profile" be and the greater will be the apparent exaggeration; when θ is 0°, there will be no "profile." There is, of course, no actual exaggeration in the presentation, since all traces are orthogonally correct, but it is apparent that a realistic appearance will be obtained by establishing θ somewhere in the middle range of the quadrant of possibilities.

There has been very little study of the problem of how much the vertical dimension should be exaggerated in relation to

Figure 8.23 A much reduced portion of a planimetrically correct terrain drawing of the Camp Hale, Colorado area. The map was drawn at a scale of 1:50,000. (Drawn by Norman J. W. Thrower. Courtesy of *The Geographical Review*.)

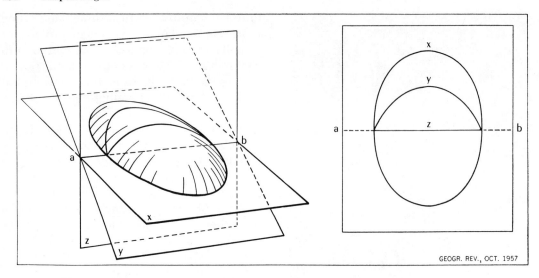

GEOGR. REV., OCT. 1957

Figure 8.24 A comparison of the appearance on a map (right) of three traces produced by passing three differently oriented planes (x, horizontal, contour; y, inclined; and z, vertical, profile) through a single landform. The drawing at left shows the relationships in perspective. (Courtesy of *The Geographcal Review*.)

the various significant factors such as relative relief and nature of the land forms, so generally the cartographer can best proceed by experimentation.* All other things being equal, the closer together the horizontal lines, the "flatter" the terrain will appear, since the nearer θ will be 90°. The relation between the angle of the planes from horizontal (θ), the contour interval in feet (h), and the scale denominator (S) of the map is

$$\frac{12h \cot \theta}{S} = D$$

where D is the distance in inches between

*In an experimental study of vertical exaggeration [George F. Jenks and Fred C. Caspall, "Vertical Exaggeration in Three-Dimensional Mapping," *Technical Report No. 2* (Project No. NR 389-146), Geography Branch, ONR, 1967], the authors related vertical exaggeration (VE) to contour interval (CI) in feet. Assuming that contour intervals reasonably reflect local relief, the equation they derived was: VE = 6.87 − 2.82 log CI.

the parallel lines on the map. For example, if a cartographer were working with a map at the scale of 1:50,000, with a contour interval of 500 ft, and wished to pass the planes through the landforms at an angle of 45° from horizontal, the equation would produce a value of 0.12 in. for D. If in a critical trial area this did not seem to provide enough relief, he could employ a smaller angle for θ, which would provide a larger value for D. When D is very small, drawing one trace for each horizontal line may be too confusing. We may then start traces on alternate lines.

Since three sets of construction lines (conventional contours, horizontal lines, and the traces of the inclined planes) are drawn on the same piece of plastic or tracing paper when utilizing this system (see Fig. 8.29, lower left), it is well to distinguish them. This can readily be done by using different colored pencils or ink. We may trace the selected contour lines from

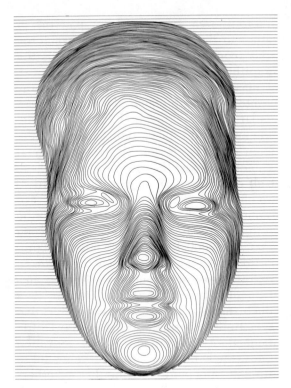

Figure 8.25 The plaster model shown in Figs. 8.8 and 8.9 shown here mapped with inclined traces from parallel planes angled 63° upward toward the rear. (Courtesy G. Fremlin.)

the topographic map in brown, for example, and as nearly as possible complete a drainage pattern in blue: then we may rule the horizontal lines in green and construct the inclined contours in red. A fresh piece of plastic or paper for the terrain rendering can be mounted over the construction sheet. Details of the landforms can then be drawn, with the framework of inclined traces as a guide. The most effective rendering is a "plastic shaded" map, produced by drawing more lines to the right and front of the features than to the left and rear (see Fig. 8.29). This will, of course, give the impression of terrain illuminated from the northwest. If the shading lines are drawn as being controlled by the pattern of

the inclined traces on the undersheet, the important attributes of unity and horizontality are preserved: that is, the landforms are realistically placed in relation to one another, and features do not seem to be "falling off" the map.

The advantages obtained by using this method for large-scale landform map drawing do not hold for very small-scale maps.

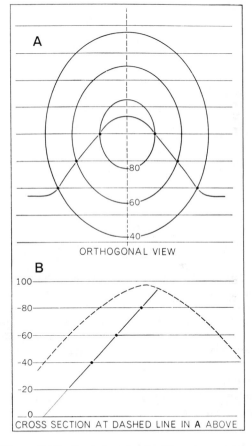

Figure 8.26 In A the circular lines are contours of the land, and the straight lines are the traces or contours of one inclined plane which slopes upward toward the top. This is seen in cross section from the side in B. In A the trace of the intersection of the land surface with the one inclined plane is shown. (Courtesy of *The Geographical Review*.)

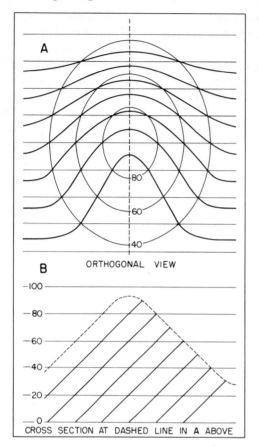

Figure 8.27 The traces of the intersections of the ground and a series of inclined planes are shown in A. The ground surface, the contour planes, and the inclined planes are shown in cross section in B. Compare with Fig. 8.26. (Courtesy of *The Geographical Review*.)

Instead, such maps should be considered more as diagrams in which the forms are being used as symbols.

Other Methods of Mapping Landform

Perhaps the most widely used method of presenting land-surface information on wall maps, in atlases, and on other "physical" maps is that called by various names, such as layer tinting, hypsometric coloring, or altitude tinting. This is the application of different area symbols (hue, pattern, or value) to the areas between the isometric lines (contours). On small-scale maps the simplification of the chosen contours must, of necessity, be large. Consequently, the "contour" lines on such maps are not particularly meaningful, and the system degenerates, so to speak, into a mere presentation of categories of surface elevation. It is an obvious fact that surface elevation is in itself of little consequence with regard to the character of third-order landforms. For example, much of the great plains of the United States lies at an elevation between 2000 and 5000 ft; yet a considerable proportion of it is as flat as any coastal plain. In contrast, most of mountainous Norway lies at an elevation of less than 5000 ft. The conclusion is inescapable, namely, that layer tinting at small scales portrays little about the land surface except elevation zones. Such information is of value to an airplane pilot and to a number of others concerned with subjects such as meteorology, wherein altitude is of some consequence. It is, however, of little value in presenting the significant differences or similarities of the land surface. It should be emphasized that the larger the scale, assuming a reasonable degree of contour simplification, the more useful is the layer system. When the scale has been increased to the point at which the character of the individual isarithms and their relationships become meaningful, then the representation graduates to being a contour map, which is most useful.

The one concept, besides elevation, that the relief map does help to portray for large areas is that of the second-order, three-dimensional structure of a region. Thus, a relief map of South America shows clearly a ridge of high land near and paralleling the western coast. This may be useful to one who is familiar with geographic interrelationships, since he can speculate with some certainty regarding the climatic, vegetational, drainage, occupance, and other possible consequences. To the geographically uninitiated, however, simple layer tinting may well be meaningless.

Colored layer tints at small scales, when combined with pictorial terrain or shading, nearly satisfy most of the landform requirements of the general map reader. They give him the major structure as well as the details of form. On the other hand, to do this is expensive and demands a skill not generally enjoyed by most cartographers. The Wenschow wall maps, previously mentioned, and the works of Richard Edes Harrison are outstanding examples of such combinations. Their reproduction is necessarily by process color (see Chapter 13) or some equally expensive method, and their reproduction cost effectively removes them from the endeavors of the average cartographer. Even the inclusion of a sample in this book is out of the question.

Because of the relative inadequacy of the layer-tinting method for showing detail, geographers and cartographers are continually searching for ways to present a more

useful representation of the land surface. Several methods have been suggested.

1. Terrain unit or descriptive landform-category method.
2. Relative-relief method.
3. Slope-value method.

With the possible exception of the first named, none has attained an acceptance for general maps that even approaches hachuring, shading, or layer tinting. The method of symbolization and presentation, which is the cartographer's major role, is straightforward; area symbols are used to reinforce either isarithms or dasymetric lines. The major problem, inherent in these methods, is the determination of what to present, not how to present it. Consequently, to utilize these methods himself, the cartographer must be essentially a landform geographer, or he must simply present the work of others.

The terrain-unit method employs descriptive terms that range from the simple "mountains," "hills," or "plains" designations to complex, structural, topographic descriptions such as "maturely dissected

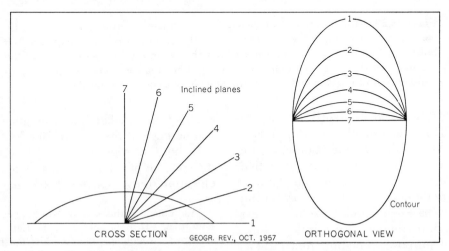

Figure 8.28 The differing orthogonal appearances of traces (right) or differently inclined planes (left). The smaller the angle between the inclined plane and horizontal, the more elongated and, therefore, more "exaggerated" the appearance will be. (Courtesy of *The Geographical Review*.)

Figure 8.29   Illustrating the successive stages in constructing a planimetrically correct pictorial map. The left top shows the contour-drainage map; the left bottom, a few traces. The right top shows the completed guide traces and the drainage; the bottom, the modelled drawing. The area shown is the upper right-hand corner of Fig. 8.23. (Courtesy of *The Geographical Review*.)

hill land, developed on gently tilted sediments." This is essentially a modified dasymetric system, and the lines bounding the area symbols have no meaning other than being zones of change from one kind of area to another. This method of presenting landforms has been found useful in textbooks and in regional descriptions for a variety of purposes, ranging from military terrain analysis to regional planning. Its basic limitations are the regional knowl-

edge of the maker and the geographical competence of the map reader.[*]

In both Europe and the United States, the concept of relative relief, as opposed to elevation above sea level, has been tried.

[*]An example of this kind of map is Plate 3 in G.T. Trewartha, A.H. Robinson, and E.H. Hammond, *Elements of Geography*. 5th ed., McGraw-Hill Book Company, New York, 1967. One also appears in the *National Atlas of the United States*, United States Geological Survey.

Relative, or local, relief is the difference between the highest and lowest elevations in a limited area, for example, a 5′ quadrangle. These values are then plotted on a map and isarithms may be drawn or the distribution may be symbolized in the simple choropleth manner. Area symbols, as in layer tinting, may also be applied. The method is of value when applied to areas of considerable size because basic landform and physiographic divisions are emphasized, but it seems to be unsuited for differentiating important terrain details too small to extend beyond the confines of the unit area chosen for statistical purposes. It is best adapted to relatively small-scale representation.

From the time the hachure became popular, the cartographer and geographer have been concerned with the representation of the slope of the land. Shading and hachuring, although not particularly commensurable, provide a graphic account of slopes on medium- and large-scale maps. The problem of presenting actual slope values on small-scale maps is not easily solved. One technique is the slope-category method. In this method, areas of similar slope are outlined and presented by means of area symbols in the dasymetric fashion (see Fig. 8.30). Other slope techniques, such as percent of flat land per unit area, have been tried, but, except for specialized teaching or research purposes, they have not been widely used.

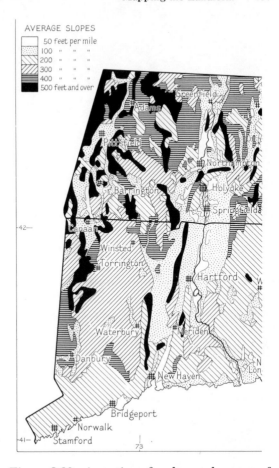

Figure 8.30 A portion of a slope-value map of southern New England by Raisz and Henry. The areas of similar slope were outlined on topographic maps by noting areas of consistent contour spacing. (Courtesy of *The Geographical Review*.)

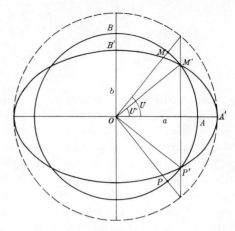

FUNDAMENTALS OF MAP PROJECTIONS

One of the major objectives of cartography is the reduction of the earth so that the spatial arrays on it can be brought into view, so to speak, and thus made comprehensible. One way of accomplishing this is the preparation of a globe map. Ignoring symbolism, when this is done, geometrically all that has been changed is the size; all other relationships (relative distances, angles, relative areas, azimuths, rhumb lines, great circles, and so on) are retained without change. A globe is, therefore, geometrically a "naturally accurate" map.

A globe map, being on a spherical surface, has, on the other hand, a considerable number of practical disadvantages, one of them being the very fact that it is a three-dimensional round body and only less than half of it can be observed at any time. In addition, it is cumbersome to handle, difficult to store, expensive to make and reproduce, and it is not easy to measure on its three-dimensional surface. Consequently, for most of the purposes for which we use a map, the globe map is less desirable than one that has included in its preparation a transformation of the spherical surface. Most of the disadvantages inherent in using the spherical form for the map are eliminated by transforming the surface to a plane; a substance in the form of a plane is easy to handle, all of it can be observed at once, it is relatively cheap to prepare and reproduce, and it is easy to measure and draw on it with ordinary instruments. Thus, the flat map is generally much more useful.

The actual process of transformation is called projection, and the term "projection" stems from the fact that many means of transformation can be accomplished by geometrically "projecting," with lines or shadows, the homologous points from the sphere to a plane surface. Actual geometric projection from the sphere to the plane includes only a few of the possibilities, however, and there are a larger number of possibilities for the retention of significant earth relationships that can be worked out mathematically. These are also called projections, and no useful purpose is served by attempting to distinguish between geometric and mathematical projections.

They are all systematic representations of the spherical surface on a plane surface, and each has specific qualities that make it more or less desirable for specific objectives.

The Cartographer and Map Projections

Ever since the ancient philosophers reasoned that the earth must be a sphere, the method to be employed in transforming the earth's spherical surface to the plane surface of a flat map has been a decision of paramount interest in cartography. The problem arises from the indisputable geometric fact that it is impossible to make the transformation without in some manner modifying the natural surface geometric relationships. But it is also a fact that there are innumerable possibilities of systematic transformation that can retain, on the plane, one or several of the spherical relationships. In addition, and of great significance, is the fact that certain systems of transformation can be employed that produce desirable geometric conditions that do not hold true on a spherical surface. For example, in one system, rhumb lines or loxodromes, which naturally take a variety of shapes on the sphere, can be changed so that they all appear as straight lines on a plane. In quite a different class, but nevertheless of great utility, is the possibility of the selective but systematic enlargement or reduction of a portion of the map area, or even the changing of its shape, in order to fit the specific needs of a presentation problem. Because a great number of systems have been devised, the cartographer does not lack for possibilities from which to choose.

The essential problem of the cartographer, then, is the analysis of the geometrical requirements of the proposed map and the *selection* of the system of transformation that will best or most nearly meet these needs. Not infrequently the cartographer is called upon to prepare a projection notwithstanding the fact that there are many base maps and projections already made that may be utilized. For this he needs to be familiar with the methods and the tabular values needed (or the mathematics of their derivation) in their plotting. Appendix G provides this information for projections commonly employed for smaller-scale maps. It should also be observed that many computer programs exist that can direct an automatic plotter to construct a projection to scale centered as desired.

It is important to reiterate, however, that the cartographer's main concern is with the selection and use of projections. Consequently, he needs to understand what happens to the geometry of the sphere when that surface is transformed to a plane.

The History of Map Projections

The earliest thought of projections of the spherical earth on a plane surface probably occurred no earlier than several centuries B.C.; but in the early centuries following the realization that the earth is a sphere, several solutions to the projection problem were presented. Probably one of the first projections was a simple representation of the spherical quadrilaterals formed by the parallels and meridians as a series of plane rectangles, and today the same kind of projection is still occasionally used. By the end of the Classical Era, enough had become known about the earth, and the problem of projection was so well understood, that Claudius Ptolemy, in his famous work on geography, was able to include a section on map projections and to devise and give directions for their construction. After Ptolemy, the western world lapsed into the Dark Ages and the knowledge of projections was one of the casualties.

Elsewhere, however, notably in the Arabic world, mathematics, geography, and cartography were kept alive. When, in the fifteenth century, Ptolemy was "rediscovered," the western world again made great strides in cartography and discovery. This period was as much a Renaissance in cartography as it was in anything else, and probably the greatest and most influential map projection ever devised, Mercator's projection, was developed during this period.

Mercator's projection is an excellent example of the proper relation that should exist between the objectives of a map and the choice of a system of transformation to meet the needs. The sixteenth century was an exciting time of exploration and much expanded sea travel. Columbus had discovered the Americas; one of Magellan's ships had succeeded in circling the globe; the earth's land areas were beginning to take shape on the world map; and ships were setting out "to all points of the compass." One of the major trials of an early navigator was that, although he had a rough idea of where lands were and had the compass to help him, he had no way of determining the bearing of a course which, with any degree of certainty, would take him to his destination.

It was pointed out earlier that the problem of the navigator is that he must sail along a rhumb line, a line of constant bearing, because he cannot readily travel any other course. The solution was to project (mathematically—not geometrically) the earth's surface in such a way that a straight line on the resulting plane, *anywhere in any direction,* was a rhumb line. Thus, if a mariner knew his starting point, he need but draw a straight line (or a series of straight lines) to his destination, and, if he made appropriate allowances for drift, winds, and compass declinations, he had a reasonably good chance of arriving somewhere near his destination. Mercator's projection suited perfectly the purpose of the map. It still does.

In the four centuries since Mercator introduced the projection that bears his name, the world of man has changed tremendously. Distances have been reduced a thousandfold; man has investigated and mapped an untold number of subjects; and all branches of science, including cartography, have progressed immensely. The development of map projections has kept pace with the developments in other fields, and as the needs arose for ways of presenting particular geographic relations, a means of transforming the spherical surface to the plane to accomplish the purpose usually became available.

Not all projections have been developed in answer to specific needs, by any means. The transformation of a spherical surface to a plane in such a manner as to maintain on the plane certain of the numerous spherical relationships is a most intriguing mathematical problem. Consequently, many projections have been devised simply as solutions to interesting problems rather than with a specific utility in mind. Also, it should be remembered that some of our common projections were originally contemplated and worked out by the ancients and were only resurrected a thousand years or more later when their utility was appreciated. Such was the case with the gnomonic, orthographic, and stereographic projections, all of which were imagined or devised before the time of Christ, but were not much employed thereafter until more than fifteen hundred years later.

The correlation between the purpose of the map and the projection used is strikingly revealed by the tremendous advances made during the last century or so. As transportation capacities have increased and social consciousness has developed, the need for maps for air navigation and other nonmarine travel (as well as for the display of population, land use, and other geographic factors) has likewise increased. Many new projections have been devised and ways of adapting many old ones worked out; thus today there is literally an unlimited number of projections from which to choose.

It may reasonably be asserted that at present cartographers need to devote little time to devising new projections but rather would do better to become more proficient in selecting from the ones available. On the other hand, if a new and particular use of maps requires a special type of projection, undeveloped as yet, such a projection might well be worth the time and effort spent in devising it.

The foregoing brief sketch of the history of map projection would not be complete without the observation that undoubtedly a great many projections have been devised without their inventors being fully aware of exactly what was involved in their transformation of the spherical system to a plane surface. That is, a given requirement, such as parallel parallels or the earth within a circle of a certain size, may be accomplished by arranging the earth's graticule by trial and error methods, without a very clear (or, for that matter, any) knowledge of the mathematical transformational processes involved. As a matter of fact, Mercator did not understand exactly the mathematical fundamentals behind the famous projection that bears his name, even though the tremendously important desired result was clearly obtained.

Obviously, a book of this sort cannot begin to go into all the theory and proof(s) involved in a subject as mathematically complex as is the subject of map projections. Nevertheless, in order that the student of cartography and the user of maps may have an understanding of the fundamentals involved in the subject of transformation, from the sphere to the plane, a relatively nonmathematical description of the more basic elements follows.

Scale and Map Projection

The understanding of map projections requires that we become fully familiar with the concept of scale variations that result from the system of projection. No problem occurs when the necessary map reduction is merely one of size, that is, when a globe map is prepared. But when the reduction is accompanied by a transformation of the spherical surface to a plane, then complexities enter the picture. Reduction to a globe is accomplished, of course, with no variation of scale from one part of the surface to another, that is the Scale Factor (SF) remains 1.0 everywhere (see Chapter 2). This is not possible, however, when transformation to a plane is involved. Consequently, the distribution of scale variation on the different projections is the key to their understanding and use.

To appreciate the problem, it is helpful to imagine the establishment of a pattern of infinitely closely spaced points on one surface, and then to establish the positions of corresponding or homologous points on another. The system employed to specify the positions of the points on the second surface constitutes the method of projection of the first surface.

If the two surfaces are applicable, that is, if one may be produced by bending the other without tearing or stretching (for example, a cone or cylinder to a plane), then the geometric relationships of angles, areas, distances, and directions on the one surface may be retained on the other. If the two surfaces are not applicable (and tearing is disallowed), then the distance relationships among homologous points on the two surfaces must be modified, that is differential enlargement or reduction must occur. Since angular and areal relationships are functions of relative distances, then alterations of these relationships are bound to occur. Consequently, it is impossible to devise a system of projection for two nonapplicable surfaces such that any figure drawn upon the one will appear exactly similar upon the other.

A spherical surface and a plane surface are not applicable; therefore, any method of map projection must include relative shrinking or stretching of portions of the spherical surface in its representation on a plane surface. Nevertheless, it is possible by properly arranging the differential enlargement and reduction involved to (1) retain some kinds of comparable angular relationships, *or* (2) to retain comparable areas of like figures from the one surface to the other. If these particular qualities are not wanted in a projection but some other geometric attribute of the spherical surface is desired, then all angular relationships will usually be changed (except perhaps with respect to particular points), and areas on the two surfaces will not have a constant ratio to each other.

In order to understand these stated facts as they apply to the projection of the spherical surface of the earth to the plane surface of a map, it is necessary to realize the following two facts.

1. Scale exists at a point.

2. The scale may be different in different directions at a point.

To demonstrate the first of these propositions, imagine an arc of 90°, as in Figure 9.1, projected orthographically to a straight line tangent at *a*. If *a*, *b*, *c*, . . . *j* are the positions of 10° divisions of the arc, their respective positions after projection to the line tangent at *a* are indicated by *a*, *b'*, *c'*, . . . *j'*. Line *aj'* therefore represents line *aj*. Whereas the intervals between successive points on the arc are equal lengths, the method of transformation has not retained this characteristic. It may be seen from the drawing that the intervals on the straight line, starting at the point of tangency (*a*), become progressively smaller as *j'* is approached. If the scale along the arc is expressed as unity (SF=1.0), it of course exists as unity anywhere along the arc; but on the projection of the arc, the SF is gradually reduced from unity at *a* to zero at *j'*. The rate of change is graphically indicated by the diminution of the spaces between the points. Since a line may theoretically be

considered as consisting of infinitely closely spaced points, every point on *aj* has its counterpart on *aj'*. But there has been a *continuous change* of scale from unity to zero along *aj'*. Since it is a continuous change, it is evident that every point on *aj'* must have a different Scale Factor.

In order to demonstrate that scale at a point may also be different in different directions, imagine a rectangle *a*, *b*, *c*, *d*, as in Figure 9.2, an orthogonal projection of which has resulted from rotating it around the axis *ad* so that side *ad* coincides in *abcd* and its projection *ab'c'd*. The perspective drawing on the left in Figure 9.2 shows how this may be done. If the RF of *abcd* is assumed to be unity, and since the length *ad* is the same in each rectangle, there has been no change in scale in that direction, that is, the SF is 1.0. Since the length *ab'* is half the length *ab*, and since it is evident from the method of projection that the change has been made in a uniform fashion, then the Scale Factor along *ab'* must be 0.5 or half the scale that obtains along *ad*. By projection, line *ac* has become line *ac'*. The ratio of lengths *ac'* to *ac* constitutes the Scale Factor along *ac'*, and it is evident that it is neither the 1.0 ratio along *ad* or the 0.5 ratio along *ab'*; it is somewhere in between. Any other diagonal from *a* to a position on the side *bc* would have its corresponding place of intersection on *b'c'*. The ratio of lengths of similar diagonals on the two rectangles would be different for each such line. Hence, the scale at point *a* in rectangle *ab'c'd* is different in every direction.

The knowledge of the characteristics of relative scale in different directions at each point and the change in scale characteristics from point to point on a projection provide the bases for analyzing, to a considerable degree, what the system of projection has accomplished for, or done to, the geometric realities of distances and directions,

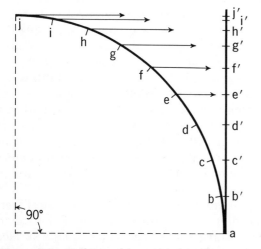

Figure 9.1   Orthographic projection of an arc to a straight line tangent.

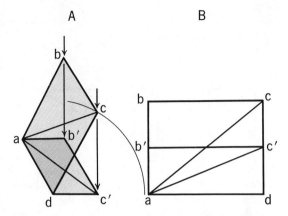

Figure 9.2   Projection of a rectangle to another rectangle with one side held constant. The reduced, perspective drawing on the left shows the geometric relation of the two rectangles, while the drawing on the right shows the relation of the two rectangles when they are each viewed orthogonally, i.e., perpendicularly to their surface.

and their functions, angles, areas, etc., on the sphere.

The Law of Deformation

At any point on the spherical surface there are, of course, an infinite number of directions and therefore an infinite number of paired perpendicular directions, such as N-S with E-W, NE-SW with NW-SE, and so on. When the spherical surface is transformed to a plane, naturally all the directions on the globe are represented by directions on the projection. Generally, however, the angular relation among the directions at each point will have been changed by the system of transformation, that is, a pair of directions that is perpendicular on the globe will not necessarily be shown as perpendicular on the projection.

If angles have been preserved by the system of projection (called conformality), there will be an infinite number of pairs of perpendicular directions at each point in

the projection in the directions of which the Scale Factor is the same. This is a special case. On the other hand, if, as is usual, conformality has not been preserved, there will be at each point only *one pair* of perpendicular directions preserved as perpendicular on the plane (except at particular points), and the Scale Factor at each point in those directions will be unequal. The law of deformation states that, whatever the system of transformation, *there are at each point of the spherical surface at least two perpendiculars (directions) which will reappear at right angles to each other on the projection*, although all the other angles at that point may be altered from their original position.

It is by an understanding of the consequences of the occurrence of equal or unequal Scale Factors in the directions of the perpendiculars that reappear at right angles in the projection that we may appreciate the kinds and amounts of deformation that can and do occur in all map projections.

Deformation in Map Projections

Regardless of the system employed to transform the spherical surface to a plane, the geometrical relationships on the sphere cannot be entirely duplicated. Angles, areas, distances, and directions are subject to a variety of changes, and there are many other specific spatial conditions that may or may not be duplicated in map projections, such as parallel parallels, converging meridians, perpendicular intersection of parallels and meridians, poles being represented as points, and so on. The major alterations, however, are those having to do with angles, areas, distances, and directions. In order to provide the student with an understanding of the basic consequences of these alterations, a brief resumé of their characteristics follows.

Angular Alteration. The compass rose appears the same everywhere on the globe surface (except at the poles); that is, at each point, the cardinal directions are always 90° apart, and each of the intervening directions is everywhere at the same angle with the cardinal directions.

It is possible to retain this property of angular relations to some extent in a map projection. When it is retained, the projection is termed *conformal* or *orthomorphic*, and both words imply "correct form" or "correct shape." It is important to understand that these terms apply to the directions or angles that obtain at infinitely small points. The property of conformality is not meant to apply to areas of any significant dimension, since no projection can provide correct shape to areas of any extent.

On the sphere, or on a globe, the scale is the same in every direction at any point and, furthermore, the scale is the same at every point. In other words, the Scale Factor is everywhere 1.0. In the projection process, stretching and compression must take place, however, and consequently, the scale may never be the same everywhere on the projection. It is possible, however, to arrange the stretching and compression so that *at each point* on a conformal projection the scale is the same in all directions, although it necessarily must vary from point to point. Thus, on all conformal projections the scale will vary from one point to another. If the condition of uniform scale in all directions at each point is maintained, then all directions around each point will be presented correctly, and the parallels and meridians will intersect at 90°. It is important to realize that just because a projection shows the parallels and meridians as crossing one another with right angles, it does not necessarily mean it has the property of conformality. The reason for this is demonstrated as follows.

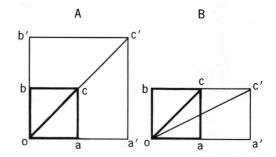

Figure 9.3 In *A*, rectangle *oacb* has been projected to rectangle *oa'c'b'* so that the ratio of lengths *oa'* and *ob'* corresponds to the ratio of lengths *oa* and *ob*. In *B* this ratio has not been preserved. In *A* all angles are preserved at *o*; in *B* they are not.

In Figure 9.3 the left-hand drawing, *A*, represents a point *o* on the sphere. Points *a* and *b* are defined as being infinitely close to *o*. Since the scale on a globe is the same everywhere in every direction, the drawing represents *oa* and *ob* as similar lengths. Line *oc* represents a diagonal. If the rectangle *oacb* is projected to rectangle *oa'b'c'* in such a fashion as to retain the relative lengths of *oa* and *ob* in the projected rectangle, then angles *coa* and *cob* will correspond to angles *c'oa'* and *c'ob'*. If, however, as in *B* in Figure 9.3, the relative lengths of *oa* and *ob* are not retained, then angles *coa* and *boc* will not be similar to angles *c'oa'* and *boc'*. But it may be seen that angles *boa* and *boa'* are the same. Therefore, just because one pair of perpendicular directions on the sphere (for example, the cardinal directions) is retained on a projection, it does not necessarily follow that all other directions will be correctly represented.

Area Alteration. It is possible to retain in a map projection the proper relative sizes of any segments of the surface of the sphere. When this characteristic of the surface of the sphere is retained, the projection is said to be *equivalent* or *equal-area*.

In Figure 9.4 the left-hand drawing, *A*, represents a point *o* on the globe. Points *a*, *a'*, *b*, and b' are infinitely close to *o* and lie in directions 90° from one another at *o*. Since on the globe the scale is the same in every direction, *aa'* and *bb'* are represented as similar lengths in Figure 9.4. Length *aa'* will be designated as *a*, and *bb'* as *b*. They constitute the bisectors of a rhombus, in this case a square. The area scale at *o* is considered unity.* The right-hand drawing, *B*, in Figure 9.4, shows a projection of point *o* in which the perpendicular directions toward *a* and *b* from *o* are retained, but the scale in the direction *a* has been doubled (SF = 2.0) and the scale in direction *b* has been halved (SF=0.5). It is evident from the conditions that the area of the rhombus in *B* remains equal to unity.† It is also clear, merely by inspection, that angular relationships from *o* to points other than *a*, *a'*, *b* or b' in *A* would be markedly changed in the projection *B*. That this must be so was demonstrated above and in Figure 9.3.

To state the proposition in general terms: if a system of projection is employed such

*The area of the rhombus is *ab*/2. If lengths *a* and *b* in the drawing are made equal to $\sqrt{2}$ then the area of the rhombus = 1.0.

$$†\frac{2\sqrt{2} \times \frac{1}{2}\sqrt{2}}{2} = 1$$

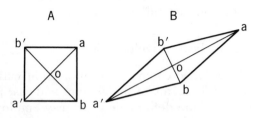

Figure 9.4 Rhombus *aba'b'* in **A** has been projected to rhombus *aba'b'* in **B**. The products of the bisectors in **A** is the same as the product of the bisectors in **B**, hence the areas of the two figures are the same.

that the *product* of the scales in directions that are perpendicular on the projection and on the globe is equal at every point, then all areas of figures on the projection will be represented in correct relative size.

Such a projection can have the scale the same in *all* directions at only one or (at the most) two points or along one or two lines. At all other places the scale will be different in different directions from each point. Hence, angles around all such points will be deformed. Since the scale requirements for conformality and for equivalence in a map projection are contradictory, it is apparent that no projection can be both conformal and equivalent. Thus all conformal projections will present similar earth areas with unequal sizes and all equivalent projections will deform most earth angles.

The alterations of areas and angles that occur when the spherical surface is represented on a plane are the most important for the majority of cartographic representations. Two other alterations need be considered, however, in order that the student may have a sound understanding of what must happen when the one surface is projected to another. One of these concerns the problem of the alteration of distances.

Distance Alteration. Needless to say, a map projection represents all distances "correctly" provided we know the scale variations involved. As generally understood, however, distance representation is a matter of maintaining consistency of scale; that is, for finite distances to be represented "correctly," the scale must be *uniform* along the extent of the appropriate line joining the points being scaled, and must be the same as the principal scale on the nominal globe from which the projection was made, that is, have a Scale Factor of 1.0. The following are possible.

1. Scale may be maintained in one direc-

tion, for example, north-south or east-west, but only in one direction. When this is done the lines (parallels or meridians, in this case) along which the Scale Factor remains 1.0, are called *standard.*

2. Scale may be maintained in all directions from *one* or *two* points, but only from those points. Such projections are called *equidistant.*

Any other scale relationship must be a compromise in order to gain better distance relationships in some or all directions in one part of the map at the expense of the representation in some other part.

Direction Alteration. Just as it is impossible to represent all earth distances with a consistent scale on a projection, so also is it impossible to represent all earth directions correctly with straight lines on a map. It is true that conformal projections represent angular relationships around each point correctly and that we can even arrange the scale so as to obtain straight rhumb lines or great circles. But no projection can show *true direction* in the proper sense that all great circles will be shown as straight lines that will have the same angular relations to the graticule of the map that they have with the earth's graticule. For example, the oft-stated assertion that the Mercator projection "shows true direction" applies only to the fact that constant bearings are shown as straight lines. Such a statement is erroneous in the sense that true direction on a sphere is along a great circle, not along a loxodrome except, of course, when the two coincide (Fig. 9.5).

When directions are defined properly as great circle bearings, and if we think of a correct direction as being that shown by a great circle as a straight line on the map having the proper azimuth reading with the local meridian, then certain representations are possible.

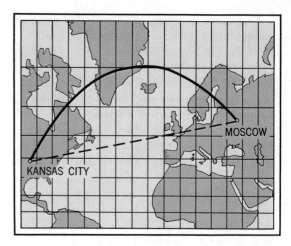

Figure 9.5   The great circle and the loxodrome from a point in the United States to a point in Russia as they appear on a Mercator projection. The great circle shown as a solid line is the "true direction," i.e., most direct, from the one point to the other, not the loxodrome.

1. The course of great circle arcs between all points may be shown as straight lines for a limited area although the angular intersections of all the great circles with the meridians (azimuths) will not be shown correctly. To do this causes such a strain, so to speak, on the transformation process that it is not possible to extend it to even an entire hemisphere.

2. Straight great circles with correct azimuths may be shown for all directions from *one* or, at the most, *two* points. Such projections are called azimuthal.

The Analysis of Deformation

In order to compare one projection with another in terms of the amount and distribution of the deformation, we may employ various approaches. Some are entirely graphic and simply show a sort of graphic index of the amount and location of the deformation. Of generally more utility is the approach that enables us to determine specific values for the angular and area exaggeration that occurs from place to place in a projection. To do this we proceed by way of the law of deformation and Tissot's indicatrix.

Tissot's Indicatrix. M. A. Tissot published in 1881 a treatise on map projections in which he developed a method for analyzing the amount and distribution of the deformation. To do this he employed a mental construct he called the indicatrix. For this purpose he imagined an infinitely small circle located at the intersection of any pair of perpendicular directions on the sphere. In projection on the plane the circle will appear either as a smaller or larger circle, if angles are preserved, or as an ellipse, if angles have not been preserved.[*] The two perpendicular diameters of the infinitely small circle on the sphere (directions) that are retained as perpendicular directions on the plane constitute the major and minor axes of the ellipse. The lengths of the axes are made proportional to the Scale Factors that occur in those directions. It is important for the student to realize that these lengths represent relative magnitudes and *not* real lengths on the sphere or projection. By an analysis of the geometric changes that result from the transformation of the original circle on the sphere to an ellipse on the plane, we can determine the amount of angular and areal deformation that has occurred at that point. To demonstrate how this may be done, an illustration is provided here without the mathematics involved in the computations or their proofs. A more complete development, with the formulas involved, is pre-

sented in Appendix H for the interested student.

Figure 9.6 is a representation of the concept of the indicatrix. It is reiterated that the lengths portrayed on the drawing are not actual distances on a map projection but are merely representative of Scale Factors. The drawing represents a point on the sphere at O. The Scale Factor in every direction at O is considered unity, so that the point O can be represented by a circle with $OM = 1 =$ the radius r. In the system of projection directions, OB and OA are the directions of the perpendiculars on the sphere that are retained as perpendiculars in the projection. The Scale Factor in the direction OA will be designated as a and that in the OB as b. Since $OB = OA = OM$, a and b in the original circle = 1.0. If the

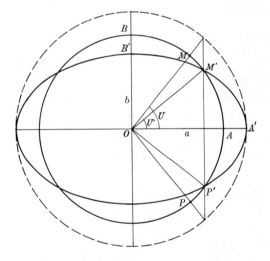

Figure 9.6 The indicatrix (the ellipse) in the above illustration has been constructed as an equal-area representation of the infinitely small circle to which the following indices apply: $OM = OB = OA = r = 1$; Scale Factors in the directions $OA' = 1.25$, and $OB' = 0.80$; $MOA = U = 51°21'40''$; $M'OA' = U' = 38°39'35''$; $U - U' = \omega = 12°42'05''$; $2\omega = 25°24'10''$. (Modified from Marschner.)

[*]If a circle is considered as a special form of an ellipse in which the major and minor axes are equal, then it can be stated that the infinitely small circle on the sphere is always transformed to an infinitely small ellipse. This will simplify subsequent wording.

system of projection is equivalent, as in Figure 9.6, the original circle will be transformed to an ellipse in which OA' is the semi-major axis and OB' is the semi-minor axis. Since, as was stated earlier, the product of the scale ratios must be equal in an equivalent projection, then in the projection (the ellipse), $a = 1.25$ and $b = 0.80$. These are the directions of maximum changes in scale in the indicatrix, and the values for a and b are all that are needed to determine for any point on any projection the amount of either the angular or areal alteration that has occurred at that point.

Angular Deformation. To analyze first the angular deformation that has taken place at O, it is necessary to understand that all points on the circumference of the circle will have their counterparts on the periphery of the ellipse. Point B has been shifted to point B', and A to A' on the ellipse. It is evident that no angular change in these directions has taken place since angle $BOA =$ angle $B'O'A'$. All other points on the arc between B and A will, when projected to the ellipse, be shifted a greater or lesser amount in their direction from O The point subjected to the greatest deflection is identified in the circumference of the circle with M, and has its counterpart in the periphery of the ellipse in point M'. The angle MOA on the sphere thus becomes $M'OA'$ in the projection. If angle $MOA = U$ and angle $M'OA' = U'$, then $U - U'$ denotes the maximum angular deformation within one quadrant. The value of $U - U'$ is designated as *omega* (ω). If an angle such as MOP were to have its sides located in two quadrants and if they were to occupy the position of maximum change in both directions, then the angle in question would be changed to $M'OP'$ and would thus incur the maximum deflection for one quadrant on both sides. Consequently, the value of 2ω denotes the possible *maximum angular* change that may occur at a point. All other angular deformations at O would be less than 2ω. Since the values of ω will range from $0°$ in the directions of the axes to a maximum somewhere between the axes, it is not possible to state an average angular deformation.

Area Deformation. Changes in the representation of areas may or may not be a corollary to the transformation of the circle into an ellipse by the projection system. If there has been a change in the surface area, its magnitude can be readily established by comparing the areal "contents" of the original circle with that of the ellipse. The area of a circle is $r^2\pi$, whereas the area of an ellipse is $ab\pi$, where a and b represent the semi-major and semi-minor axes, respectively. Therefore, since the axes of the ellipse are based upon the original circle whose radius was unity, and since π is constant, the product of ab compared to unity expresses how much the areas have been changed. The product of ab is designated as S.

For the purpose of comparing projections, only the values of 2ω and S are needed. On conformal projections the Scale Factor, by definition, is the same in every direction at a point. It will differ, however, from point to point, but it will always be the same in every direction around a point. Therefore, on all conformal projections, $a = b$ everywhere on the projection. When $a = b$ the value of 2ω is $0°$. Hence there is no angular deformation at points on a conformal projection, but because the values of a and b vary from place to place, the product of ab, that is, S, will also vary from place to place. Consequently, all conformal projections exaggerate or reduce relative areas, and S at various points provides an index of the degree of areal change.

On equivalent projections the scale relationships at each point are such that the product of ab always equals 1.0. Any difference between the values of a and b will produce a value of 2ω, that is, greater than $0°$. Consequently, all equivalent projections deform angles and the value of 2ω at various points provides an index of the degree of angular deformation.

On all projections that are neither conformal nor equivalent, a will neither equal b nor will the product of $ab = 1.0$. Therefore, on such projections both the values of S

and 2ω will vary from place to place. Their relative magnitudes will provide an index of the degree of areal change and of the angular deformation.

The values of S *and* 2ω that occur at various points may be plotted on a projection. The distribution of intermediate values may be shown by drawing isarithms of equal values. In this manner, a diagram showing the pattern of the distribution of angular deformation and areal change over the entire projection may be prepared. Most of the projections illustrated in the next chapter have such distributions shown on them. For some, data have not been determined. It is also possible from such a diagram of the angular or areal deformation on a projection to derive the value of the mean deformation for either the entire projection or for only a portion, such as the land area included, by the use of a cumulative frequence graph. A comparison of mean values for several similar projections is helpful in evaluating the relative qualities of different systems of projection.

Graphic Representation of Deformation

Tissot's indicatrix is limited, in its analytical function, to values at a point. It does not provide much help in depicting another kind of deformation, which exists in all projections, that of changes in the distance and angular relation between widely spaced points or areas such as continents. To date, this kind of deformation has not been found to be commensurable, and consequently, graphic means have been employed to help show the alteration that takes place with respect to the larger spatial relations.

Various devices have been employed to this end, such as a man's head plotted on different projections to illustrate elongation, compression, and shearing of larger areas (Fig. 9.7). The deformation of directions is effectively demonstrated by plotting various

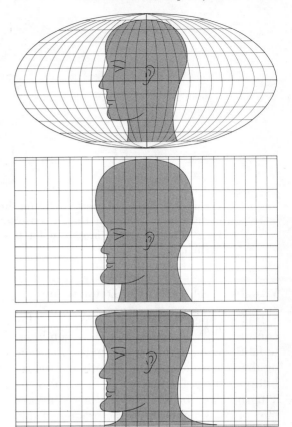

Figure 9.7   A profile plotted with the same latitude and longitude on (top) Mollweide's projection, (center) Mercator's projection; and (bottom) the cylindrical equal-area projection with standard parallels at 30°, to illustrate deformations. Because the profile looks most natural on Mollweide's projection does not mean that projection is "better." The natural profile could be drawn on any one and then plotted on the others.

great circles in different parts of a projection (Fig. 9.8). Another device has been the covering of the globe with equilateral triangles and then reproducing the same triangles on the different projections.*

*This device appears to be particularly helpful, and the reader is referred to Fisher and Miller, *World Maps and Globes*, Essential Books, New York, 1944, for excellent illustrations of this use of triangles.

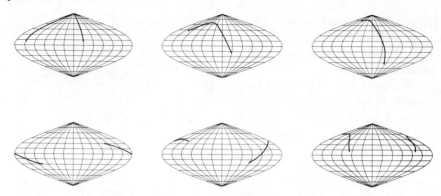

Figure 9.8   Selected great circles on the sinusoidal projection showing the departure from a straight line. Each uninterrupted and interrupted arc is 150° long. (From a report produced for the Office of Naval Research, Geography Branch, by W. Tobler.)

Visual-Logical Analysis of Deformation

It is not uncommon for a map reader to be confronted by a projection with which he is unfamiliar. A visual-logical method of recognizing to some extent the kinds and distribution of the deformation may be employed. This involves becoming familiar with certain of the geometric characteristics of the graticule, since this supplies a kind of index of the spacing and location of points, and then comparing these same characteristics as they are portrayed by the graticule of the projection.

The visual characteristics of the graticule, some of which have been discussed previously, are given in the following list.*

1. Parallels are parallel.
2. Parallels are spaced equally on meridians.
3. Meridians and other great circle arcs are straight lines (if looked at perpendicular to the earth's surface, as is true on a map).
4. Meridians converge toward the poles or diverge toward the equator.
5. Meridians are equally spaced on the parallels, but their distance apart decreases from the equator to the pole.
6. Meridians at the equator are spaced the same as parallels.
7. Meridians at 60° are half as far apart as parallels.
8. Parallels and meridians cross one another at right angles. (Therefore, with 10 below, the compass rose is the same everywhere, except at the poles.)
9. The area of the surface bounded by any two parallels and two meridians (a given distance apart) is the same anywhere between the same two parallels.
10. The Scale Factor at each point is the same in any direction.

To illustrate the mental processes the cartographer should employ when analyzing a projection, Figures 9.9 and 9.10, together with their analyses, are given. The numbers of the visual characteristics in the foregoing list, referred to in the legends, are placed in parentheses.

*In this list several slight variations have been approximated in order that the principles may be more clearly grasped. In 2, parallels actually vary in their spacing by about 0.7 mile; in 6, the discrepancy is negligible; and in 7, there is a difference of about 0.1 mile. None of these approximations would be of significance in a general-use map with a scale smaller than 1:2,000,000.

NORTH   POLE

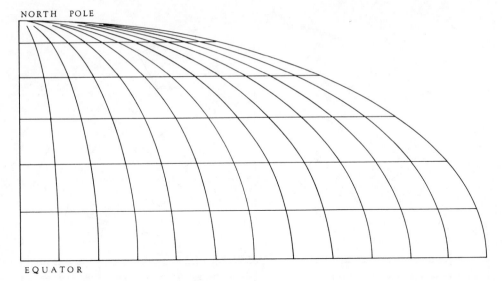

EQUATOR

Figure 9.9   One quadrant of Mollweide's projection with a 15° spacing of the graticule. Since the scale is not the same in every direction (10), as shown by the disproportionate length of the meridian sections at the equator (6), and the compass rose is not the same because parallels and meridians do not cross one another at right angles (8), the projection obviously is not conformal. Because numbers nine (9) and five (5) appear to be satisfied, it is likely that the projection approaches equal-area. Because the parallels are not equally spaced (2) and because they do not always cross meridians at right angles (8), it is evident that angular relations are greatly distorted. (Courtesy *Annals of the Association of American Geographers.*)

The Arrangement of the Graticule

Repeated operations over a long period of time usually result in the establishment of conventions. These are generally the consequence of the realization that one way of doing something is more convenient or efficient than another. Map projections have not escaped this process.

Since the earth's surface is that of a sphere, the surface is the same everywhere insofar as its general geometry is concerned. At any point, the surface curves away in all directions at the same rate. One segment is just like any other segment. In order to have something to provide positive identification of location on such a limitless, uniform space, man devised the earth's coordinate system. It is so useful that we find it difficult to think of the earth's surface without automatically including the graticule. Consequently, we tend to conceive of a map projection as a representation of the graticule. Furthermore, because the concept of cardinal directions is so important to man in his thinking about the earth, it has become conventional to represent the surface in such a way as to present these directions to good advantage. This is commonly done by making significant directions appear as straight lines or arcs, such as east-west parallels or north-south meridians. This necessitates organizing the chosen system of projection in a particular or conventional way.

It is important for the student to realize

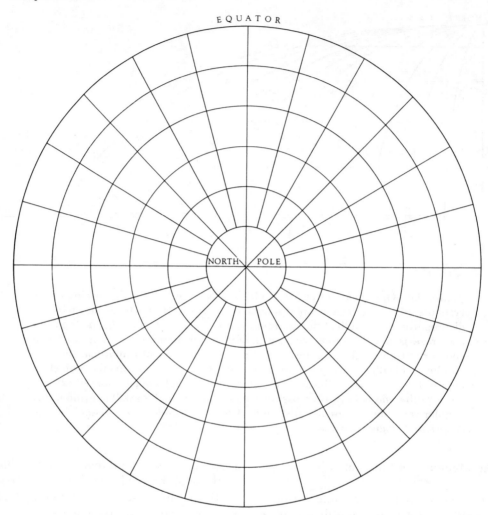

Figure 9.10 One hemisphere of the azimuthal equidistant projection. Since a series of straightline meridians (3), which are, of course, arcs of great circles, converge to a point and are properly arranged around the point as shown by their equal spacing on each parallel (5), the projection must be azimuthal from the point of convergence. The foregoing plus the fact that the parallels are equally spaced on the meridians (2) make it evident that there is no scale change along the meridians; hence the projection must be equidistant from the point of convergence. Because, however, on the projection the meridians at the equator are not spaced the same as parallels (6) the scale cannot be equal in every direction from each point. Hence the projection cannot be conformal. Also, since the scale along the meridians has been shown to be correct by the spacing of the parallels (2) and because of the fact that azimuths from the pole are correct, it is likely that the central (polar) area is close to truth. If that is true, then the equatorial areas are disproportionately large because of the excessive distances between the meridians (6). Therefore, the projection cannot be equivalent. (Courtesy *Annals of the Association of American Geographers.*)

that, however the system of projection may be arranged or oriented with respect to the graticule, any system of projection is merely one of transforming a spherical surface to a plane. The graticule has nothing to do with it except to provide a handy series of reference points. For example, a projection that shows the meridians and parallels as straight lines intersecting at 90° is in reality merely showing one set of great circles that intersect at two points and a series of small circles concentric to those points as parallel orthogonal straight lines. There are an infinite number of such arrangements of great and small circles on a sphere, and whichever set the system of projection displayed, the characteristics of the projection would still be the same. The pattern of deformation and the amounts of deformation would not change. Figure 9.11 shows how the graticule appears when the sinusoidal projection is centered at various places. They are all the same system of projection.

On the other hand, in working with a spherical surface, we must have reference points. The coordinate system provides these. Therefore, since the graticule is useful to establish convenient reference points (because it is well known and because the cardinal directions are important), it is common for a transformation system to be applied to the sphere in such a way that the graticule is displayed in a simple fashion. This may be accomplished by arranging the system so that deformation is symmetrical around well-known lines, such as a meridian or one or two parallels. Such lines, when the scale is held constant along them and is made equal to that on the nominal globe of the same scale, are standard lines. They constitute reference lines that define the scale employed and from which the scale departs in other parts of the map.

The concept of standard parallels and standard meridians will be used in the next chapter to help describe the conventional form of projections. It should be remem-

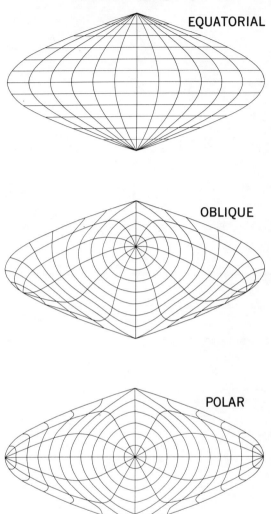

Figure 9.11 **Different centerings of the sinusoidal projections produce different appearing graticules. Nevertheless, the distribution pattern of the deformation is the same on all, since the same system of transformation is employed; therefore, they are all the same projection. (From a report produced for the Office of Naval Research, Geography Branch, by W. Tobler.)**

bered, however, that the arrangement of the projection system so that standard great circles coincide with the equator or meridians, or so that the standard small circles coincide with parallels, is merely a great convenience; it is not a necessity.

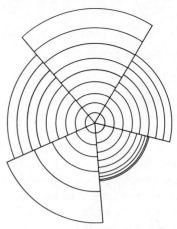

THE EMPLOYMENT OF MAP PROJECTIONS

10

The number of possible systems of projection is unlimited, and so is the variety of objectives for which maps may be made. Fortunately, in practice, the combination of purposes that recur frequently and the fact that a number of systems of projection combine several useful characteristics result in relatively few projections being commonly used. Anyone who uses or constructs maps should have a working knowledge of the character of those projections that are frequently employed or encountered.

The number of diverse factors that may influence the choice of a map projection is surprising. The geographer, historian, and ecologist are likely to be concerned with sizes of areas. The navigator, meteorologist, astronaut, and engineer are generally concerned with angles and distances. The reference map maker generally wants a compromise, the illustrator is often limited by the prescribed format, and the maker of a series of maps is interested in how sheets may be made to fit together. To balance the sometimes contradictory desires against the projection possibilities is always an operation that requires ingenuity and imagination.

Because the selection of a projection involves such diverse elements as expediency, publication format, geometric properties, and many others, there is likely to be a tendency, common among those who do not quite appreciate the nature of projections, to think of them as but poor representations of an actual globe surface. The selective process then proceeds on the basis of "the least of the evils." Such an attitude leads to the conviction that one projection is intrinsically better than another. Nothing could be farther from the truth. Projections are commonly advantageous for reasons other than the fact that it is cheaper to make a flat map than a globe map. The majority of projections enable us to map distributions and derive and convey concepts that would be either impossible or at least undesirable on a globe. The truth of the matter is that regardless of how much we are concerned with "deformation," "alteration," "distortion," and all the other concepts used in analyzing, comparing, and describing projections, a projection is a triumph of ingenuity and a positive, useful device.

The notion that one projection is by nature better than another is as unfounded as saying pliers are better than wrenches. Each (tool or projection) is a device to use for a particular purpose, and some will be good for one purpose and poor for another. There are some projections for which no useful purpose is known, but there is no such thing as a bad projection—there are only poor choices.

The Choice of Map Projections

There are many factors to be kept in mind when choosing a system of projection. No specific formulas can be given that will lead to the right selection because each map is a complex compound of objectives and constraints. A few generalizations can be drawn that will exemplify the complex nature of this fundamental choice that the cartographer must make.

Probably the largest class of maps (and map users) is concerned not with the portrayal of some one or combination of the facts of spherical geometry but more with the widest possible uses of the map. In this class, for example, would be atlas makers, whose reference maps serve a wide clientele. One of the more important factors involved in selecting projections to be used for general maps is the manner in which the inherent deformation is arranged with respect to the area covered by the graticule. Certain general classes of projections have specific arrangements of the deformation values, and the knowledge of these patterns helps considerably in both choosing and using a particular system.

Maps to be made in series, such as sets for atlases or even topographic series, have different requirements from those made as individual maps. For example, a most useful attribute of some aspects of some projections is the fact that any portion can be cut from the whole and provide a segment that is in itself a relatively good selection for a smaller area, with good symmetry and deformation characteristics. Any projection in which the meridians are straight lines that meet the parallels at right angles satisfies this requirement. With such a projection it is possible, for example, to make a map of the entire United States and then make from it separate, overlapping maps of various regions, each centered in the middle of a neat projection.

Consideration of geometrical needs, expediency, methods of subdivision, etc., all play a part in the selection of a system of projection that will give the best result.

A great many maps demand more from the map projection than one or a combination of the special properties of projections (equivalence, conformality, and azimuthality). Such projection attributes as parallel parallels, localized area deformation, and rectangular coordinates frequently become greatly significant to the success of a map. For example, a map of some sort of distribution that does not require equivalence may have a concentration of the information in the middle latitudes. In a case of this kind, a projection which to some extent expanded the areas of the middle latitudes would be a great help by allowing relative detail in the significant areas. Any small-scale map of temperature distributions over large areas is made more expressive if the parallels are parallel and even more so if they are straight lines allowing for easy north-south comparisons. A map for which indexing of places is contemplated is more easily done with rectangular coordinates than with any other kind.

The overall shape of an area on a projection is likewise of great importance. Many times the shape and size (format) of the page or sheet on which the map is to be made is prescribed. On one map projection the area to be mapped may fit this format, and on another it may not; but each may have the desirable properties. By using a projection that fits a format most efficiently, a considerable increase in scale can often be effected, which may be a real asset to a crowded map.

The Classification of Projections

The usual categorization of projections is based on general geometric characteristics. Theoretically, the spherical surface is transformed to a "developable surface," which is a geometric form capable of being flattened, such as a cone or a cylinder (both of which may be cut and laid out flat) or a plane (which is already flat). Conventionally, the axis of the earth is aligned with the axes of the cylinder and cone (see Fig. 10.1) so that the graticule lines will be simplified. In a projection based upon a cone, meridians converge in one direction and diverge in the other, and on the opened-up cylinder, meridians are straight parallel lines. Projections on a plane are not so conventionally aligned, and no generalizations can be made about their appearance. Such a constructional grouping of projections results in categories called

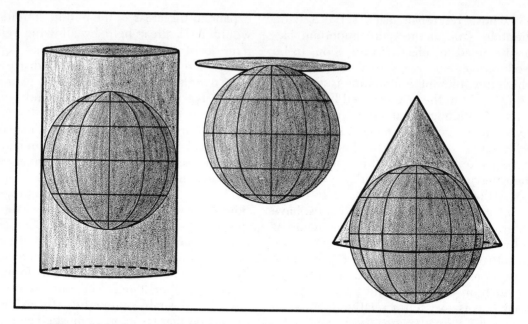

Figure 10.1   Some of the theoretical developable surfaces on which the earth's surface may be projected. The origin of the projecting rays may assume various positions. For example, it may be at the center of the earth or at the antipode of the point of tangency of the plane.

cylindrical, conic, azimuthal (plane), pseudocylindrical, and miscellaneous (those based on no geometric form). Whatever the terminology employed, the grouping is, strictly speaking, not a classification but a listing. Other classifications or listings of projections include the parametric, the appearance of the graticule, and the relation of the spherical surface to the plane (secant, tangent, transverse, or oblique, for example).

The subject of classification of projections has been much investigated and general proposals for its accomplishment have been made, none of which seems to have gained general acceptance. The terminology of the subject has been described as being "in a state little short of chaotic," and the treatment here is not intended to set a pattern. The topic is so large and compli-

cated that a detailed treatment of it is out of the question in a book of this kind.*

The primary elements in the choice of projections are generally their major property or properties. Each property (for example, equivalence) is generally of sufficient significance so that it is usually the first distinguishing characteristic with which the cartographer begins to make his choice. The properties of projections are not entirely mutually exclusive, however, since there are some azimuthal projections that are also conformal or equivalent. Hence, properties cannot easily be used as a basis

*The interested reader is referred to two important studies: Waldo R. Tobler, "A Classification of Map Projections," *Annals Association of American Geographers,* **52,** 167–175 (1962); and D. H. Maling, "The Terminology of Map Projections," *International Yearbook of Cartography,* **8,** 11–64 (1968).

for classification. The inherent qualities of the projections in a group that has a particular property are, however, relatively distinctive, and when one such projection is chosen on such a utilitarian basis, the fact that it has other properties is usually of minor significance.

The Terminology of Projections

A variety of special terms are used in describing and analyzing map projections. Some have already been referred to in the previous chapter, but for the sake of completeness they have been included here. Many will be much clearer if the student will visualize the projection process as consisting of two steps: (1) the reduction of the earth to a nominal globe, and (2) the projection of this reduced globe to the plane. The inevitable scale departures from the nominal scale of the globe can be thought of in terms of the Scale Factor.

1. *Equivalence* is that quality of a projection in which the product of the quantities *a* and *b* is everywhere unity on the projection. Hence, the sizes of all areas will be represented on the projection in correct proportion to one another.

2. *Conformality* is that quality of a projection in which the quantities *a* and *b* are at every point equal but which differ from point to point. Angles around every point are thus correctly represented.

3. *Azimuthality* is that quality of projection in which the angle between any straight line from *A* to *B* on the earth and the meridian at *A* is shown as the same angle as that which would occur on the globe between a meridian at *A* and the great circle arc from point *A* to point *B*.

4. *Linear scale* refers to the ratio of distance in some direction on the map compared to distance along that same direction

on the earth. Where the two RF values (map and globe) are the same, i.e., where the SF = 1.0, the scale is said to be "correct" or "true."

5. *Area scale* is the ratio of areas on the map to areas on the earth. Relative area differences from one place to another on the map are obtained by computing the values of S.

6. *Standard lines* are those great or small circles on the projection that have along them a uniform true linear scale as defined above. In the usual or conventional arrangement of the projection system with respect to the globe surface the lines of true scale are often represented by parallels or meridians of the earth coordinate system. Hence it is usual to refer to *standard parallels* or *standard meridians*. It is not, however, necessary that standard lines coincide with elements of the graticule.

7. *The central meridian* of the map or projection is the meridian that divides it in equal parts. In many conventional projections it is the straight line, which may or may not be a standard line, around which the projection is symmetrical, that is, the projection on one side of the central meridian is the mirror image of the other side.

8. *Angular deformation* (ω) is the change in angular relationships that can occur at a point. Since the amount of deviation that takes place varies from 0° to a maximum and then back to zero within any quadrant at any point, we cannot state that there is an average value at a point. The maximum angular deformation that could occur at a point is twice the amount that could occur in one quadrant. It is symbolized by 2ω, which is stated in degrees.

9. *Pattern of deformation* is the arrangement of either S or 2ω values on a projection. It is most easily symbolized and visualized by thinking of the quantities as representing the Z values of an S or 2ω third

dimension above the projection. Then isarithms drawn for this surface will show the arrangement of the relative values and, by their closeness, the gradient or rate of change. Certain classes of projections have similar patterns of deformation.

(a) A *cylindrical pattern* occurs on all projections that in fact or in principle are developed by first transforming the spherical surface to a tangent or an intersecting cylinder. In all cases the lines of equal deformation are straight lines parallel to the standard lines, the least deformation being along the line of tangency or intersection (see Fig. 10.2).

(b) A *conic pattern* results if the initial transformation is made to the surface of a true cone tangent at a small circle or intersecting at two small circles on the sphere. Lines of equal deformation parallel the standard small circles (Fig. 10.3).

(c) An *azimuthal pattern* occurs if the

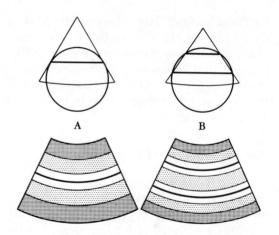

Figure 10.3 Conic patterns of deformation. Diagram A shows the pattern when the cone is tangent to one small circle; B, when it intersects along two small circles.

transformation takes place from the sphere to a tangent or an intersecting plane.* The trace of the intersecting plane and sphere will of course be circular. Lines of equal deformation are concentric arcs around the point of tangency or the center of the circle of intersection (see Fig. 10.4).

Deformation will increase in all these instances away from the standard lines (or point). The greatest gradient will usually be along the line normal (perpendicular) to the standard line.

10. *Mean deformation,* either maximum angular or area (2ω or S), is the weighted arithmetic mean of the values that occur over the projection. When derived for similar areas on different projections, a comparison of the mean deformation values provides an index of the relative efficiency of the forms of projection.

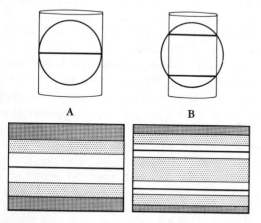

Figure 10.2 Cylindrical patterns of deformation. Diagram A shows the pattern when the cylinder is tangent to a great circle; B, when the cylinder is secant and the standard lines are parallel small circles (they need not be parallels of the graticule). The darker the shading the more the deformation.

*When an intersecting form (cylinder, cone, or plane) is employed in the transformation process, the resulting projection is commonly termed a secant projection. The term is usually applied to the conic forms.

MAJOR MAP PROJECTIONS

The number of map projections that have been devised and described in the literature is very large. No attempt is made here to be at all inclusive in naming or describing this vast list. Instead, only those that are commonly seen are included.*

Equivalent Projections

For general reference map use, the quality of equivalence is perhaps more necessary than any other. For research tools and for the presentation of results in the geographic, economic, historical, and political fields, the map is more than merely a graphic record. The framework of boundaries, rivers, and other base data provides a kind of correlative background for study, and of the general elements needed for this

*A useful list of named map projections with a short description of each is included as Appendix I, Named Map Projections, in *Glossary of Technical Terms in Cartography*, London, The Royal Society, 1966.

background, proper delineation of relative sizes probably takes precedence.

In the presentation of many kinds of distributional data, the property of equivalence is more than a passive factor. The mapping of some types of data specifically requires that the reader be given a correct visual impression of the relative sizes of the areas involved; otherwise he will receive an erroneous impression of relative densities, which would clearly be contrary to the purpose of the map. An example of how this can come about is illustrated in Figure 10.5.

Many of our general impressions of the relative extent of various regions are gained subconsciously through frequent experience. Because nonequivalent projections have been so frequently used for general maps in the past, most people think Greenland is considerably larger than Mexico (nearly the same size) and that Africa is smaller than North America (Africa is more than 2 million square miles larger). It is, of course, obvious that if any area measurement, (for example, by planimeter) is contemplated, the projection should have a uniform area scale.

The choice among equivalent projections depends upon two important considerations:

1. The size of the area involved.

2. The distribution of the angular deformation.

There are a great many possibilities from which to choose, and if the cartographer will but keep these two elements in mind, he will rarely make a bad choice.

As a general rule, the smaller the section of the earth to be represented, the less significant is the choice of projection for a general map. Any equal-area projection

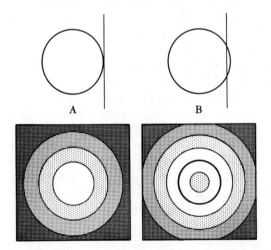

Figure 10.4 Azimuthal patterns of deformation. Diagram *A* shows the pattern when the plane is tangent to the sphere at a point; *B*, when it intersects along a circle on the earth.

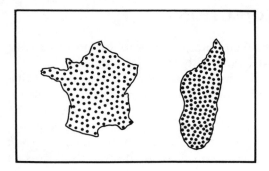

Figure 10.5 These two areas (France, left; Madagascar, right) are nearly the same size on the earth. On the nonequal-area projection from which these two outlines were traced, France appeared larger than it should by comparison to Madagascar. The same number of dots has been placed within each outline, but the apparent densities are not the same.

will contain areas of relatively little angular deformation. Consequently, if the area is not large, a projection may be chosen with fortuitous deformation relationships for the area involved. For large areas or the whole earth the distribution of the deformation becomes of paramount significance. The regions of topical importance on the map should be represented in the best fashion possible by the choice of a projection with an advantageous distribution of deformation.

A few representative types of common equivalent projections are shown, along with brief notes on their employment. Most of the illustrations show isarithms of 2ω (angular deformation) to show the overall pattern of deformation. Since the values of the lines are not always the same on the different illustrations, it is necessary for the reader to note the values carefully if he wishes to compare the various projections. The areas shown with lighter shading on the drawings are the areas of lesser deformation.

Because equivalent projections are often employed for world maps, it is convenient

to consider this class separately. Consequently, equivalent projections appropriate for areas of lesser extent are dealt with first, and then those appropriate for world maps are considered as a separate class. This is not entirely consistent; some parts of projections appropriate for world maps are also useful for specific types of smaller areas.

Albers' projection (Fig. 10.6) has two standard parallel small circles (conventionally these are made parallels of the graticule) along which there is no angular deformation. Because it is conically derived, deformation zones are arranged parallel to the standard lines. Any two small circles in one hemisphere may be chosen as standard, but the closer together they are, of course, the better will be the representation in their immediate vicinity. Because of the low deformation value and its neat appearance with straight meridians and concentric arc parallels that meet the meridians at right angles in its conventional form, this is a good choice for a middle-latitude area of greater east-west extent and a lesser north-south extent. Outside the standard parallels, the Scale Factor along the meridians is progressively reduced. Parallel curvature ordinarily becomes excessive if the projection is extended for much over 100° longitude. In recent decades Albers' projection has replaced other projections (notably the Polyconic) as the common choice for many maps of the United States. Its obvious superiority for maps on which to plot and study geographical distributions has led to its selection as the standard base map by many governmental agencies such as the Bureau of the Census or the Bureau of Agricultural Economics.

Bonne's projection (Fig. 10.7) has a central meridian and a central parallel along neither of which is there any deformation. Any parallel may be chosen as central and

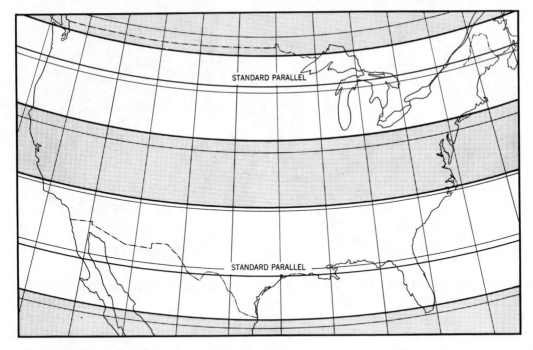

Figure 10.6 Albers' conic projection. Values of lines of equal maximum angular deformation are 1°.

all the others will parallel it. The representation decreases in angular quality outward from the central meridian and away from the central parallel. Although it is still seen in general atlas maps, its use is decreasing.

Lambert's equal-area projection (Fig. 10.8) is azimuthal as well as equivalent. Its azimuthal properties are described later in the chapter. Since deformation is symmetrical around the central point, which can be located anywhere, the projection is useful for areas that have nearly equal east-west and north-south dimensions. Consequently, areas of continental proportions are well represented on this projection, It is limited to hemispheres. As a base for general maps it is fast replacing the formerly much used Bonne's projection.

World Projections. The entire surface area of the earth may be plotted in equivalent fashion within the bounding lines of a plane figure of almost any shape. Most of these pseudocylindrical projections have more theoretical interest than practical util-

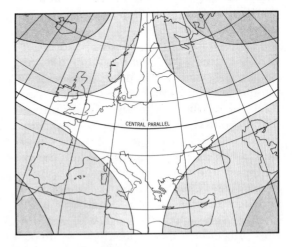

Figure 10.7 Bonne's projection. Values of lines of equal maximum angular deformation are 1° and 5°.

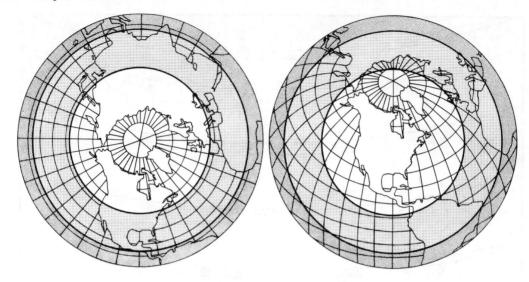

Figure 10.8 Lambert's azimuthal equal-area projection. Values of lines of equal maximum angular deformation are 10° and 25°.

ity; only the commonly used few can be treated here.

Any world equal-area projection will necessarily have considerable angular deformation, for the modification of angular relationships necessary to maintain the relationship of $ab = 1.0$ over the entire projection is great. The pattern of arrangement of the deformation is an important quality of these projections.

The cylindrical equal-area projection (Fig. 10.9) is capable of variation like Albers' conic, that is, the projection has two parallel standard small circles, usually parallels of latitude. The two small circles (parallels) may "coincide," so to speak, and be a great circle (the equator), or they may be any two others so long as they are homolatitudes (the same parallels in opposite hemispheres). Deformation is arranged, of course, parallel to the standard small circles. Although for a variety of reasons this projection "looks peculiar" to many people, it does in fact provide, when standard parallels just under 30° are chosen, the least *overall* mean deformation of any equal-area world projection.

The sinusoidal projection (Fig. 10.10), when conventionally oriented, has a straight central meridian and equator, along both of which there is no angular deformation. All parallels are standard and a merit of this projection is that they are equally spaced, giving the illusion of proper spacing so that it is useful for representations where latitudinal relations are significant.[*] The sinusoidal is particularly suitable when properly centered for maps of less-than-world areas such as South America for which area the deformation distribution is especially fortuitous.

Mollweide's projection (Fig. 10.11) does not have the excessively pointed polar areas of the sinusoidal, and thus it appears a bit more realistic. In order to attain equivalence within its oval shape, it is necessary to decrease the north-south scale in the high latitudes and increase it in the low latitudes. The opposite is true in the east-west direction. Shapes are modified accord-

[*]It is an illusion because the distance between parallels is properly measured along a meridian, not simply perpendicular to the lines representing the parallels on the map.

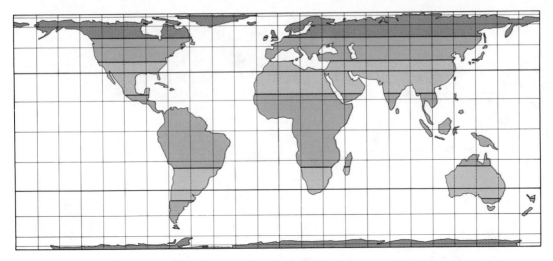

Figure 10.9   The cylindrical equal-area projection with standard parallels at 30°. Values of lines of equal maximum angular deformation are 10° and 40°.

ingly. The two areas of least deformation in the middle latitudes make the projection useful for world distributions when interest is concentrated in those areas.

Eckert's IV (Fig. 10.12) projection, when conventionally arranged, has a pole represented by a line half the length of the equator instead of by a point. Accordingly, the polar areas are not so compressed in the east-west direction as on the preceding two

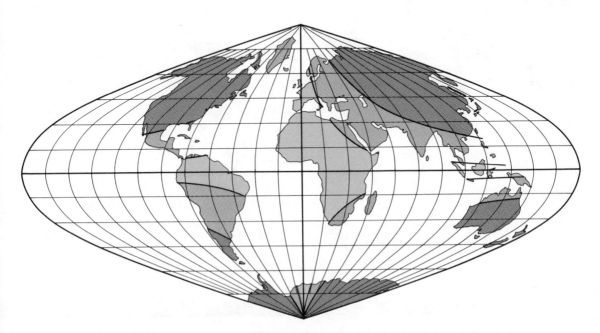

Figure 10.10   The sinusoidal projection. Values of lines of equal maximum angular deformation are 10° and 40°.

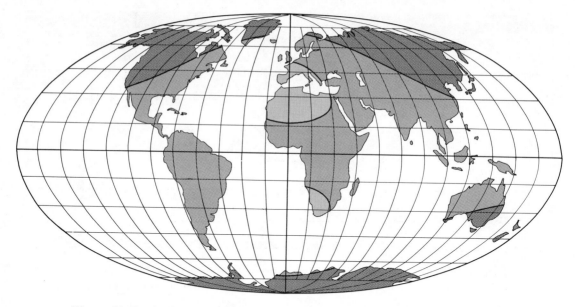

Figure 10.11 Mollweide's projection. Values of lines of equal maximum angular deformation are 10° and 40°.

projections. This takes place, however, at the expense of their north-south representation. As in Mollweide's, the equatorial areas are stretched in the north-south direction. Deformation distribution is similar to that of Mollweide's.

The flat polar quartic projection (Fig. 10.13) has a line one-third the length of the

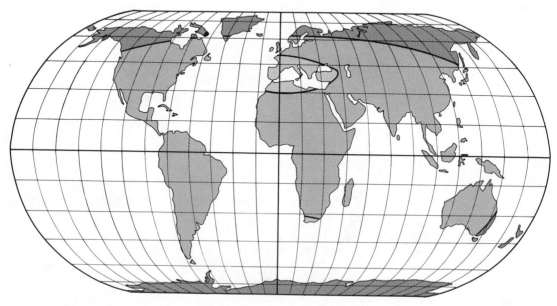

Figure 10.12 Eckert's IV projection. Values of lines of equal maximum angular deformation are 10° and 40°.

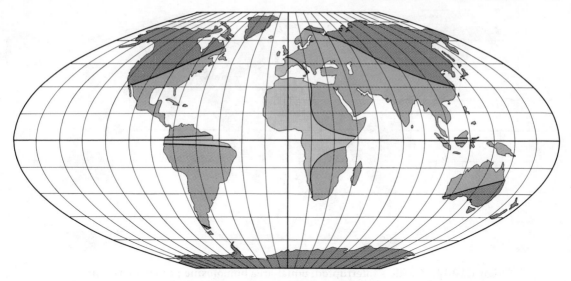

Figure 10-13   The flat polar quartic equal-area projection. Values of lines of equal maximum angular deformation are 10° and 40°.

equator for the pole (when conventionally arranged). This provides a better appearance for the polar areas than either the sinusoidal or the Mollweide. It is becoming increasingly popular for world maps.

Several projections for world maps have been prepared by combining the better parts of two. The best known of these is Goode's Homolosine (Fig. 10.14), which is a combination of the equatorial section of the sinusoidal and the poleward sections of Mollweide's, thus, it is equal-area.° The two projections, when constructed to the same area scale, have one parallel of identical length (approximately 40°) along which they may be joined. It is usually used in interrupted form (see below), and has been widely employed in the United States. Its overall quality, as shown by a comparison of mean values of deformation, is not appreciably better than Mollweide's alone.

°Mollweide's projection is sometimes called the *homolographic*, hence the combined form *homolo + sine.*

Interruption and Condensing. In order to display the land areas of the earth to better advantage on an equal-area world projection, it is possible to interrupt and recenter the projection, that is, construct the projection in such a way as to repeat the zones of lesser deformation.

All that is necessary to interrupt and recenter a projection constructed in the conventional manner is that the parallels be uniformly subdivided by the meridians, or, to put it another way, that the linear scale along each parallel be uniform. Any number of central meridians may be chosen and the grid constructed around each; no central meridian need extend any particular latitudinal distance; the central meridians in opposite hemispheres need not match; and, as a matter of fact, a "central meridian" may even be displaced from latitude to latitude.

When, through interruption and recentering, the zones of lesser deformation are duplicated, then no land area need be far removed from the area of least angular alteration (Fig. 10.15). Thus, the deformation

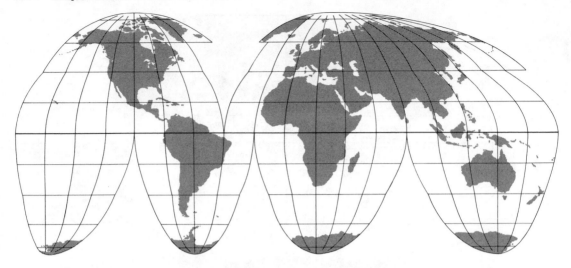

Figure 10.14   Goode's interrupted, equal-area homolosine projection is a union of Mollweide's in the areas poleward of approximately 40° and the sinusoidal equatorward of 40°. See Figs. 10.10 and 10.11 for the patterns of deformation.

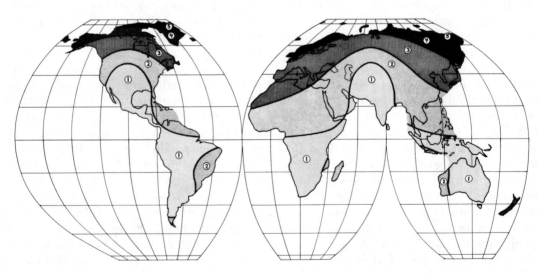

Figure 10.15   An interrupted flat polar quartic equal-area projection. The values of the zones of maximum angular deformation are (1) 0° − 10°, (2) 10° − 20°, (3) 20° − 40°, (4) 40° − 60°, and (5) over 60°. Note how much more land area is included in the low deformation zones compared to Fig. 10.13. (Courtesy *Annals of Association of American Geographers*.)

of the significant parts of the map, usually the land, may be considerably reduced.*

Another operation that is commonly applied to world equal-area map projections is to condense them, that is, cut away the unnecessary water areas (if the land is the

*Arthur H. Robinson, "Interrupting a Map Projection: A Partial Analysis of its Value," *Annals of the Association of American Geographers*, **43**, 216–225. (1953).

concern). This does not improve the deformation aspects, but it obviously does make it possible to employ a larger scale within a given format (Fig. 10.16).

Conformal Projections

Maps that are to be used for analyzing, guiding, or recording motion and angular

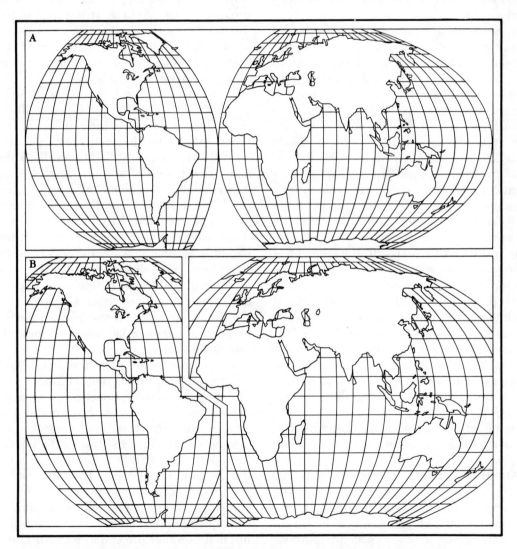

Figure 10.16 An interrupted, flat polar quartic equal-area projection (*A*) compared to one which has been condensed as well as interrupted (*B*).

relationships require the employment of conformal projections. In these categories fall the navigational charts of the mariner or aviator, the plotting and analysis charts of the meteorologist, and the general class of topographic maps. For topographic maps, the property of conformality is not quite so indispensable as it is for maps to be used for the other purposes, since topographic maps serve many needs that do not require conformality.

Because there is no angular deformation at any point on a conformal projection, the notion is widespread that shapes of countries and continents are well presented on projections having this quality. Although it is correct to state that very small areas on conformal projections are practically perfect, it is also true that in order to retain angular representation it is necessary to alter the area relationships. Thus, on conformal projections the area scale varies from point to point, and consequently large areas are imperfectly represented with respect to shape.

It is difficult to deal with the concept of deformation on a conformal projection because, in a sense, there is nothing deformed since all angular relationships at each point are retained. All that changes is the Scale Factor, and one point is as "accurate" as another; only the scales are different. Thus we may refer to the principal scale as "correct," and the other areas will be merely relatively exaggerated or reduced.

Some of the conformal projections together with some notes on their qualities follow.

Mercator's projection (Fig. 10.17) is one of the most famous projections ever devised. It was introduced in 1569 by the famous Dutch cartographer as a device for navigation, and it has served this purpose well. In its conventional form it has the property that all loxodromes are represented as straight lines, an obvious advantage to someone trying to proceed along a compass course. Except for the meridians and the equator, directions (great-circle courses) are not straight lines, so this projection does not show "true direction"; but such courses can be easily transferred from a projection (gnomonic) that does so, and a series of straight rhumb lines can thus approximate a great circle. It is apparent that the projection enlarges (not distorts) areas at an increasing rate as we move toward the higher latitudes, so it is of little use for purposes other than navigation.

Lambert's conic projection (Fig. 10.18) is very similar in appearance to Albers' equivalent projection, since in its conventional form it too has concentric parallels and equally spaced, straight meridians that meet the parallels at right angles. Like Albers' it has two standard parallels, but the Scale Factor is less than 1.0 between them and more than 1.0 outside them. Area deformation between and near the standard parallels is relatively small, and thus the projection provides exceptionally good directional and shape relationships for an east-west latitudinal zone. Consequently, the projection is much used for air navigation in intermediate latitudes and for meteorological charts.

The stereographic projection (Fig. 10.19) belongs also in the azimuthal group. Like the other azimuthal projections, the deformation (in this case, area exaggeration) increases outward from the central point symmetrically. As in the case of the equivalent azimuthal, this is an advantage when the area to be represented is more or less square or of continental proportions. In addition to being conformal and azimuthal, the stereographic has an additional attribute that no other projection has. All circles on the earth remain as circles on the

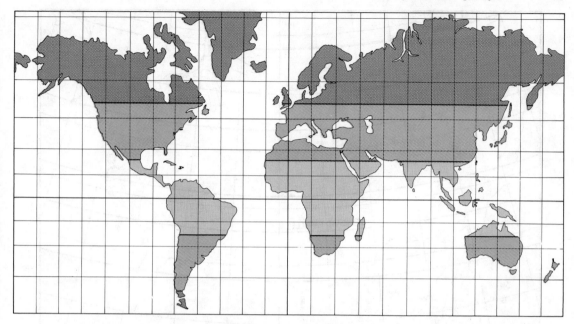

Figure 10.17   Mercator's projection. Values of lines of equal area exaggeration over the principal scale at the equator are 25% and 250%.

projection.* It is possible therefore to plot the ranges of radiating objects, from radio waves to airplanes, merely with a compass. This projection, centered on a pole, is much used for navigation in the very high latitudes.

It bears repeating that recent practice has turned to the selection of conformal projections for topographic maps. Each of the mentioned three projections is so used in one form or another. The most widely used are Lambert's conic and a transverse form

of the Mercator. The latter is of considerable significance because it has been chosen as the projection for at least twenty of the modern topographic map series. In addition, the universal transverse Mercator (UTM) projection is employed for the rectangular grid system, described in Chapter 2. This transverse form is also called the Gauss-Krüger projection. The manner in which a projection is "transversed" is considered below.

Azimuthal Projections

As a group, the azimuthal projections (sometimes called zenithal) have increased in prominence in recent years. The advent of common air travel, the development of radio electronics, and the general increase in scientific activity all have contributed to this development. It has occurred because azimuthal projections have a number of

*Since all great circles through the center of the projection are straight lines, we must define these as circles with a radius of infinity to make this statement not open to argument. Any small circle on the earth can truly be drawn with a compass on the projection. It may be done by locating the ends of a diameter along a straight line radial (great circle) through the center, and then finding the *construction* center of the circle by halving the diameter. The actual center on the earth and the construction center for the circle do not coincide except at the center of the projection.

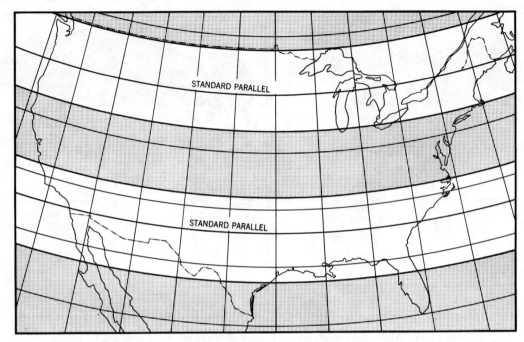

Figure 10.18   Lambert's conformal conic projection. Values of lines of equal areal exaggeration are 2%.

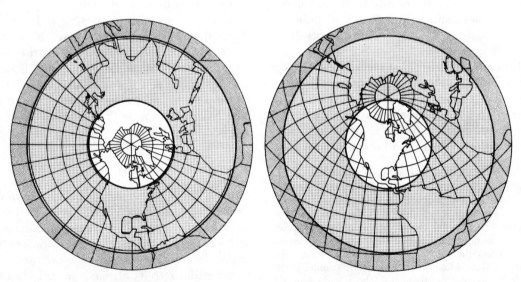

Figure 10.19   The stereographic projection. Values of lines of equal areal exaggeration are 30% and 200% (approximately).

qualities not shared by other classes of projections.

All azimuthal projections are theoretically "projected" upon a plane perpendicular to a line passing through the center of the sphere. Such a line may, of course, intersect the surface of the sphere at any point; consequently, these projections are symmetrical around a chosen center. The variation of the Scale Factor in all cases radiates from the center at the same rate in every direction. If the plane is made tangent to the sphere, there is no deformation of any kind at the center. Furthermore, since all these projections are "projected" on a plane parallel to a tangent to the sphere, all great circles passing through the center point will be straight lines on the projection and will show the correct azimuths from and to the center in relation to any point. It should be emphasized that only azimuths (directions) from and to the *center* are correct on an azimuthal projection.

At the center point all azimuthal projections with the same principal scale are identical, and the variation among them is merely a matter of the scale differences along the straight great circles that radiate from the center. Figure 10.20 illustrates this relationship. Any azimuthal projection may be changed to any other one by changing the scale relations along the azimuths. The fact that the deformation radiates symmetrically makes this class of projections useful for areas having more or less equal dimensions in each direction, or for maps in which interest is not localized in one dimension.

Because any azimuthal projection can be centered anywhere and still present a reasonable appearing graticule, the class is rather more versatile than others. We frequently see an azimuthal projection with the north pole at the center, for it is easy to draw and provides an illusion of reality

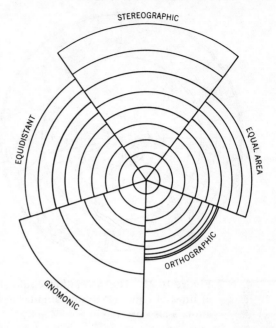

Figure 10.20 Comparison of common azimuthal projections centered at the pole. In all cases meridians would simply be straight lines radiating from the center. Note that the only variation is in the spacing of the parallels; in other words, the only difference among them is the scale away from the center. This relationship between graticules and small circles around the center obtains wherever the projections may be centered.

because of the regularity of the graticule. The several qualities of azimuthal projections so centered are somewhat wasted.

The above characteristics of azimuthal projections are in addition to whatever other attributes they may have, and unlike equivalence and conformality, which are mutually exclusive, there is an equivalent azimuthal projection (Lambert's equal-area) and a conformal member of the class (stereographic). The employment of these two was considered earlier in the chapter. There are three other important members of the azimuthal family among an infinite

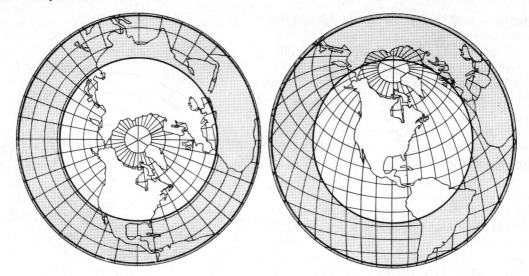

Figure 10.21 The azimuthal equidistant projection of a hemisphere. Values of lines of equal maximum angular deformation are 10° and 25°. Areal exaggeration is also present.

number of possibilities. These are the azimuthal equidistant, the orthographic, and the gnomonic.

The azimuthal equidistant projection (Fig. 10.21) has become popular in recent years. It has the unique quality that the linear scale does not vary along the radiating straight lines through the center. Therefore, the position of every place is shown in consistent relative position and distance in relation to the *center*. Directions and distances between points whose connection does not pass through the center are not shown correctly. It is apparent, then, that unless movement along a straight line through the center is of major significance, the azimuthal and equidistant qualities may be wasted and some other projection might be better choice.

Any kind of movement that is directed toward or away from a center is well shown on this projection, such as radio impulses and seismic waves. The projection has an advantage over many of the other azimu-

thal projections in that it is possible to show the entire earth on the projection (Fig. 10.22). Most azimuthals are limited to presenting a hemisphere or less. Since the bounding circle is the antipode of the center point, shape and area deformation in the periphery are excessive.

The orthographic projection (Fig. 10.23) looks like a view of a globe or the earth's graticule from a considerable distance, although it is not quite the same. For this reason it might almost be called a visual projection in that the deformation of areas and angles, although great around the edges, is not particularly apparent to the viewer since it appears the same as though he were looking at a portion of the globe. On this account it is useful for presenting some kinds of directional concepts (for example, Europe as seen from the south), illustrative maps, and for those maps wherein the sphericity of the earth is of major significance.

The gnomonic projection (Fig. 10.24),

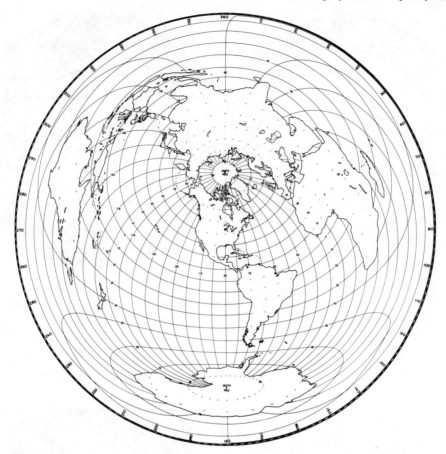

Figure 10.22   An azimuthal projection centered on Madison, Wisconsin, prepared for geophysical uses. Note the deformation near the outer edge. (University of Wisconsin Cartographic Laboratory.)

like Mercator's, is one of the best known projections of any class, but it is useful only for special purposes. It has the unique property that all great-circle arcs are represented anywhere on the map as straight lines.* The deformation, both angular and areal,

*For example, the marine navigator need but join the points of departure and destination with a straight line and the location of his course is determined. Because compass directions constantly change along most great circles, the navigator transfers the course from the gnomonic graticule to the graticule of Mercator's projection and then approximates it with a series of rhumbs, which are straight lines on the Mercator.

increases very rapidly away from the center, so that the transformation is not much good for any purpose other than projecting great circles as straight lines. Because projection is from the center of the earth to the tangent plane, less than a hemisphere can be constructed.

As was observed earlier, an infinite number of azimuthal projections are possible and many have been devised but not widely used. A vertical air photograph is a "perspective" azimuthal projection, as is a photograph of the heavens (gnomonic).

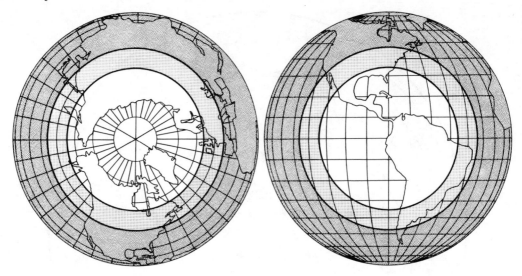

Figure 10.23   The orthographic projection in equatorial and polar form. Values of lines of equal maximum angular deformation are 10° and 25°. Areal deformation is also present.

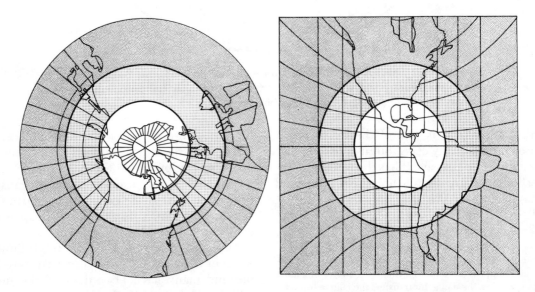

Figure 10.24   The gnomonic projection. Values of lines of equal angular deformation are 10° and 25°. Areal exaggeration, extreme toward the peripheries, is also present.

Other Systems of Projection

There are any number of projections possible that have none of the precise relationships that make possible equivalence, conformality, or azimuthality. Some of them are very useful indeed. In this group would fall those projections that have extremely little linear, angular, or areal scale error for small areas but that can be quickly constructed. Similarly, a projection system that can be developed for individual sections of the earth in such a manner that indivudual sheets will fit precisely but that have the same pattern of deformation on each sheet is very useful. In this class are the polyconic, the modified polyconic, and a number of others.

The polyconic projection (Fig. 10.25) was first widely used in the United States by the Coast and Geodetic Survey, and it was subsequently adopted as the projection for the standard topographic series of the United States Geological Survey. It has a straight, standard central meridian. The parallels are arcs of circles, and each is a standard parallel; that is, each is truly subdivided by the meridians, and each is drawn with the proper radius for its cone and hence with its own center. Thus the parallels are not concentric. Although the Scale Factor along each parallel is 1.0, the scale along the curved meridians is greater and increases with increasing distance from the central meridian. The projection is, therefore, neither conformal nor equivalent. On the other hand, when it is used for a small area, bisected by a central meridian, both these qualities are so closely approached that the departure is very small.

For the mapping of a large area on a large scale, the development of each small section on its own polyconic projection is therefore an acceptable solution to the projection problem. Each small section will fit perfectly with the adjacent ones to the north or south, but on account of the curved meridians they will not fit together east-west. The variations of the Scale Factor within the 7-12' and 15' quadrangles of the standard topographic maps of the United States is insignificant and usually less than that which results from paper shrinkage or expansion. Although it is admirably suited for individual maps covering small areas, it is clearly not suited for maps of larger areas. A number of others are definitely preferable for an area even the size of a midwestern state in the United States.

A modified polyconic projection is used for the so-called Millionth Map or International Map of the World. It was changed by making the meridians straight instead of curved, and by making two of them, instead of one, standard on each sheet. Whereas in the ordinary polyconic all parallels are standard, in the modified polyconic only the bottom and top one in each sheet are standard. This makes it possible to fit sheets together east-west as well as north-south at the slight expense of some linear and area scale variation. The diagonal sheets cannot fit. On the ordinary map sheet, 4° of latitude by 6° of longitude, the scale errors are considerably less than 1 in 1000.

The plane chart, sometimes called the equirectangular projection, is one of the oldest and simplest of map projections (Fig. 10.26). It is useful for city plans or base maps of small areas. It is easily constructed and for a limited area has small deformation. All meridians and any chosen central parallel are standard. The projection may be centered anywhere.

There are several possible arrangements of scale that may be used to produce a conic projection with two standard parallels,

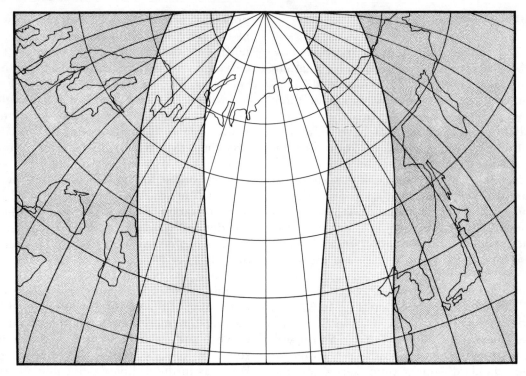

Figure 10.25 The polyconic projection of a large area. Values of lines of equal maximum angular deformation are 1° and 5°. Areal deformation is also present. Only a narrow zone along the central meridian is used for topographic maps.

an example of one being Figure 10.27. These projections are similar in appearance to Albers' and Lambert's conic projections, but they do not have the special properties of those two. Such a projection does not distort either areas or angles to a very great degree if the standard parallels are placed close together and provided the projection is not extended far north and south of the standard parallels. This kind of projection is frequently chosen for areas in middle latitudes of too large an extent for an equirectangular projection, and for maps not requiring the precise properties of equivalence or conformality.

Miller's cylindrical projection (Fig. 10.28) employs a spacing of parallels that

allows the poles to be shown (impossible on Mercator's projection). It has neither the excessive angular deformation characteristic of the cylindrical equal-area nor the excessive rate of change in the representation of areas that occurs on Mercator's. It has no special properties other than the rectangularity of the graticule in the conventional form.

There are scores of other map projections that have no special properties. Many of these have been devised for special purposes or simply represent the result of some mathematician's interest in the problem of transformation. Most projections that are widely used have been included in this survey.

Transverse and Oblique Projections

As was observed earlier, it is conventional to arrange the system of transformation and the earth so that the graticule will be represented by relatively simple lines. In order to obtain these regular smooth lines in the projection, the axis of the theoretical cone or cylinder is made to coincide with the earth's axis, the center of an azimuthal projection is located at the equator or a pole, or the pseudo-cylindrical projections are constructed around a straight-line equator and central meridian. Although this is conventional and usually desirable, it is obviously not necessary. Projections are systems of transformation of the spherical surface to a plane surface, not systems of transformation of the graticule to a plane. That this is so is most easily demonstrated by reference to the azimuthal class of projections (and to their illustrations shown), which may be centered anywhere. Although the graticule will appear differently, depending upon the location of the center, the pattern of angular and/or areal deformation of the projected surface will not vary because the surface of a sphere does not vary and neither does the surface of the plane on which it is projected.

One way of visualizing these relationships is to think of the coordinate system as being loose on the sphere while the land and water bodies retain their locations relative to one another. Then it would be possible to slide or shift around the graticule in any way desired. We might even go so far as to slide it around a full 90° so that the equator of the graticule coincides with the former location of a meridian and runs through the Arctic and Antarctic regions, whereas the poles of the loose graticule, being antipodal points, would then be located at opposite points on the former equator, as in Figure 10.29.

Projections of the earth that are not conventionally oriented with respect to the earth grid are usually called oblique (Fig. 10.30) if the aspect has been shifted less than 90° from the conventional arrange-

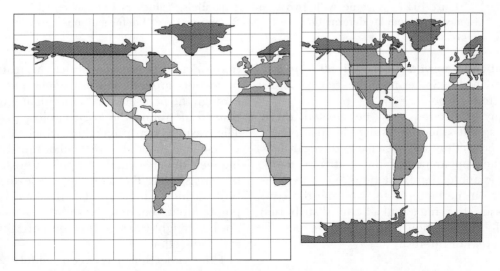

Figure 10.26 The plane chart or equirectangular projection. On the left the standard parallel is the equator; on the right, 45°. Values of lines of equal maximum angular deformation are 10° and 40°. Areal deformation is also present.

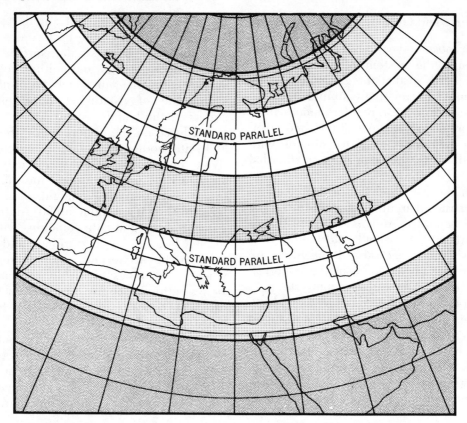

Figure 10.27 A conic with two standard parallels. Values of lines of equal maximum angular deformation are 1° and 2° (approx.). Areal deformation is also present.

ment and transverse if the full 90° shift has been made. Sometimes the transverse is also termed polar.

Any projection may be treated these ways, and since the structural relations of the projection are not changed, regardless of the appearance of the graticule, many cartographic representations of earth relations can be bettered by "shifting the land masses" to bring about a better distribution of the deformation pattern with respect to the areas of interest.

Transverse Mercator Projection . One of the commonly used projections of this type is the transverse form of Mercator's projec-

tion. For this we construct Mercator's projection around a great circle passing through the poles, that is, a meridian circle, instead of around the great circle equator. The result is a projection which is, of course, conformal, but instead of representing well the areas in the tropics, it presents the areas along the meridian circle in the best fashion (Fig. 10.31). It should be pointed out that most loxodromes would not be straight lines on this form of Mercator's projection.

This projection has been widely employed to provide a conformal projection base for topographic maps. Because the Scale Factors are uniform along lines that

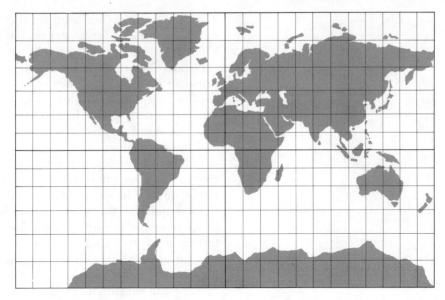

Figure 10.28 Miller's cylindrical projection. (University of Wisconsin, Cartographic Laboratory.)

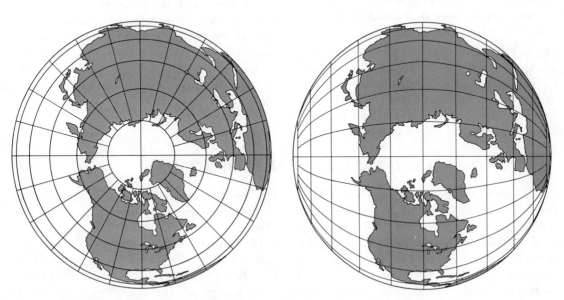

Figure 10.29 (Left) A pole centered orthographic projection. (Right) The graticule shifted 90° on the earth. Since the earth is spherical, the graticule will "fit" it in any position.

Figure 10.30 A portion of an oblique Moll-weide projection. This is exactly the same projection system as is shown in Fig. 10.11. North Atlantic relations are well presented on this representation, which gives somewhat the appearance of a portion of a globe, but is, of course, equal-area. Note that the pattern of deformation is unchanged from that shown on Fig. 10.11.

parallel the standard great circle, it is possible to construct over the map area a perfectly rectangular coordinate reference grid that has the merit that a given X value will have the same Scale Factor at any Y value.

This is the projection used for the Universal Transverse Mercator grid system or UTM referred to in Chapter 2.

Because Mercator's projection is cylindrical in concept, the lines of equal deformation are small circles paralleling the standard great circle. In the conventional form the standard great circle is the equator and the parallel small circles are, of course, parallels of latitude. In the transverse Mercator a meridian circle is the standard great circle and the small circles parallel to it are represented by the vertical parallel lines of the rectangular grid system. The horizontal lines of the grid system are equally spaced great circles that cross the central meridian at right angles. They correspond to the earth meridians on the conventional Mercator. To keep to a practical minimum the inevitable increase in scale change perpendicular to the central meridian in the UTM, a new central meridian is employed every 6° of longitude.

In order to improve the scale characteristics in the UTM, the theoretical cylinder of the transverse Mercator is assumed to intersect the earth's surface along two of the small circles instead of being tangent at the central great circle. This distributes the deformation more evenly over the map, as shown in Figure 10.32.

CONSTRUCTION OF PROJECTIONS

The cartographer occasionally finds it necessary to construct a projection for a specific purpose. Often, however, many are available in one form or other that may be copied or machine programs can be employed in which a computer will direct a plotter in drawing a graticule. In any case, a cartographer should know how projections are constructed because many times he needs to modify, shift, recenter, or fit

ready-made projections, and a knowledge of their construction aids in their understanding.

In the brief sections that follow, only the general characteristics relating to the construction of projections are treated. Directions for the actual construction of the more commonly used projections are given in Appendix G, together with the requisite tables.

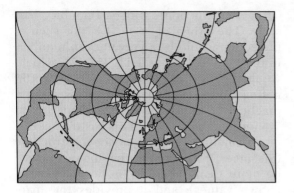

Figure 10.31 The transverse Mercator projection. This form of Mercator's projection gives a conformal representation with the least deformation along the meridian chosen as the "equator," i.e., the tangent standard line, in this example the meridian of 0°–180°.

Basic Aspects of Construction

In order to transform the spherical surface to a plane surface, it is necessary to establish a series of points on the spherical surface and then to transfer these points to the plane surface according to the particular transformation system being employed. Thus, by locating the points properly on the plane, the system is "defined" by the points and all intervening spaces will conform. Because the earth's coordinate system is convenient for this purpose, the parallel and meridian system provides the medium for the location of homologous points on the two surfaces.

As was observed earlier, many projections may be visualized as being the result

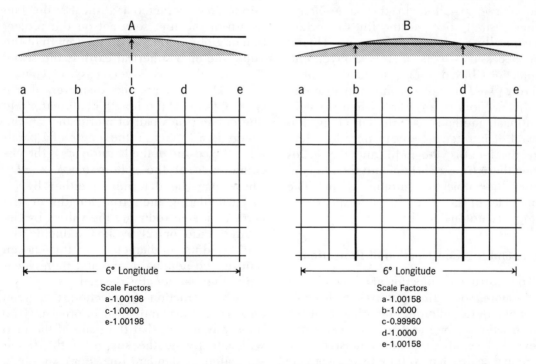

Figure 10.32 Instead of a single line being standard as in A, two are made standard, as in B, on the transverse Mercator projection system used for UTM rectangular grid reference purposes. In the UTM system each grid zone covers 6° of longitude and extends about 400 kilometers on each side of the central meridian. The standard lines (small circles) are 180 kilometers east and west of the central meridian.

of a kind of geometric transfer of the graticule from the spherical form to cylindrical, conic, or plane surfaces that may then be developed into a plane. The lines of tangency or intersection of the sphere with the cylinder, cone, or plane are the *lines* or *points of origin*. Because such lines will in all cases be either great or small circles, it is also convenient to arrange the system of projection so that these circles will correspond to great or small circles of the graticule. This ordinarily results when the axes of the cylinder and cone and a perpendicular to the plane are made to coincide with the earth's axis of rotation.

The equator or a pair of parallels north and south of it are usually the lines of origin for cylindrically based projections, and one or two parallels in the same hemisphere usually become the lines of origin in conically conceived projections. This is not necessary—it is merely convenient. Projections based upon a tangent or intersecting plane are not so commonly oriented in this fashion. In any case, when the conventional method of orientation is followed, it is often necessary only to calculate the lengths, the radii, and the scales along these lines of origin, and then to construct the projection around them. The remainder of the area will, so to speak, fall into place automatically.

Scale and the Construction of Projections

To construct a projection at a given linear scale along some line requires, at least in theory, a quite different operation from the construction of one at a given areal scale. Nevertheless, all projections must have a principal scale, that is, they bear a specific relation to a certain size globe on which their construction is theoretically based.

A number of the common projections used for small-scale maps may be graphically derived to fit a format and provided (after construction) with a graphic bar scale. The principal scale (RF) for such a projection would be a very uneven number in most cases, such as 1:11, 453,421, which would be inconvenient and hardly worth noting. Whenever possible, it is better practice to construct the graticule to a "round number" scale. This has the merits of promoting accuracy of construction (distances, for example, on the map may be more easily checked for accuracy with known values); and it provides the map user and reader with an even, readily understood, and usable fractional scale.

A projection constructed so that the scale along a standard line bears the given scale relationship to that line on the sphere, or equals its length on the nominal globe, is said to be constructed at a linear scale. The length of the line segment on the projection or globe in relation to the length of the same line on a spherical earth is the scale. To determine this is very easy, of course, for if D represents the diameter of the sphere, then (1) the length of a great circle (the equator or a pair of meridians) on the sphere is πD, and (2) the length of a parallel is the cosine of the latitude (ϕ) × the circumference of the sphere ($\cos\phi \times \pi D$). These may be determined either by (1) first calculating their true lengths on the earth and then reducing the values by the desired ratio or (2) by first reducing the earth (radius or diameter) by the chosen ratio to the nominal globe and then calculating the map lengths desired.

The construction of projections to a given area scale is accomplished in precisely the same way except that the scale of the map will actually be the square of the linear scale along a standard line. For example, a globe having a diameter at a scale of 1:40,000,000 would have an area in relation to the earth as 1 is to 40,000,000². Care must be exercised in working with the pseudocylindrical equivalent world projections, since most of them are constructed by using the equator and one meridian as axes of construction. In most cases these are *not* standard lines but instead bear a specified length ratio to the radius (R) or the diameter (D) of the nominal globe.

Methods of Construction

Projections may be mechanically constructed in a number of ways: (1) those that can be derived geometrically can be constructed by working from an elevation of the globe drawn to the proper scale, and then the graticule may be derived by transfer methods; (2) by calculating the radii of curves and spacings of parallels and meridians; (3) by consulting tables showing the X and Y plane coordinates of the intersections of given parallels and meridians, and then joining the points thus established by smooth lines to form the graticule.

In many cases it is convenient to begin the construction by first drawing one or two of the graticule lines and then using them as lines of reference. Most conventional projections are drawn around a straight *central meridian*.

The directions that are given in Appendix G are meant to be used only for the construction of moderately sized projections generally suitable for small-scale maps. Large-scale maps wherein considerable precision is desired should be prepared on projections that have been derived from more extensive tables readily available in the standard treatises on map projections.

Mention should also be made that for precise mapping of the earth it must be treated as a spheroid. To simplify the calculations in constructing equal-area and conformal projections, the geodetic latitudes on the spheroid have been determined for true spheres such that on the one they are in equal-area (or authalic) relation-

ship and on the other in conformal (or isometric) relationship. These latitudes, corresponding to the geodetic latitudes on the spheroid, may then be used with spherical formulas and the resulting projections will "automatically" be of the spheroid.[*]

Frequently, the tables that are available for the construction of map projections are given in some unit of distance (such as meters, miles, and minutes of longitude at 0° latitude) calculated at a scale of 1:1, that is, actual earth size. To construct to scale it is necessary (1) to reduce each unit by the scale ratio, and (2) convert the tabular unit of measure to a convenient unit for plotting, for example, centimeters or inches. It should be remembered that the scale relationship is simply an arithmetic linear relationship between the actual radius of the earth and the radius of the nominal globe of desired scale that is being projected.

Oblique and Transverse Construction

There are several ways of changing the "viewpoint" of a projection, that is to say, of centering it anywhere on the sphere. The general procedure involves locating the positions of the graticule intersections on the sphere when the earth grid has been "shifted." This is relatively easy when working with the azimuthal projections, and it is possible to do this by employing an ingenious nomographic process described in detail in Appendix G. The nomographic method can be used with other classes of projections such as the cylindrical, but it is not quite as simple a procedure as it is with the azimuthals.

Another method of shifting the center of a projection is by calculation of coordinates. For example, if we wished to "shift"

---

[*]Tables of authalic and isometric latitudes for every 30 minutes on the spheroid are available in United States Coast and Geodetic Survey, *Special Publication No. 67*, pp. 102-113.

the graticule of a Mollweide projection so that the equator of the conventional projection coincided with some other great circle, it would be the same as shifting the earth "inside" its coordinate system, so to speak. If the "loose" graticule were thus shifted on the earth, all the intersections of the new graticule could be located and expressed in terms of latitude and longitude in the old graticule. These new positions may be calculated. The procedure is as follows, assuming the central meridian is 0° longitude in both old and new graticule:

θ = The number of degrees shift in the new system (90° minus the position of the new pole)
ϕ = latitude in new system
λ = longitude in new system
ϕ' = latitude in old system
λ' = longitude in old system

The problem is to find the latitude (ϕ') and longitude (λ') on the conventional projection of the *same* latitude (ϕ) and longitude (λ) in the new system. The formulas are

$$\sin\phi' = \sin\phi \, \cos\theta - \sin\theta \, \cos\phi \, \cos\lambda$$

$$\sin\lambda' = \frac{\sin\lambda \, \cos\phi}{\cos\phi'}$$

The calculations are not difficult, and their appearance should frighten no one. Nothing but arithmetic is necessary, since tables of sines and cosines are available.°

Interrupting and Recentering

Interruption involves using several central meridians in place of one, and it results in a lobate kind of projection with the continental masses (or oceans) being shown separately on either side of a single equator. Recentering merely means not placing central meridians in opposite hemispheres opposite one another. Even those projections with a line for the pole instead of a point may be interrupted. It should not be inferred that the conventional form (straight-line equator, for example is necessary). After interruption any projection could then be "tilted" in the manner described in the previous section.

The mechanics of interruption involves constructing the chosen central meridians and then duplicating the projection around each as far to each side as is necessary. This provides several "points or axes of strength" in place of the one or two on the uninterrupted projection. In practice, only the minimum necessary section of the conventional projection need be constructed, and then the appropriate sections or lobes may each be traced in their proper position.

°Tables of the position of ϕ' and λ' have been computed for every 5° of shift of the graticule and are available in E. Hammer, *Über die geographisch wichtigsten Kartenprojektionen. . .*, Stuttgart, 1889.

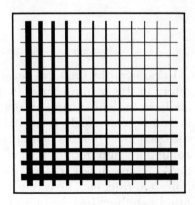

11 CARTOGRAPHIC DESIGN

In cartography the term *design* ordinarily refers to the planning of a map and especially to the choice and arrangement of its graphic elements. It is a vital part of the cartographic process because effective communication requires that the various marks, lines, tones, colors, patterns, symbols, lettering, etc., be carefully chosen and fitted together. Just as an author (that is, a "literary designer") must employ words with due regard for many important structural elements of the written language, such as grammar, syntax, and spelling, in order to produce a first-class written communication, in parallel fashion the cartographer or "map designer" must pay attention to the principles of graphic communication.

The great majority of maps are made with the objective of conveying some geographical information to a reader, and the processes of compilation, symbolization, choice of scale, and projection all are focused to that end. Although the substantive aspects of a map can be quite complex, the designing of it as a communication is equally important and complicated. The manner of presentation of the many map components so that together they appear as an integrated whole, devised systematically to fit the objectives, includes elements ranging from mathematics to art. Regardless of the essential accuracy or appropriateness of the map data, if the map has not been properly designed it will be a cartographic failure.

It is not necessary to have the latent talents of an artist to learn map design, any more than it is necessary to be a potential literary giant to learn to write clearly. Like the study of written composition, the basic elements of graphic composition lend themselves to systematic analysis, and the principles can be learned.* A fundamental requirement, however, is a willingness to think in visual terms, uninhibited by prejudices resulting from previous experience, or, to put it another way, a willingness to exercise imagination. The range of imagination must, of course, be disciplined to some extent, since, like many fields, cartography has developed traditions and conventions; to disregard them completely would inconvenience the user of the maps, which would in itself be proof of poor design.

Cartography is not aesthetic art in the sense that we may have complete freedom with techniques and media. On the other hand, the possibilities for variation of shapes, sizes, forms, and other perceptual relationships of the map components seem practically to be unlimited. The aim of cartographic design is to present the geographical data in such a fashion that the map, as a whole, appears as an integrated unit and so that each item included is clear, legible, and neither more nor less prominent than it should be.

*This is not meant to suggest that every individual can become highly creative merely through study. As has been often repeated, one can be taught his craft (i.e., the characteristics of the materials with which he works) but cannot be taught his art.

GRAPHIC ELEMENTS OF MAP DESIGN

The planning of a cartographic communication involves working with a variety of components ranging from the symbolism treated in earlier chapters to the lettering and map reproduction methods covered in the following chapters. Because cartography is a visual technique, there are various graphic elements involved that are as indispensable to good visual composition as are such things as syntax, paragraphing, organization, phrasing, and so on, to good literary composition. The most important of these elements are contrast, color, figure-ground, and balance. Each of these elements of visual composition will be treated below in summary fashion, and it should be remembered that the abbreviated attention that can be payed to them in this book is not a proper measure of their importance. It is worth observing that cartography involves the combination of many kinds of elements, and that it is simply not possible to provide instructions "from the ground up" in topics that are fundamental to it, whether it be mathematics, geography, or visual composition. Nevertheless, the basic principles involved can be treated.

Clarity and Legibility

The transmission of information by means of the coding built into marks of various sorts (lines, letters, tones, etc.) requires that the marks be clear and legible. Although the various graphic elements treated in this chapter have other functions to perform in a visual composition, one of their basic objectives is to promote clarity and legibility.

Clarity and legibility are broad terms, and many of the techniques and principles considered in other parts of this book are important factors in obtaining these qualities in a presentation. Furthermore, a considerable portion of the task of achieving clarity and legibility will have been accomplished if the map maker has made sure that the intellectual aspects of his map are not open to doubt or misinterpretation. In writing or speaking, the aim is to state the thought with the right words, properly spelled or pronounced, and clearly written or enunciated. In cartography, the symbols take the place of the written or spoken words; their form and arrangement substitute for broad organization and the syntax; and their delineation takes the place of writing or enunciation. It is apparent, then, that no matter what the form of a presentation may be, the principles behind clarity and legibility are much alike; only the "vocabulary" and the principles of composition vary.

If it is assumed that the geographical concepts underlying the purpose and data of a map are clear and correct, then legibility and clarity in the presentation can be obtained by the proper choice of lines, shapes, and colors and by their precise and correct delineation. Lines must be clear, sharp, and uniform; colors, patterns, and shading must be easily distinguishable and properly registered (fitted to one another); and the shapes and other characteristics of the various symbols must not be confusing.

One important element of legibility is size, for no matter how nicely a line or symbol may be produced, if it is too small to be seen it is useless. There is a lower size limit below which an unfamiliar shape or symbol cannot be identified. Although there is some disagreement as to the exact measure of this threshold, a practical application sets this limit as being a size that subtends an angle of about 1 minute at the eye. That is to say, no matter how far away the object may be, it must be at least that size to be identifiable. It is well to point out that this limit sets rather an ideal, since it assumes perfect vision and perfect conditions of viewing. Because of the unreasonableness of these assumptions, it is wise for the cartographer to establish his minimum size somewhat higher, and it may be assumed that 2 minutes is more likely to be a realistic measure for average (not "normal") vision and average viewing conditions. Table 11.1 (based upon 2 minutes) is useful in setting bottom values of visibility. It should be remembered that some map symbols have length as well as width, and in such cases, as for example with lines, the width may be reduced considerably since the length will promote the visibility. In similar fashion, other elements, such as contrasting colors or shapes, may enhance visibility and legibility, but even though the existence of a symbol on the map is made visible by such devices, if it does not stand at or above the sizes given in Table 11.1 it will not be legible. In other words, it might be seen (visibility), but it might not be read or recognized (legibility).

Table 11.1

Viewing Distance	Size (Width)
18 in.	0.01 in.
5 ft	0.03 in.
10 ft	0.07 in.
20 ft	0.14 in.
40 ft	0.28 in.
60 ft	0.42 in.
80 ft	0.56 in.
100 ft	0.70 in.

A second element regarding legibility is also operative in cartography. As a general rule, it is easier to recognize something we are familiar with than something that is new to us. Thus, for example, we may see a name in a particular place on a map, and, although it is much too small to read, we can tell from its position and the general shape of the whole word what it is.

Contrast

The fact that symbols, lines, and the other elements of a map are large enough to be seen does not in itself provide clarity and legibility. An additional element, that of contrast, is necessary.

No element of the cartographic technique is so important as contrast. Assuming that each component of the map is large enough to be seen, then the manner and the way in which it is contrasted with its surroundings determine its visibility. The degree to which a map appears precise and "sharp" is dependent on the contrast structure of the map.

Contrast is a subtle visual element in some ways, and in others it is blatant. The character of a line, and the way its curves or points are formed, may set it completely apart from another line of the same thickness. The thickness of one line in comparison to another may accomplish the same thing. The relative darkness of tonal areas or the differences in hue of colored sections may be made similar or contrasting. The shapes of letters may blend into the background complex of lines and other shapes, or the opposite may be true. If one element of the design is varied as to hue, darkness, thickness, or shape, then the relationship of all other components will likewise be changed. It requires careful juggling of the lines, shapes, and color characteristics, on a kind of "trial and error" basis, to arrive at the "right" combination.

Contrast is achieved by varying the visual characteristics of lines and shapes, pattern, value (relative darkness), and the other characteristics of color. Each of these will be examined below.

Contrast of Lines and Shapes

Most maps require the use of several kinds of lines, each symbolizing some geographical element or concept, such as coastlines, rivers, railroads, various political boundaries, and so on. In order to make each clearly distinct from the next, it is necessary to vary their character, design, or size in some way. Figure 11.1 shows some of the many possible variations in line width. Note that the upper left and lower right quadrants are "more interesting." Figure 11.2 shows only a few of the various possibilities for variation of line design. Figure 11.3 shows some possible variations

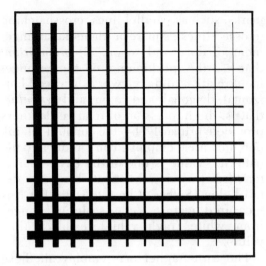

Figure 11.1 Size contrast of lines. Uniformity produces unpleasant monotony. Note that a clear visual difference between sizes of lines requires considerable actual difference.

in character of line. Only on the largest-scale maps does accuracy require an exact position for every part of a line. On smaller-scale maps the lines, if large enough to be seen, cover much more area than the element they represent on the earth. Therefore, they may be drawn precisely and firmly. A wobbly, wavering line looks weak and indecisive and should be avoided.

Just as lines may have an almost infinite variety, so also may the shapes of other graphic elements vary. Oftentimes shape is given by nothing but the bounding line, and there are numerous components of a map, such as the point symbols, legend or title boxes, insets, labels, and so on, that may have definite shapes without benefit of a geographic delimitation. Part of the ability of the eye to perceive and take note of the existence (and contents) of such shapes is because of the way they contrast with their surroundings.

It is impossible, of course, to catalog all the ways in which lines and shapes may be

varied and contrasted. In other portions of the book some specific elements, such as the shapes and lines of letter forms, the delineation of coastlines, and so on, have been considered. It is up to the student, however, to let his imagination roam and to consider critically those maps on which the line and shape structure appears well designed in order to become familiar with the range of possibilities.

Contrast of Pattern

Because many kinds of data cannot effectively be shown by tonal values alone, and because the nature of the data may dictate that other devices be used, it is common for the cartographer to rely also upon various kinds of patterns. These patterns are composed of dots (stippling), lines, or combinations. The possibilities are unlimited.

Very little study has been directed toward the understanding of patterns, their

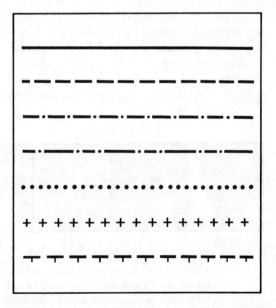

Figure 11.2 A few examples of design variations of lines. Many others are possible, of course.

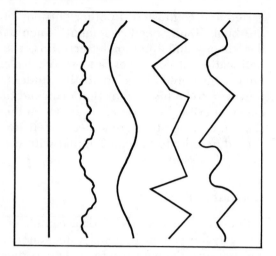

Figure 11.3 Variation of shape or character of lines. Irregular "wiggliness," such as the line second from the left, produces an impression of weakness.

marks, such as parallel lines, crosshatching (crossed lines), square or triangular arrangement of dots, and so on; (3) *orientation*, that is, the positioning of the arrangement relative to the viewer as vertical, horizontal, diagonal, and so on; and (4) *value*, that is, the impression of relative darkness provided primarily by the amount of ink per unit area employed in producing the pattern.° Value is one of the basic characteristics of color and is treated in more detail below. Suffice it here to state that it is measured against a gray scale and ranges from white (high value) to black (low value).

The cartographer may employ all these characteristics of pattern to obtain contrast. Dot patterns having textures finer than 75 lines per inch will usually be perceived as gray value areas without any pattern, while

perceptual consequences, and the ability to discriminate among them. Some basic terminology is necessary in discussing them. Every pattern, whether composed of dots, lines, or a combination, has several characteristics: (1) *texture,* that is, the spacing of the marks, usually specified by the number of lines per inch; the greater the number the "finer" the texture; (2) *arrangement,* that is, the relation of the positioning of the

°These four basic attributes of pattern are perceived in integrated fashion in an impression that can be called *pattern-value*. It is impossible to describe this total impression by a single measure (just as it cannot be done for color); instead, we must rely on the listed characteristics, texture, arrangement, orientation, and value. See Henry W. Castner and Arthur H. Robinson, *Dot Area Symbols in Cartography: The Influence of Pattern on their Perception,* A.C.S.M. Monographs in Cartography, No. 1, American Congress on Surveying and Mapping, Washington, D.C., 1969.

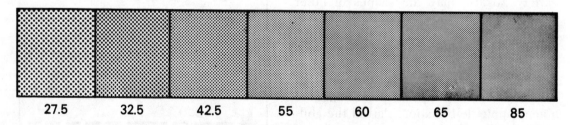

| 27.5 | 32.5 | 42.5 | 55 | 60 | 65 | 85 |

Figure 11.4 A series of dot area symbols (Artype) in which arrangement, orientation, and value (percent area inked) are held essentially constant but vary in texture. The numbers below the boxes show the numbers of lines per inch in each dot screen. Note that textures coarser than about 40 are primarily seen as patterns without value while a texture as fine as 85 appears primarily as a value area without visible pattern.

those with textures coarser than 40 lines per inch will probably not convey much impression of value to most readers (see Fig. 11.4). So long as the marks making up the patterns are bold and the textures relatively coarse, differences in both arrangement and orientation are quite noticeable. The most significant contrasts in these characteristics are the results of inducing visual lines or direction in the eyes of the viewer as detailed below.

Contrasts in value among patterns are obtained by varying the percent area inked, but it is not possible to specify any simple lower threshold for discrimination. From the meager data now available it appears that a difference of 10 to 12% will usually be noticeable. The important point in obtaining contrasts among patterns is varying as many of the characteristics as possible, rather than trying to hold some constant. The more characteristics that are held constant, the greater the difference must be in the quality that is varied.

Area symbols employing patterns may be classified broadly into two contrasting groups, those composed of lines and those composed of dots. The general visual characteristics of these two groups are quite different. Any line anywhere has, in the eye of the viewer, an orientation, and he will tend to move his eyes in the direction of the line. If irregular areas are differentiated by line patterns, as in Figure 11.5 the reader's eyes will be forced to change direction frequently. Consequently, he will experience considerable difficulty in noting the positions of the boundaries.

If the line patterns of Figure 11.5 are replaced with dot patterns, as in Figure 11.6, the map is seen to become much more stable, the eye no longer jumps, and the positions of the boundaries are much easier to distinguish. Lettering is also easier to read against a dot background than against a line background. If, however, the parallel lines are fine enough and closely

enough spaced, the perceptual effect is largely one of value and it will have but a slight suggestion of direction.

Many parallel-line patterns are definitely irritating to the eye. Figure 11.7 is an exaggerated example of the irritation that can occur from using parallel lines. It is probably because the eyes are unable quite to focus upon one line. Whatever the reason, the effect is somewhat reduced if the lines, regardless of their width, are separated by white spaces greater than the thickness of the lines. Generally, a cartographer should be very wary of using parallel line patterns.

Contrast of Value

It was stated above that contrast is of fundamental importance in the cartographic technique. It may be further asserted that the variation of light and dark (whether colored or uncolored) is the most important of all contrast elements. The quality of relative lightness is termed brightness by the physicist, tone by the photographer, and

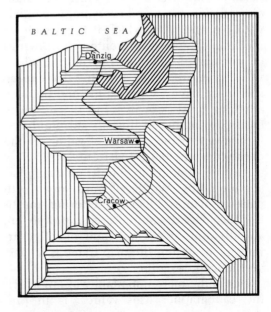

Figure 11.5 A map employing line patterns.

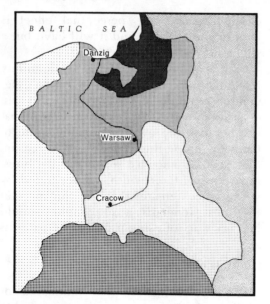

Figure 11.6 Same data as Fig. 11.5 but employing dot patterns.

senting a graded series of information. Thus, for example, geographical phenomena, such as variations in rainfall, depth of oceans, elevation of land, density of population, intensity of land use, and so on, are usually depicted by some technique that depends for a large portion of its effectiveness upon classes as differentiated by value contrasts.

The precise ratios of black to white in a series of grays that will result in a progression of equal visual steps have been a question under study for some time. The problem is very complex and, even if a theoretically based answer is ever found, its utility in cartography will be somewhat limited. The fact that most such value steps in mapping must be created by patterns of dots or lines introduces the complication that the reader reacts to the stimulus of pattern as well as to the value (value-pattern)

value by the psychophysicist and artist. The last will be used here. Lightness is termed high value, and white would be the maximum attainable.

Value contrasts are the most important element of seeing, and everyone is familiar with the ease with which it is possible to recognize objects represented in drawings or photographs merely by their tonal or value structure. Since anything that can be seen must have a value rating, and because anything must vary in value if it is to be easily distinguished from its surroundings, it follows that the contrast of values is one of the fundamentals of visibility.

Any object or group of units on a map has a value rating. Widths of lines, areas with patterns, names, blocks of lettering, the title, the legend, colored areas, and so on, are all value areas, and their effective arrangement within the map frame is a basic part of map designing.

One of the more important ways in which the cartographer uses values, is in pre-

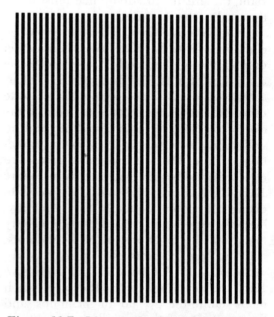

Figure 11.7 Line patterns can hurt the eyes and the map on which they are used, if they are not chosen with due regard to their potential as irritants.

and thus upsets the progression. The relation between physical black/white ratios of printed dot patterns and their perceptual effects, empirically derived by Williams, and shown in Figure 11.8, is probably as good a statement as can be derived of this complex question.° Note that it requires less difference in black/white ratios to make an equal difference in the higher (lighter) values than in the lower.

Of fundamental importance is the fact that the human eye is not particularly sensitive when it comes to distinguishing value differences. Six shades of gray, between a black and white at the ends of the value scale, is about the limit. Consequently, the cartographer must be relatively restrained in this respect. If a greater number of divisions in a series must be shown, it is neces-

°Robert L. Williams, "Map Symbols: Equal Appearing Intervals for Printed Screens," *Annals Association of American Geographers,* **48,** 132-139 (1958). See also George F. Jenks and Duane S. Knos, "The Use of Shaded Patterns in Graded Series," *ibid,* **51,** 316-334 (1961).

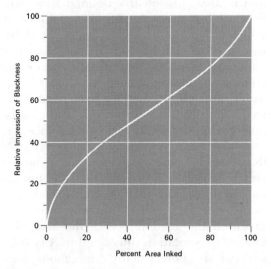

Figure 11.8 **The relation derived by Williams between the percent area inked of a series of printed dot patterns and perceptual reactions to them.**

sary to add hue or some pattern of dots or lines to the areas to aid the user to distinguish among them.

Another characteristic of vision and its relation to color in general, which the cartographer must bear in mind, is a general phenomenon called simultaneous contrast. In value contrasts it is particularly important. The basic generalization is that any color characteristic is markedly modified by its environment. With respect to value, a dark area next to a light area will make the dark appear darker and the light appear lighter. This specific effect, called induction, makes it difficult for a reader to recognize a given value in various parts of a map when it is surrounded by or is adjacent to different values. Figure 11.9 illustrates one aspect of induction. The "wavy" appearance of the value blocks causes the recognition of values to be difficult, and the difficulty increases in direct ratio to the similarity or closeness of the values. In Figure 11.10 the gray spots are the same, but the one on the dark block appears much lighter, and the reverse is true of the other. It is obvious that a reader would have difficulty in recognizing values under such a situation. The effects of induction may be largely removed by separating adjacent values with a white space or by outlining the areas with black lines.

Figure-Ground

In small-scale thematic cartography and, to a lesser extent, even in larger-scale reference cartography, the immediate perception of the fundamental elements in the map is of primary importance. The reader must be able to tell land from water, recognize and separate the outline of the city, island, or harbor from its surroundings; in short, he should be able to focus immediately on the characteristics that the cartographer had as his objectives without visu-

Figure 11.9   A value scale in which the steps are equal in terms of differences in black/white ratios. Note the apparent "waviness" or induction. Note also that a middle gray does not appear to be half way between black and white. (From a *Kodak Gray Scale*, courtesy Eastman Kodak Company.)

ally fumbling and groping to find what he is supposed to be looking at. The most important element in graphic composition that leads to such quick recognition is called the *figure-ground* relationship.

The eye and the mind are a team, and they react spontaneously to any complex of visual stimuli. They tend immediately to organize any display into two basically contrasting perceptual impressions, a figure upon which the eye settles and sees clearly as distinct from the amorphous, more or less formless ground. The figure is perceived as a coherent shape or form with clear outlines, and it is not confused with the ground that surrounds it. While this spontaneous visual organization is occurring, the previous experience of the reader is also brought into play in terms of his mental constructs of what the particular or several geographical form relationships are. Both occur at the same time, and it is the cartog-

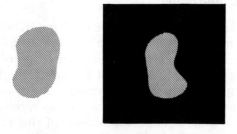

Figure 11.10   How environment modifies appearance. The two gray areas are exactly the same gray.

rapher's responsibility to design his display graphically so that the intended outlines, forms, or shapes will immediately and without confusion emerge as the dominant figure or figures.

A simple illustration of this phenomenon of perception is provided by Figure 11.11. In *A* the simple line, wandering across the field, merely separates two areas and is quite ambiguous. Assuming the map reader knows he is supposed to see an area separated into familiar shapes, he is confused as to which is land and water, and unless he is unusually familiar with the area, he will continue to be perceptually confused. In *B* and *C* areas are clearly defined from one another but there is still the problem of which is land and which is water. In *D*, by the use of land-based names, familiar boundary symbols that usually occur on land areas, the graticule to "lower" the water, and slight shading to separate the basic land from the water, the land clearly emerges on the figure. Of course, once we have clarified the display by observing *D*, the others become much easier. It is important to observe, however, that the fundamental perceptual relationships are always there, and regardless of how well acquainted the reader is with the area, the kind of presentation illustrated in *A* is subconsciously perceptually ambiguous and therefore confusing and irritating.

It is impossible to develop in depth in this book the variety of visual stimuli that

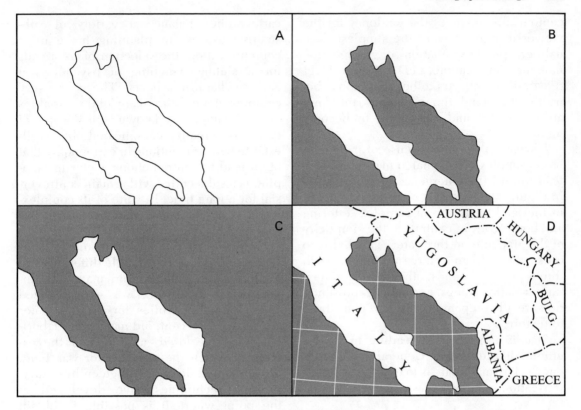

Figure 11.11   Four simple sketch maps to illustrate various aspects of the figure-ground relationship. See the text for explanation.

lead toward the production of visual form or figures. This is a complex field of study still under intensive investigation by students of perception. Nevertheless, some of the basic principles can be outlined.

Differentiation must be present in order for one area to emerge as the figure. The figure area must be homogeneous and the homogeneity of the entire visual field (the entire map) must not be greater than each of the two areas taken separately. Such differentiation may be obtained in a variety of ways, such as by color, value, and texture.

Brightness or high value tends to lead toward figure differentiation on a map. In *B*

and *C* in Figure 11.11 it is relatively easy to see the light area as the figure; it is more difficult to make the light areas recede and become ground.

Closed forms, such as islands, entire peninsulas or countries, and other complete entities are more likely to be seen as figures than if they are only partially shown. If all of the peninsula of Italy had been shown in *A* in Figure 11.11, it would have appeared more easily as figure.

Good contour is a catchall term that includes the characteristics of "continuous-appearing," "symmetrical," and "simplicity" of visual forms, all of which lead toward figure-ground differentiation. Good contour also involves the tendency for the viewer to move toward the simplest "visual perceptual explanation" of graphic phenomena. For example, in *D* in Figure 11.11 the graticule appears to be continuous "beneath" the land thus "raising" the land area above it and helping it to become figure.

Texture, in the broad sense of it being a basic complex of articulated marks, tends to lead toward the emergence of figure. In *D* in Figure 11.11 there is considerable detail in the land, made up of complex lettering and boundaries in contrast to the simplicity of the graticule on the water. This helps to cause the land to emerge as figure. Such things as city symbols, other names, rivers, transportation routes, and relief representation are all components that can contribute to texture.

Area is important in leading to figure-ground differentiation. Generally, the tendency is for smaller areas to emerge as figure in relation to larger areas.

In composing his graphic communication, the cartographer must work with the elements of the figure-ground relationship with great care. Some geographical relationships have "bad contour," are large, cannot be closed, and so on, and in these cases he must lean more heavily on those factors such as brightness or texture. Each composition is a new problem and standard specifications cannot be made any more than we can say that a well-designed paragraph will contain a given number of words or sentences.

Color

Of the various elements of cartographic design, color is probably the most interesting and yet also the most frustrating to the cartographer. Just as the process of generalization in cartography requires that the cartographer balance a variety of substantive aspects in planning his graphic communication, the selection of color calls for a similar weighing of psychological and mechanical aspects. These are very complex; not only are many of the variables not yet completely known, but the use of color is greatly complicated by well-established conventions in cartography that often lead to contradictions. Color in mapping is a subject that will remain controversial for a long time. In spite of its complexity, it is much sought after because it has several advantages.

Even the use of a small amount of color seems to produce remarkable differences in legibility and emphasis on maps. Its importance in this respect was early realized and, although facilities for printing color are relatively recent, old maps were laboriously hand colored—an index of the esteem in which the use of color was held. Since the middle of the nineteenth century, printing techniques have developed to the point where it is possible to obtain almost any desired effect. Unfortunately, as the techniques have increased, so has the relative cost. Nevertheless, the use of color will no doubt continue to increase, and the cartographer must know the basic principles of its use in cartographic design.

The importance of color in cartography arises for several reasons. One of the more important of these is that it is a remarkable simplifying and clarifying element. Even a small amount of color used as a light tint tends to subdue the visual clutter of lines, point symbols, and lettering. It is an important element in the figure-ground relationship, and it acts as a unifying agent in the visual composition or organization of a map. For example, a tint on the land clearly helps to identify its shape and distinguish it from the water.

A second reason why color is so important is that it leads to remarkable subjective effects on the part of map readers. It is an important aesthetic element, and the cartographer must take care not to use inadvertently colors that are generally unpleasant—unless, of course, that is his objective.

Color has important effects on the clarity and legibility characteristics of a map, since it affects the reader's ability to distinguish fine detail, to read the lettering, and to see the boundaries defined by colored areas. Naturally, color also is significant in that various kinds of construction and reproduction problems are associated with its use.

The terminology of color is very complex because its characteristics may be considered in several ways—such as those that arise from the chemistry and physics of pigments and light or those that have to do with their perception by man. The last are by far the most important in cartography. In order to deal with this subject, specific color terms are necessary, and terminology has been developed in each of the above fields. Color can be described by terms that refer to its three main dimensions: (1) *hue,* (2) *value,* percent reflectance, lightness, or brightness, and (3) *chroma,* saturation, purity, or intensity. The three terms, hue, value, and chroma, are those employed in the Munsell system of notation.*

Hue is a sensation the reasons for which are not understood. When we speak of red, green, blue-gray, or chartreuse, we are attempting to describe a particular hue. Millions of hues are, of course, possible and, except in a general way, words are quite

*Color specification is very complicated. There are two main systems: the Munsell system devised in 1915 and since defined with great precision, and the C.I.E. system recommended in 1931 by the Commission Internationale de l'Éclairage (International Commission on Illumination). The Munsell system is commonly used in the United States. (*A Color Notation,* 10th ed., Munsell Color Company, Baltimore, 1946; see also *Journal Optical Society of America,* July 1943, for its relation to the C.I.E. System.)

incapable of communicating exact meanings of hue. When passed through a prism, "white" light is changed, forming the familiar spectrum of the rainbow. Many of the descriptive terms use these rainbow colors as a base ingredient and then attempt to qualify them as to what kind of blue, green, yellow, orange, or red is under consideration. Those hues that come relatively close to the colors of the spectrum are called spectral colors, and a progression of hues used as area symbols, when arranged in the same order as they appear in the spectrum, is called a spectral progression.

Value is the sensation of lightness or darkness of a color; for example, a red is usually darker than a yellow. The tones of the value scale range from black to white and are rated by reference to a gray scale as previously described. A gray scale precisely constructed on the basis of equal changes in black/white ratios is widely used in photography and printing (see Fig. 11.9). All colors, modified or pure spectral hues, have a value rating and can be matched in tonal value with one of the tones on a gray scale.

Chroma is the most difficult to describe of the reactions of the eye to color. It is somewhat analogous to the brightness of the physicist, and bright, brilliant, or intense colors have high chroma as contrasted with dull colors or lighter pastels. More precisely, chroma is varied by changing the ratio of gray and hue while maintaining a constant value rating. Thus, chroma also refers to saturation, that is, the "amount" of hue in a color. A scale of chromas ranges from the pure hue at one end to a neutral gray at the other. At no place would it vary in value.

Much of the color printing in cartography is accomplished by combining various proportions of inks of three primary colors, magenta, cyan, and yellow (called process colors), plus black and the white of the printing surface. From these an almost unlimited number of hues (with associated values and chromas) can be obtained. The cartographer who can work with color has, therefore, a very large "palette" with which to work, and it is necessary that he be fully familiar with the variety of reactions map readers have to these dimensions of color. The more important of these are indicated below.[*]

Hue. Although man can see the difference between colors side by side, it is difficult for him to recall colors, and we must conclude that in that respect he is not very sensitive. An important rule is that the smaller the area (point symbol or line width) the more difficult it is to distinguish a color. Relatively, man appears to be most sensitive to red, followed in order by green, yellow, blue, and purple. In terms of its effect on the reader's ability to distinguish fine detail, monochromatic light is superior to polychromatic, while yellow is the best background color and blue the poorest.

Hues have many subjective aspects. Each individual (and even each culture) has its preferences. In the western world blue and red are greatly preferred over yellow. Some colors appear to be more individual than others. For example, red is red, but orange seems to be composed of both red and yellow, and purple is made up of blue and red. Many colors are named for their apparent components, such as yellowish green, greenish blue, and blue-violet. Normally only blue, green, yellow, red, brown, white, and black appear as individual colors. The reason for this is not definitely known, but nevertheless, all authorities agree that the phenomena of "pure hues" together with the "intermediate hues" are of considerable significance in color use. Their importance in cartography for showing interrelationships or mixtures of distributions is apparent.

Convention plays a large role in the employment of hues in cartography. One of the most firmly established is the spectral progression of colors for layer tinting of elevations, with blue for water and greens in the lowlands ranging upward through yellows and browns to reds in the higher elevations. Because of the light and dark values associated with the hues, the relative clarity and legibility of the various areas is quite unequal, with the highest visibility being in the intermediate elevations. Such a progression should never be used for a graded series other than elevation. The use of blue for water is probably the oldest of the color conventions.

There are many connotations associated with hues: red, yellow, and orange with warm temperatures; blue, green and gray with cool; yellow and tan with dryness and sparse vegetation; brown with land forms and contours; green with vegetation; and so on. Clarity and legibility are of fundamental importance in cartographic design, and when color conventions conflict with them, clarity and legibility are, of course, more important.

Value. Of the three dimensions of color, value is probably the most important in terms of the fundamental perceptual aspects of map design. As was observed in connection with contrast, value is basic to clarity and legibility because of the effect of value contrast in definition. The basic precept is that the greater the value contrast the greater the definition and the greater the clarity and legibility.

As we saw in connection with area symbols in Chapter 7, representation of a

---

[*]This section is simplified from Arthur H. Robinson, "The Psychological Aspects of Color in Cartography," *International Yearbook of Cartography,* **7**, 50-61 (1967).

graded series should be accomplished by value changes rather than changes in pattern. The fundamental rule is that the darker the value the greater the magnitude. For example, we would not think of making a thematic map of rainfall distribution on which the heaviest rainfall was shown by the lightest tone and the least rainfall by the darkest tone. If there is to be a gradation of amounts on a map, we can approach the presentation problem by thinking in terms of a graph, on which value (or brilliance) is plotted as the abscissa and the phenomenon being presented is plotted as the ordinate, as in Figure 11.12.

In terms of graphic composition, it is well to use areas of extreme values with caution since extreme values tend to dominate. Dark blues for oceans or deep colors on political areas will generally overshadow the other parts of a map.

The inability of the eye to distinguish among values and the important effects of induction have been referred to earlier.

Chroma. Chroma is the least significant of the perceptual dimensions of color as applied to cartographic design. Probably the most important aspect of chroma is that the apparent intensity or brilliance of a color varies directly with the size of the area. A small section of color in a legend may be quite acceptable, but when used to cover a large area on a map, it may be overpowering in its intensity. It is especially important to keep this in mind when choos-

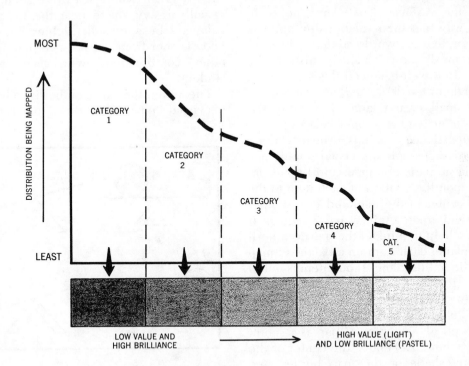

Figure 11.12 Theoretical graph of a graded interval or ordinal map distribution showing the basis upon which one chooses a scale of values or brilliance. The graph could be constructed in a number of ways, but the curve of the progression of value or brilliance should correspond with the progression of the data.

ing colors from the small chips usually found in sample charts.

Differences in chroma carry the same connotation of relative amounts that value differences do, that is, the greater the intensity or saturation the greater the implication of magnitude.

Balance

Designing a visual composition requires a number of preliminary decisions involving what is called by the general term balance. These involve problems of layout concerning the general arrangement of the basic shapes of the presentation. The basic shapes may include the land-water masses, titles, legend, boxes, color areas, and so on.

Balance in graphic design is the positioning of the various visual components in such a way that their relationship appears logical or, in other words, so that it does not unconsciously or consciously disturb the viewer. In a well-balanced design nothing is too light or too dark, too long or too short, or too small or too large. Layout is the process of arriving at proper balance.

Visual balance depends primarily upon the relative position and visual importance of the basic parts of a map, and thus it depends upon the relation of each item to the optical center of the map and to the other items, and upon their visual weights. It may also help to think of the map as a horizontal plane; each item on a balanced map would lie in this plane. If one item is out of balance it may be prominent and lie "above" the plane visually, or if weak it will recede "below" the balance plane.

The optical center of a map is a point slightly (about 5%) above the center of the bounding shape or the map border (see Fig. 11.13). Size, value, brilliance, contrast, and to some extent, a few other factors influence the weight of a shape. The balancing of the various items about the opti-

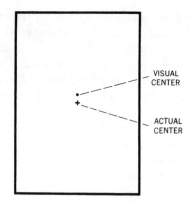

Figure 11.13   The visual as opposed to the actual center of a rectangle. Balancing is accomplished around the visual center.

cal center is akin to the balance of a beam or lever on a fulcrum. This is illustrated in Figure 11.14, where it can be seen that a visually heavy shape near the fulcrum is balanced by a visually lighter but larger body farther from the balance point. Many other combinations will occur to the student.

The aim of the cartographer is to balance his visual items so that they "look right" or

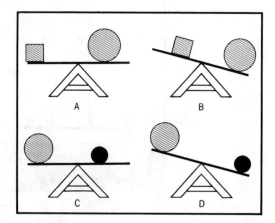

Figure 11.14   Visual balance. A, B, C, and D show relationships of balance. A and B are analogous to a child and an adult on a "teeter-totter"; C and D introduce relative density or visual weight, darker masses being heavier than lighter.

appear natural for the purpose of the map. The easiest way to accomplish this is to prepare thumbnail sketches of the main shapes, and then arrange them in various ways within or around the map frame until a combination is obtained that will present the items in the fashion desired. Figure 11.15 shows some rough sketches of a hypothetical map in which the various shapes, that is, land, water, title, legend, and shaded area, have been arranged within the border in various ways. Many reference map series utilize the sheet margins for titles, legends, scales, acknowledgements, etc. These masses must also be appropriately balanced.

The format, or the size and shape of the paper or page on which a small-scale map is to be placed, is of considerable importance in the problem of balance and layout. Shapes of land areas vary to a surprising degree on different projections, and in many cases the desire for the greatest possible scale within a prescribed format may suggest a projection that produces an undesirable fit for the area involved. Likewise, the necessity for fitting various shapes, such as large legends, complex titles, captions, and so on, around the margins and within the border makes the format a limiting factor of more than ordinary concern. Generally speaking, a rectangle with sides having a proportion of about 3:5 seems to be the most pleasing shape (see Fig. 11.16).

Often in thematic cartography, the cartographer wishes to emphasize one portion of the map or a particular relationship thereon. For example, his objective may be to show territories that changed hands in Europe since 1900 and he would like to make them appear above the background base data so that the eye will focus upon these areas and will only incidentally look at the locational base material "beneath" them. In cases like this the map is out of balance intentionally, that is, all elements of the map do not lie in the same plane (see Fig. 11.17). To accomplish this adjustment of "vertical balance," he draws upon the precepts involved in the figure-ground relationship.

The possibilities of varying balance relationships to suit the objective of the map are legion. The cartographer will do well to analyze every visual presentation he sees, from posters to advertisements, in order to become more versatile and competent in working with this important factor of graphic design.

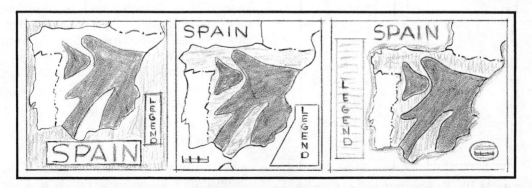

Figure 11.15 Thumbnail sketches of a map made in the preliminary stages to work toward a desirable layout and balance.

PLANNING ASPECTS OF MAP DESIGN

When graphic means are employed for communication, the display must be as carefully organized and planned as any other sort of communication. Just as when we plan to write something we first prepare an outline, so it is also necessary to outline a graphic communication. Each visual element of the map should be evaluated in combination with the other elements in terms of its probable effect on the reader. To do this requires a full and complete understanding of the objectives of the map to be made. We can scarcely imagine writing an article or planning a lecture without first arriving at a reasonably clear decision concerning (1) the audience to which it is to be presented and (2) the scope of the subject matter. With these well in mind, the writer uses them as a framework upon which to plan and as a yardstick against which to measure the significance of the items to be included.

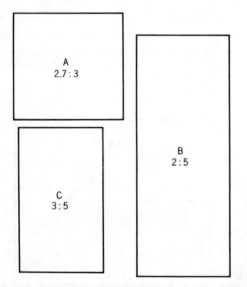

Figure 11.16 Various rectangles. *C*, with the ratio of its sides 3:5, is considered to be more stable and pleasing than the others.

It is impossible to categorize the kinds of maps that can be made in such a way that the rubrics will help provide more than a very general guide to design planning. It is true, however, that many maps generally fall into a class of "reference maps." These maps, such as topographic maps, charts, and reference atlas maps, are to be used rather like a dictionary; they are for the reader to find many kinds of information. Generally, few of the represented items are more important than the others. Consequently, clarity and legibility are paramount, and contrast, color, and figure-ground relationships are employed to enhance visibility, balance the small against the large, and in general to produce a graphic presentation that can serve many purposes. Naturally, even in some kinds of reference maps, some items are more important than others. In navigational charts, soundings and isobaths (depth contours) are more important than most items on the land; in air charts, flying hazards must be emphasized; and even in topographic maps, specific user requirements can be scaled from the more important to the lesser.

Thematic maps, on the other hand, are essentially graphic essays dealing with geographical subjects, and by their very nature they require a structuring or organization of emphasis. Organizing and outlining a graphic presentation is always an important part of the cartographic design, but it is of paramount significance in thematic cartography.

The Graphic Outline

As an illustration of how a graphic communication may be outlined, *A*, *B*, *C*, and *D* in Figure 11.18 have been prepared. The assumption is made that the planned

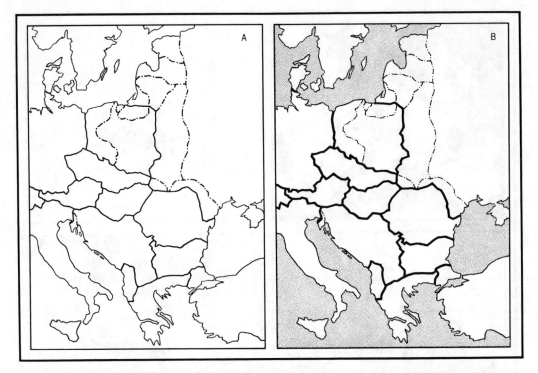

Figure 11.17   All elements in map *A* lie generally in the same visual plane. In map *B* the land has been made to appear above the water, and the more prominent boundaries have been made to rise above the visual plane of the land. Lines of the graticule on the water only would also tend to make the land appear above the water level.

thematic map is to show two related (hypothetical) distributions in Europe. The fundamental organizational parts of the communication objectives and graphic elements are as follows.

1. The place—Europe.
2. The data—the two distributions to be shown.
3. The position of the data with respect to Europe.
4. The relative position of the two distributions.

Any one of these four elements may be placed at the head of the graphic outline, and the order of the others following it may be varied in any way the cartographer de-

sires. In *A* in Figure 11.18 the design places the items in the general order of 1-2-3-4; in *B*, 2-3-4-1; in *C*, 3-1-4-2; and in *D*, 4-2-3-1. Other combinations are, of course, possible. It should not be inferred that the positioning of elements in the graphic outline can be as exact and precise as in a written outline. In the latter, the position is reasonably assured since the reader traditionally starts at one point and proceeds systematically in one direction to the end, whereas in a graphic display, the viewer, at least potentially, sees the items all at once. It is up to the cartographic designer to attempt to lead him by applying the various principles involved in the elements of graphic design.

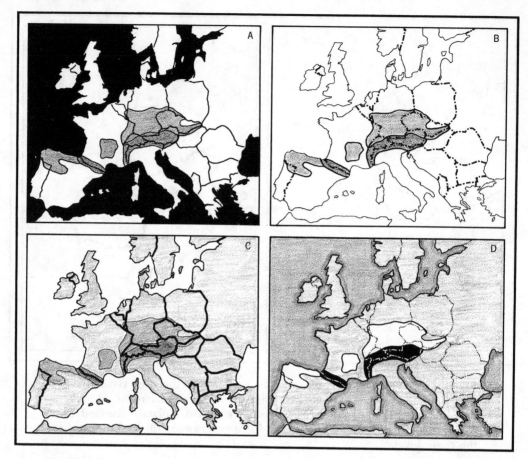

Figure 11.18   Examples of variations in the primary visual outline. See text for explanation.

It is appropriate at this point to digress slightly in order to emphasize one of the more difficult complicating factors a cartographer must face in cartographic design. Any component of a map has, of course, an intellectual connotation as well as a visual meaning in the design sense. It is difficult to remove the former in order to evaluate the latter, but many times it is not only necessary but definitely desirable. Graphic designers turn their works upside down; advertising layout artists "rough in" outlines, and even basic lettering, as design units without "spelling anything out"; and because the intellectual connotations cannot always be predicted, cartographers, for their own purposes, will obtain better design if they do likewise, except for obvious, well-known shapes, such as continents and countries.

After the structuring of the major components has been decided upon, attention must be shifted to the second stage of graphic outlining. Just as is the case with a written outline, the major items, or the primary outline, are first determined, then the position of the subject matter within each major rubric is decided. In the case of a map, the visual presentation of the "detail" is primarily a matter of clarity, legibility, and relative contrast of the detail items.

Often, the problems of cartographic design for a series of maps, whether reference or thematic, call for the preparation of one or more "trial maps" in order that primary decisions may be made regarding various forms of symbolism, lettering styles, area patterns, consistency of layout, and many other important design elements. Not infrequently, the decision making in such instances includes opinions of noncartographers. Such trial maps should be made with great care to the end that the substantive aspects are as correct as they can be. Even though such accuracy may confuse the design issues involved, all names should be properly spelled, lines placed in their correct planimetric position, etc. Unfortunately, it appears to be difficult, if not downright impossible, for those inexperienced in design problems to ignore the substantive aspects of a trial design. The noncartographer, with seeming perversity, is likely to dwell upon any planned "inaccuracies" and fail to give the needed attention to the design problems. One experience in such a situation serves clearly to exemplify the fact that any map component has both an intellectual and a visual connotation; but it is an experience the cartographer can well avoid.

Titles, Legends, and Scales

These standard elements serve two major purposes in cartography. Naturally, they have a denotative function in identifying the place, subject matter, symbolization, etc., but they also serve as graphic masses that can be positioned to provide the graphic organization of a map (Fig. 11.19). Generally, fragmentation of the map should be avoided as much as possible.

A title serves a variety of functions. Sometimes it informs the reader of the subject or area on the map and is therefore as important as a label on a medicine bottle. But this is not always the case, since some maps are obvious in their subject matter or area and in reality need no title. In these instances the title may yet be useful to the designer as a shape that he may use to help balance the composition.

It is impossible to generalize as to the form a title should take; it depends entirely upon the map, its subject, and its purpose. Suppose, for example, a map had been made showing the 1968 density of population per square mile of arable land in

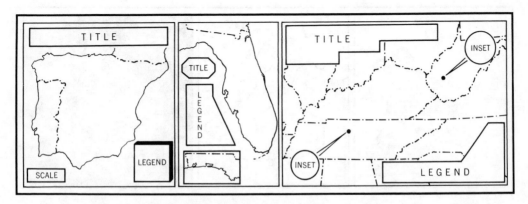

Figure 11.19 Titles, legends, scales, and insets may be arranged in various ways in the graphic organization of a map.

France. The following situations might apply.

1. If the map appeared in a textbook devoted to the general world-wide conditions at that time with respect to the subject matter, then only

FRANCE

would be appropriate because the time and subject would be known.

2. If the map appeared in a study of the current food situation in Europe, and if it were an important piece of evidence for some thesis, then

France
POPULATION PER SQUARE MILE
ARABLE LAND

would be appropriate.

3. If the map appeared in a publication devoted to the changes in population in France, then

POPULATION PER SQUARE MILE
ARABLE LAND
1968

would be appropriate, since the area would be known but the date would be significant.

Many other combinations could be worked out, but there is no need to belabor the fact that the title must be tailored to the occasion. Similarly, the degree of prominence and visual interest displayed by the title, through the style, size, and the boldness of the lettering employed, must be fitted into the whole design and purpose of the map.

Legends or keys are naturally indispensable to most maps, since they provide the explanation of the various symbols used. It should be a cardinal rule of the cartographer that no symbol that is not self-explanatory should be used on a map unless it is explained in a legend. Furthermore, any symbol explained should appear in the legend *exactly* as it appears on the map, drawn in precisely the same size and manner. Legend boxes can be emphasized or subordinated by varying the shape, size, or value relationship. Figure 11.20 illustrates several variations. In the past it was the custom to enclose legends in fancy, ornate outlines, called *cartouches*, which, by their intricate workmanship, called attention to their presence. Today it is generally conceded that the contents of the legend are more important than its outline, so the outline, if any, is usually kept simple, and the

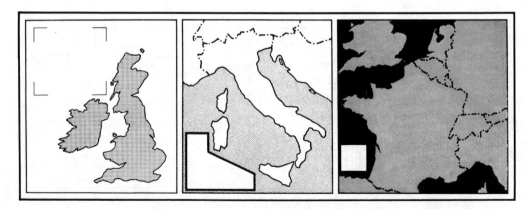

Figure 11.20   Variations in the prominence of legend boxes. Note the operation of the principles of the figure-ground relationship.

visual importance is regulated in other ways.

The scale of a map also varies in importance from map to map. On maps showing road or rail lines, air routes, or any other phenomenon or relationship that involves distance concepts, the scale is an important factor in making the map useful. In such cases the scale must be placed in a position of prominence, and it should be designed in such fashion that it can be easily used by the reader.

The method of presenting the scale may vary. For many maps, especially those of larger scale, the Representative Fraction, the RF, is useful because it tells the experienced map reader a great deal about the amount of generalization and selection that probably went into the preparation of the map. It should, however, be borne in mind

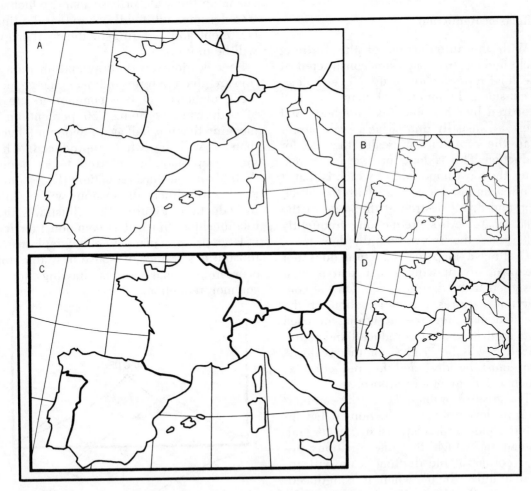

Figure 11.21 Effects of reduction. (*A*) When artwork is designed for proper line contrasts at drawing scale, then (*B*) reduction will decrease the contrasts too much. (*C*) When artwork is designed for reduction, then (*D*) reduction produces proper line contrast relationships.

that if a map is to be changed in size by reduction, this will not change the printed numbers of the RF. On a map designed for reduction, the RF must be that of the final scale, not the construction scale. A graphic scale is much more common on small-scale maps, not only because it simplifies the user's employment of it, but because an RF in the smaller scales is not so meaningful. Furthermore, enlargement or reduction have no effect on a graphic scale.

Effects of Reduction

With the introduction of the scribing technique, many maps are constructed at the final reproduction scale. Nevertheless, a considerable number of maps are still prepared by "pen and ink" methods and these are usually drafted at a scale larger than the reproduction scale, that is, for reduction. This is done for a variety of reasons, the most important of which is that it is often impossible to draft with the precision and detail desired at the scale of the final map. Also, reduction frequently "sharpens" the line work of the drawing.

Drafting a map for reduction does not mean merely drawing a map that is well designed at the drafting scale. On the contrary, it requires the anticipation of the finished map and the designing of each item so that when it is reduced and reproduced it will be "right" for that scale. A map must be designed for reduction as much as for any other purpose.

The greatest problem facing the cartographer in designing for reduction is that involving line-width relationships. In general, a map on which the line relations appear correct at the drafting scale will appear a little "light" when it is reduced. Consequently, the map maker must make his map overly "heavy" in order to avoid its appearing too light after reduction. It is necessary for the cartographer to "overdo" his lettering somewhat, just as it is necessary

for him to make lines and symbols a little too large and dark on the drawing. Figure 11.21 illustrates these relationships. The use of a reducing glass will aid in visualizing how the map will appear when reduced.

Maps of a series should appear comparable and, if they are drawn for reduction, they should be drafted for the same amount so that similar lettering, lines, etc., will appear comparable. This may necessitate changing the scale of base maps, which is troublesome, but it will insure that the line treatment and lettering in the final maps will be uniform.

Specifications for drafting and for reduction are given in terms of linear change, not areal relationships. It is common to speak of a drawing as being "50 percent up," meaning that it is half again wider or longer than it will be when reproduced. The same map may be referred to as being drafted for one-third reduction; that is, one-third of the linear dimensions will be lost in reduction. Figure 11.22 illustrates the relationships. Since it is common practice in large printing plants to photograph many illustrations at once, it is also desirable, for economy, to make series drawings for a common reduction.

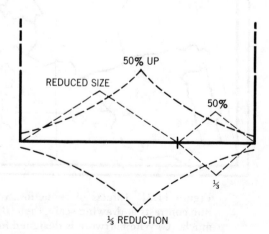

Figure 11.22 Relation of enlargement to reduction.

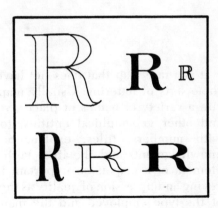

CARTOGRAPHIC TYPOGRAPHY AND LETTERING

It is a rare map that does not have on it some sorts of lettering. Usually maps contain a variety of names of places, regions, and other geographical entities, together with numbers, titles, explanations, acknowledgments and the like. Perhaps no element of a map is so important in conveying an impression of quality as the styles of the type employed and the manner in which the map has been lettered. Uncongenial type or poorly designed letter forms badly positioned mark a map as having come from the hand of an incompetent cartographer, and the reader is likely unconsciously to put little trust in it. The cartographic quality of a map, like the attributes of literary quality in writing and artistic style in painting, is heavily dependent on its lettering, and because of this we can often recognize at a glance the author or cartographic establishment from which a map comes.

The utility of a reference map depends to a great extent on the characteristics of the type and its positioning. The recognition of the feature to which a name applies, the "search time" necessary to find names, and the ease with which the lettering can be read are all important to the function of the map. Like other aspects of cartographic design, we must approach the problem of cartographic typography and lettering by settling clearly on the objectives of the map and then fitting the selection of type and the positioning of the lettering to them.

The Function of Map Lettering

Like all other marks on a map, the type functions as a symbol, but it is considerably more complicated in its function this way than are most symbols. Its most straightforward service is as a literal symbol in that the individual letters of the alphabet, when arrayed, encode sounds that are the names of the symbolized elements of the map. Generally, this is the most important role of the lettering in the communication system that is the map.

In several ways the type on a map is a graphic symbol. By its position within the structural framework, it helps to indicate the location of points (for example, cities); it more directly shows by its spacing and array such things as linear or areal extent (for example, mountain ranges and national areas); and by its arrangement with respect to the graticule, it can clearly indicate orientation. By systematically employing distinctions of style, form, and color, the type on a map may be used as a means of showing nominal classes to which the labeled feature belongs. For example, we can identify all hydrographic features in blue type, and within that general class, can further show open water by all capital letters, and running water by capitals and lower-case letter forms. By variations of size, it can portray the ordinal characteristics of geographical phenomena, ranking them in terms of relative area, importance, and so on.

In a more subtle way, the type on a map serves as an indication of scale. Primarily by its size contrasts with respect to other factors such as line width, symbol size, and relative chroma in color, we can often tell "by looking" that one map is a larger or smaller scale than another. Of course, if the designer has not been careful, this important impression can be lost or even reversed.

The complexity of the ways type performs as a map symbol sets this subject quite apart from the other kinds of cartographic symbolism.

The History of Map Lettering

Before the development of printing, all manuscript lettering was of course done freehand, and each copy of a map had to be laboriously hand lettered by the calligrapher. The styles were varied and ranged from the cramped and severe to the free flowing.

After printing and engraving became the accepted methods of map reproduction, the lettering on maps was a task of the engraver, who cut his letters with a burin or graver in reverse on the copper plate. The great Dutch atlas makers were wont to include many pictures of animals, ships, and wondrous other things for, as Hondius explained, "adornment and for entertainment," but their lettering was generally well planned in the classic style and well executed, as is that illustrated in Figure 12.1. The problems of lettering were complex even then, and some maps were printed from wood cuts and engravings with metal type inserted in holes cut in the blocks.

Figure 12.1   The well-lettered Hondius map of America. This map was included in the Hondius-Mercator Atlas of 1606. Note also the ornate cartouches. The original is in the Newberry Library, Chicago. (Courtesy Rand McNally and Company.)

As might have been expected, when hand lettering was done by those more interested in its execution than in its use, it tended to become excessively ornate. The trend toward poor lettering design seemed to accelerate after the development of lithography and continued well into the Victorian era. A process called cerography or wax engraving came into use in the 1840's and later made lettering a map quite easy. This led to considerable "over-lettering."

Lettering and type styles in general became so bad by the close of the nineteenth century that there was a general revolt against them, which caused a return to the classic styles and greater simplicity. Figure 12.2 is an example of ornate lettering in the title of a nineteenth century geological map. The fancy lettering of this and earlier periods provide good examples of manual dexterity, being intricate and difficult to execute; but they are examples of poor communications design because they are difficult to read and call undue attention to themselves.

In the past half century, and especially during the past several decades, many changes have taken place in cartographic lettering practice. No longer does the engraver do the lettering; instead, maps are "engraved" by means of photography, and it is entirely up to the cartographer to plan the lettering.

The introduction of the stick-up process whereby type is printed on thin plastic, which is then wax-backed for adhesion to the map, has all but made freehand lettering a thing of the past as far as map construction is concerned. The draftsman now need only cut out words or the individual letters and place them on the map in correct position. In the larger map-producing establishments there are machines for automatically positioning photographic "type," and it is safe to say that the cartographer now has an almost unlimited array of type styles and sizes from which to choose.

Although relatively few maps are hand lettered for reproduction in their construc-

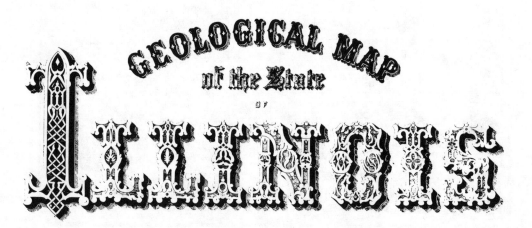

Figure 12.2   Ornate lettering on a lithographed map of 1875. The name "Illinois" is more than 1 foot long on the original.

tion stage, it is well to point out that even if the cartographer never executes freehand the ink lettering on a map, he will find the ability to letter neatly a most useful accomplishment. To be "sloppy" in almost anything usually requires more time in the long run than it does to be neat. In the cartographer's pencil compilations it is usually necessary to insert many names and to pay careful attention to positionings and spellings. If he hurries and is messy, he will make mistakes of one kind or another and these will cause trouble later. For this reason, if for no other, it is well for the student of cartography to study and practice freehand lettering. It merely involves learning principles and practicing execution; almost anyone can learn to do neat freehand lettering—especially with pencil—if he will but try.

Planning for Lettering

Lettering a map means the preparation of this aspect of the artwork, which includes all the names, numbers, and other typographic material. It involves the entire process from the beginning to the end and requires careful consideration of many factors.

The more elaborate and complex the map, of course, the more elements must be considered; but, in general, there are at least seven major headings to the planning "check list." The complexities of the map and its purposes will add subheads to the following major elements.

1. The style of the type.
2. The form of the type.
3. The size of the type.
4. The contrast between the lettering and its ground.
5. The method of lettering.
6. The positioning of the lettering.
7. The relation of the lettering to reproduction.

The style refers to the design character of the type, and it includes such elements as thickness of line and serifs. The form refers to whether it is composed of capitals, lowercase, slant, upright, or combinations of these and other similar elements. The methods of lettering include the mechanical means whereby the type or lettering is affixed to the map. The positioning of the lettering involves the placement of the type. Also important is a consideration of when and where on the map, and in the construction schedule, the lettering is done. Different methods of reproduction require variations in parts of the above processes, and this is especially so when special effects are to be gained. This topic will be treated in Chapter 14.

Regardless of the kind of map, the lettering is there to be seen and read. Consequently, the elements of visibility and legibility are among the major yardsticks against which the choices and possibilities are to be measured.

ELEMENTS OF TYPOGRAPHIC DESIGN

Type Style

The cartographer is faced with an imposing array of possible choices when he sets out to plan the lettering for a map. There is a surprising number of different alphabet designs from which to select, but he must also settle upon the wanted combinations of capital letters, lowercase letters, small capitals, italic, slant, and upright forms of each alphabet. There is no other technique in cartography that provides such opportunity for individualistic treatment, and this is especially true with respect to the monochrome map. The cartographer who becomes well acquainted with type styles and their uses finds that every map or map series presents an interesting challenge. When this attitude prevails lettering the map ceases to be the mechanical chore some, whose maps reflect their disinterest, consider it to be.

Type styles have had a complex evolution since Roman Times.* The immediate ancestors of our present-day alphabets include such grandparents as the capital letters the Romans carved in stone and the manuscript writing of the long period prior to the discovery of printing. Subsequent to the development of printing, the types were copies of the manuscript writing, but it was not long until designers went to work to improve them. Using the classic Roman letters as models for capitals and the manuscript writing for the small letters, there evolved the alphabets of uppercase and lowercase letters that it is our custom to use today. There are three basic classes.

One class of designers kept much of the free-flowing, graceful appearance of some freehand calligraphy so that their letters carry something of the impression of having been formed with a brush. The distinction between thick and thin lines is not great, and the serifs with which the strokes are ended are smooth and easily attached. Such letters are known as Classic or Old Style. They appear "dignified" and have about them an air of quality and good taste that they tend to impart to the maps on which they are used. This style has a neat

---

CHELTENHAM WIDE

Cheltenham Wide

CHELTENHAM WIDE ITALIC

Cheltenham Wide Italic

GOUDY BOLD

Goudy Bold

GOUDY BOLD ITALIC

Goudy Bold Italic

CASLON OPEN

Figure 12.3 Some Classic or Old Style type faces (Courtesy Monsen Typographers, Inc.)

appearance, but at the same time it lacks any pretense of the geometric. (see Fig. 12.3).

A radically different kind of face was devised later, and for that reason it (unfortunately) is termed Modern. Actually the Modern faces were tried out more than two centuries ago, although we think of them as coming into frequent use around 1800. These type faces look precise and geometric, as if they had been drawn with a straightedge and a compass, which they have. The difference between thick and thin lines is great and sometimes excessive in Modern Styles (see Fig. 12.4). Both Old

---

BODONI BOLD

Bodoni Bold

BODONI BOLD ITALIC

Bodoni Bold Italic

Figure 12.4 Some Modern letter forms. (Courtesy Monsen Typographers, Inc.)

---

*A very well illustrated, interesting treatment of the development of lettering and type styles is Alexander Nesbitt, *The History and Technique of Lettering*, Dover Publications, New York, 1957.

Style and Modern types have serifs, and for that reason are sometimes loosely classed together as Roman.

A third style class includes some varieties that are definitely modern in time but not in name, as well as some of older origin. This class, increasingly important in modern cartography, is called Sans Serif (without serifs), and has about it an up-to-date, clean-cut, new, and nontraditional appearance. The forms are angular or of perfect roundness. There is nothing subtle about most Sans Serif forms. There are many variants in this class, some of which include variations in the thickness of the strokes (see Fig. 12.5). Sans Serif forms are sometimes called Gothic as opposed to Roman.

There are several other styles that are not common but that are occasionally used on maps. These are Text, Script or italic, and Square Serif. Script and italic in Old Style and Modern types are similar to handwriting with flowing lines. They have been traditionally replaced in map work by the Sans Serif or Gothic slant letter, which is also called italic (Fig. 12.5). Text, or black letter is dark, heavy, and difficult to read. Square Serif is rarely seen any more, but

MONSEN MEDIUM GOTHIC

Monsen Medium Gothic Italic

COPPERPLATE GOTHIC ITALIC

FUTURA MEDIUM

LYDIAN BOLD

DRAFTSMANS ITALIC

Figure 12.5 Some Sans Serif letter forms. (Courtesy Monsen Typographers, Inc.)

𝕮𝖑𝖔𝖎𝖘𝖙𝖊𝖗 𝕭𝖑𝖆𝖈𝖐

𝕮𝖑𝖔𝖎𝖘𝖙𝖊𝖗 𝕭𝖑𝖆𝖈𝖐

STYMIE MEDIUM

Stymie Medium

Figure 12.6 Examples of Text and Square Serif letter forms. (Courtesy Monsen Typographers, Inc.)

was popular during the last century (see Fig. 12.6).

This listing by no means exhausts the possibilities. There are literally hundreds of variations and modifications possible, such as the open letter, light or heavy face, expanded or condensed, and so on.

In the selection of type, the cartographer must be guided by certain general principles that have resulted from a considerable amount of research by the students of alphanumeric symbolism as well as from the evolved artistic principles of the typographer.

Recognition depends upon the occurrence of familiar forms and upon the distinctiveness of those forms from one another. For this reason, "fancy" lettering or ornate letter forms are hard to read and text lettering is particularly difficult. Conversely, well-designed Classic, Modern, and Sans Serif forms stand at the top of the list, and apparently they rate about equally. Ease of recognition also depends, to some extent, upon the thickness of the lines forming the lettering. The thinner the lines relative to the size of the lettering, the harder it is to read. The cartographer must, therefore, be careful in his selection, since, although the bold lettering may be more easily seen, the thicker lines may overshadow or mask other equally important data. The type on a map is not always the most important element in the visual outline; rather, it may even be desirable that the type recede into the background. If so, light-line type may be an effective choice.

The problem of the positions of the lettering in the visual outline is significant. For example, the title may be of great importance, while the balance of the type may be of value only as a secondary reference. Size is usually much more significant, in determining the relative prominence, than style, but the general design of the type may also play an important part. For example, rounded lettering may be lost along a rounded, complex coastline, whereas in the same situation angular lettering of the same size may be sufficiently prominent.

It is the convention in cartography to utilize different styles of lettering for different nominal classes of features, but this may be easily overdone. Although there is a paucity of evidence, there does seem to be some indication that the average map reader is not nearly so discriminating in his reactions to type differences as cartographers have hitherto thought. The use of many subtle distinctions in type form for fine classificational distinctions is probably a waste of effort.

As a general rule, the fewer the styles, the better harmony there will be. Most common type faces are available in several variants; it is better practice to utilize these as much as possible (see Fig. 12.7). If styles must be combined for emphasis or other reasons, good typographic practice allows Sans Serif to be used with either Classic or Modern. Classic and Modern are normally not combined. Differences in style are best utilized for nominal differentiation; size and boldness are more appropriate for ordinal and interval distinctions.

The cartographer is usually less concerned with what has been called the "congeniality" of type style, that is, the correspondence between the subjective impression one gains from a type design in

Futura Light

Futura Light Italic

Futura Medium

Futura Medium Condensed

Futura Medium Italic

Futura Demibold

Futura Demibold Italic

Futura Bold

Futura Bold Italic

Futura Bold Condensed

Futura Bold Condensed Italic

Figure 12.7   Variants of a single face. This face has a larger number of variants than is usual, but the list is representative except that expanded (opposite of condensed) is missing. (Courtesy Monsen Typographers, Inc.)

relation to its use. There is no question that readers respond to various styles with reactions such as "authoritative," "masculine," "feminine," "arty," and "clean." Sans Serif seems to be assuming a dominant position in cartography while the Old Style and Modern types are most used in literary composition. It is not uncommon to find a Sans Serif used in the body of a map with a contrasting style, usually Old Style, employed for the title and legend materials.

Type Form

Alphabets consist of two quite different letter forms called capitals and lowercase letters. These two forms are used together in a systematic fashion in writing, but conventions as to their use are not so well established in cartography. In general, past practice has been to put more important names and titles in capitals, and less important names and places in lowercase letters. Names requiring considerable separation of the letters are commonly placed in capitals.

Most tests have shown that capitals are more difficult to recognize or read than are lowercase letters, since the latter contain more clues to letter form. A greater use of well-formed lowercase letters probably will improve the legibility of a map.

Most styles of type can be had in either upright or italic form. The tendency in cartography is for hydrography, landform, and other natural features to be labeled in slant or italic, and for cultural features (man-made) to be identified in upright forms. This can hardly be called a tradition, since departure from it is frequent, except in the case of water features where this convention is very strong. The slant or italic form seems to suggest the fluidity of water.

There is a fundamental difference between slant and italic, although the terms are sometimes used synonymously (see Fig. 12.8). True italic in the Classic or Modern faces is a cursive form similar to script or handwriting. Gothic slant and Sans Serif italic are simply upright letters with the upper portions tilted forward. Italic forms are considerably harder to read than their upright counterparts; however, it is doubtful that there is much difference

Kennerley—an upright Classic

Kennerley in the Italic form

Monsen Medium Gothic—upright

Monsen Medium Gothic Italic

Figure 12.8 Differences between italic and slant forms. (Courtesy Monsen Typographers, Inc.

between the upright and slant letters of Gothic or Sans Serif insofar as recognition is concerned.

Type Size

The size of type is commonly designated by *points*, 1 point being nearly equal to 1/72 in. With the development of photo lettering machines, sizes may be expressed also in the height of letters. Lettering that is 1/4 in. high is roughly equal to 18-point type although not exactly, since designation by points refers to the body height of type and not to the letter face on it (Fig. 12.9). Type (or the print from a photo matrix in a photo lettering machine) that is enlarged or reduced will be slightly different from the designed type at that size, since the line thickness of one font (one size of one style) is usually designed for that size only.

Perhaps the commonest kinds of decisions concern the various sizes to be used

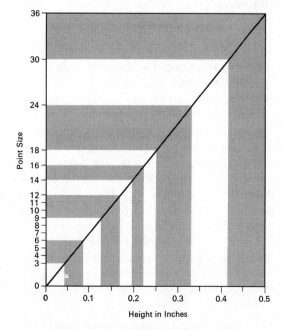

Figure 12.9 The relation between letter height and nominal point size.

for the great variety of items that must be named on maps. Traditionally, specifications for lettering are usually based on the magnitude of the object being named or the space to be filled, but the type must also be graded with respect to the total design and intellectual content of the map. Much of the criticism, however, conscious or unconscious, that is leveled at map type is aimed specifically at size—or lack of it. There seems to be a widespread tendency among amateur (and even some professional) cartographers to overestimate the ability of the eye.

As was observed in the last chapter, and assuming no other complications (the assumption is a bit unreal), the eye reacts to size in relation to the angle the object subtends at the eye. Generally speaking, with normal vision an object that subtends an angle of 1 minute can just be recognized. Letter forms are complex, however, and it has been determined that about 3-point type is the smallest, just recognizable type at usual reading distance. Normal vision is, however, a misnomer, for it certainly is not average vision. It is safer to generalize that probably 4-point or 5-point type comes closer to the lower limit of visibility for the average person.

A common employment of type size variations in cartography is for the purpose of ordinal or interval ranking of places, usually settlements, in terms of their relative magnitude as rated by their populations. The ability of map readers to recognize differences in size has not been extensively studied, but there is evidence that children generally require a height difference of at least 20%. This would be approximately 2 points in the range from 6 to 12 point.° The average adult probably reacts in similar fashion. Furthermore, the number of cate-

gories of similar sizes that will normally be noted appears to be three, simply, small, medium, and large. For a greater number of categories we probably should introduce significant differences in form. In the interest of promoting ease of recognition ("findability" when looking for a name), the cartographer should use the largest type sizes possible, consistent with good design.

Frequently, cartographers are called upon to prepare presentations for groups. They must face such questions as, "What is the minimum type size that may be used, under normal conditions, on a wall map or chart from the back or middle of a 40-foot room?" Figure 12.10 is a nomograph, the construction of which is based on the assumption that if a particular size, at normal reading distance from the eye (18 in.), subtends a certain angle at the eye, then any size lettering if viewed at such a distance that it subtends the same angle is, for practical purposes, the same size. Thus, 144-point type at 30 ft from the observer is the same as 8-point type at normal reading distance, since each circumstance results in the same angle subtended at the eye. It will be seen from the graph that legibility diminishes very rapidly with distance. For example, any type of 16-point size or smaller cannot be read even at 10 ft from the chart or map, and letters 1 in. high can be read from a distance of 40 ft only by a person with above-average vision. To those with average (not normal) vision, such letters are likely to be illegible from 25 ft.°

Type Color

In typography, "color" of type frequently refers to the relative overall blackness of

---

°Barbara S. Bartz, *Map Type, Form and Function*, Field Enterprises Educational Corporation, Chicago, 1966.

°For the preparation of type for maps and charts that are to be made into slides for projection, see Arthur H. Robinson, "The Size of Lettering for Maps and Charts," *Surveying and Mapping*, **19**, 37–44 (1950).

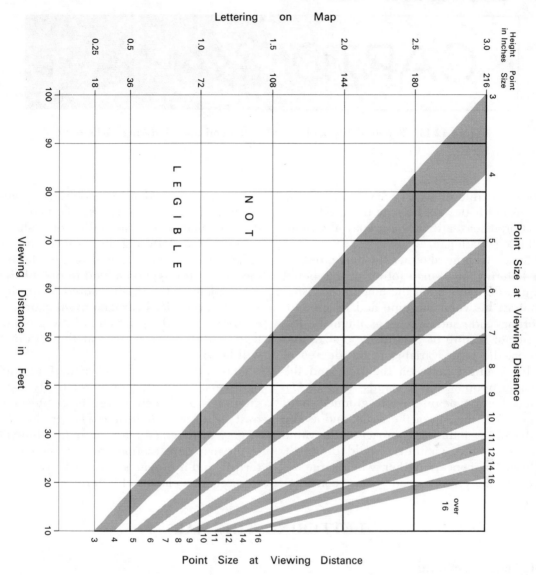

Figure 12.10 A nomograph showing the approximate apparent point size of lettering when viewed from various distances. Equivalent sizes less than 3 point are not legible.

pages set with different faces. Here it is restricted to the actual hue (such as black, blue, gray or white) of the letters and the relation between the color of the type and that of the ground on which it appears. The perceptibility of lettering depends upon the contrast between the lettering and the ground. Commonly, lettering of equal intrinsic importance does not appear as equal in various parts of the map because of ground differences. Even when the same type face must be used in various graphic situations because of other design requirements, the cartographer should be aware of

CARTOGRAPHY

Figure 12.11   Perceptibility and legibility depend upon lettering-background contrast.

these possible effects. Perhaps, within the limits of his design, he may be able to correct or at least alleviate a graphically inequitable situation.

Stated in general terms, the perceptibility of lettering on a map (other effects being equal) depends upon the amount of visual contrast between the type and its ground. Putting aside such effects as might be the result of texture of ground, type size, and so on, the basic variable is the degree of value contrast between the type and the ground upon which it stands (Fig. 12.11). Thus, black type on a white ground would stand near the top of the scale, and as the value of the lettering approached the value of the ground, visibility would diminish. This is of concern either when large regional names are "spread" over a considerable area composed of units colored or shaded differently or when names of equal rank must be placed on areas of different values.

The usual lettering in cartography is either black or open (called reverse lettering), but often type is added to one of the color plates of a multicolored map, for example, blue for hydrography. Regardless of the color of the print and of the ground, if the value contrast is great the lettering will be visible.

Gray lettering on a light or dark ground or reversed lettering on a relatively dark ground is an effective way to create contrast in the map lettering. The production of gray or reversed lettering is not difficult in the map construction process (see Chapter 14), and it provides a very effective way to categorize classes of lettering.

LETTERING THE MAP

Methods of Lettering

Until less than fifty years ago, the lettering for most maps was done either by the draftsman or the engraver. To be sure, the cerographic technique ("wax engraving") employed printing type pressed into a film of wax, which was then electroplated to produce a printing plate; but the majority of map lettering was either freehand on an ink drawing or was freehand in metal by the engraver. The desire for lower cost, speed, uniformity, and standardization led to the development of mechanical lettering devices and other means to escape from the restraints imposed by freehand methods. Today almost all map-making organizations use some method other than freehand to letter their products. These methods may be grouped in two categories: (1) stick-up or preprinted lettering and (2) mechanical aids for ink lettering. Freehand lettering, which will always be the most versatile of the ways to letter a map, is largely restricted to the occasional special map produced for illustrative purposes or to the products of the cartographic "do it yourselfer." The art is still employed to some extent in traditional fashion in some old-world mapping houses, but it is fast dying out.

Stick-up Lettering. Stick-up lettering has a number of advantages over other methods of lettering, in addition to the fact that it is generally faster, requires less skill, etc. Any of the thousands of type styles and sizes used in printing can be selected. The letters may be used as composed in a straight unit, or they may be cut apart and applied separately to fit curves. If the position first selected for the name turns out not to be suitable, the name may be relocated with no attendant problem of erasing.

The type used in this process is ordinarily printed on a thin plastic sheet, the back of which is then given a slight coating of colorless wax. The words, numbers, or letters may then be cut from the sheet, positioned carefully, and then burnished to the art work by rubbing. Rubbing is usually done with a smooth tool, not directly on the film, but on a sheet laid over it in order to protect it. Light rubbing will hold it in place but makes it possible to reposition it. Final burnishing is postponed until all positioning decisions have been made.

It is possible, of course, to print (or even typewrite) the letters on other materials and to affix them to the map in other ways. Gum-backed paper, thin tissue, etc. have been employed, and liquid adhesives, self-adhesive materials, and so on have all been tried. The standard process, however, uses thin plastic film or cellophane with a wax backing.

There are two ways of producing such stick-up materials, by photography and by letterpress printing. There are a number of photographic devices manufactured for the purpose that may be used in the map-making establishment, and each has its special characteristics. They will not be detailed here. Suffice it to say that some provide several advantages. These include the ability to vary the size of the lettering. This can be accomplished because in these devices the "type" appears as a photographic negative (matrix) and, by enlargement or reduction, the print can be scaled to the desired size. In others, a separate set of matrices must be obtained for each size desired.

For letterpress printing, done commercially, the cartographer must make for the printer a complete list of all names and other lettering that are to appear on the map or series of maps for which he plans to use stick-up* Spellings must, of course, be reviewed carefully. Since the printer will supply a number of copies of the composition, it is only necessary to repeat names, such as "river" or "lake," as many times as the total number of map occurrences divided by the number of copies of the composition to be obtained. Type style, point size, and capital and lowercase requirements must be carefully designated. The printer will compose the names as listed, and since possibilities of inadvertent omission are relatively high, it is good practice to include a request for alphabets of the styles and sizes being used. Prices vary considerably, depending upon whether the type to be used must be hand set or machine set, so careful planning is desirable.

One disadvantage of stick-up is that,

*The major commercial supplier of letterpress stick-up is Monsen Typographers, Inc., 22 East Illinois Street, Chicago, Ill. 60611, and 960 West 12th Street, Los Angeles, Cal. 90015. Their Trans-Adhesive impressions are available on glossy acetate, matte-finish acetate, and flexible film (Vinylite) for slight curves without cutting apart; and all have a wax backing. The same sort of material is available also in black ink on a white backing, which automatically masks the area around the letters, or in white ink for reverse lettering.

generally speaking, the type faces used for both photo and letterpress stick-up were originally designed for printing books. Such type faces have not been designed for any position except horizontal. When such letters are closely spaced in a curved position, the design is less than perfection. Furthermore, each size of a type style has been designed separately for use at that size. Consequently, when used on maps planned for reduction, there is a tendency for the lettering to appear somewhat light, since 18-point lettering reduced one-half is made of slightly thinner strokes than is the 9-point size of the same style. It is generally necessary, when using stick-up lettering for maps to be reduced, to choose a somewhat larger size than would ordinarily be warranted.

Mechanical Lettering. Mechanical devices that enable someone unskilled in freehand lettering to produce acceptable ink lettering are available. With them we can obtain neat lettering, but it should be emphasized that the lettering produced with most of these aids appears rather mechanical and gives the impression of looking at a building blueprint. They are most used for just this purpose—engineering drawings, wherein variations in the character and orientation of letter forms are not particularly needed. Some of the complexities of good quality map lettering have been detailed in previous sections, and it is not to be expected that any relatively inexpensive mechanical device can approach their attainment. Nevertheless, mechanical lettering devices will continue to find a place in "do it yourself" cartography, especially for the production of graphs, charts, and diagrammatic maps.

The best-known devices require a special pen, feeding through a small tube, while the pen is guided either mechanically or by hand with the aid of a template.

Leroy is the patented name of a lettering system involving templates, a scriber, and special pens. A different template is necessary for each size lettering. The template is moved along the T-square or steel straightedge, and the scriber traces the depressed letters of the template and reproduces them with the pen beyond the template. A variety of letter weights and sizes in capitals and lowercase is possible by interchanging templates and pens.

Wrico is the patented name of a lettering system involving perforated templates or guides and special pens. The lettering guides are placed directly over the area to be lettered and are moved back and forth to form the various parts of a letter. The pen is held in the hand and is moved around the stencil cut in the guide. The guide rests on small blocks which hold it above the paper surface to prevent smearing. A considerable variety of letter forms are possible, including condensed and extended. A different guide is necessary for each size although pens may be varied.

Varigraph is the patented name of a versatile lettering device also involving a template with depressed letters and a stylus. The device is actually a sort of small, adjustable pantograph that fits over a template. The letters are traced from the template and are drawn by a pen at the other end of the pantograph-like assembly. Adjustments may be made to make large or small, extended or condensed, lettering from a single template. Templates of a variety of letter styles are available.

Freehand Lettering. As was observed earlier, the freehand method is the most versatile of the ways of applying the lettering to a map. With this technique, names may be designed, oriented, scaled, curved, and inserted to the best graphic advantage. Unless one is unusually experienced, however, stick-up will lead to a better overall result.* It is important, however, that all cartographers become reasonably adept at freehand lettering with pencil so that the large amount of lettering necessary in compilation work will be neatly executed. Furthermore, the cartographer, in designing his graphic composition, will be called upon to lay out titles, legends, etc., and to work out the spaces and sizes needed, prior to obtaining stick-up. Consequently, for executing this kind of lettering, the cartographer needs to gain facility in three basic aspects of freehand lettering: spacing, use of guide lines, and stroking. The term "freehand lettering" is rather a misnomer; such lettering well done is done with great care.

Spacing refers to the distance between letters. In letterpress type the spacing is mechanical and, as a consequence, words often appear ragged and uneven. Visual spacing in which the same amount of *apparent* total space appears between each letter and its neighbor is far better (Fig. 12.12). For this reason, the letters of large names and especially titles done by stick-up need to be positioned one at a time. The beginner will soon learn that there are

*The learning of freehand lettering requires much practice, patience, and familiarity with and an ability to use the basic tools. It also requires a thorough appreciation of the principles of letter formation and design. These elements require much more treatment than can be given in a book of this sort, and the reader is referred to Nesbitt, *op. cit.* Very helpful to beginners is Ross F. George, *Speedball Textbook* (various editions), Hunt Pen Company, Camden, New Jersey.

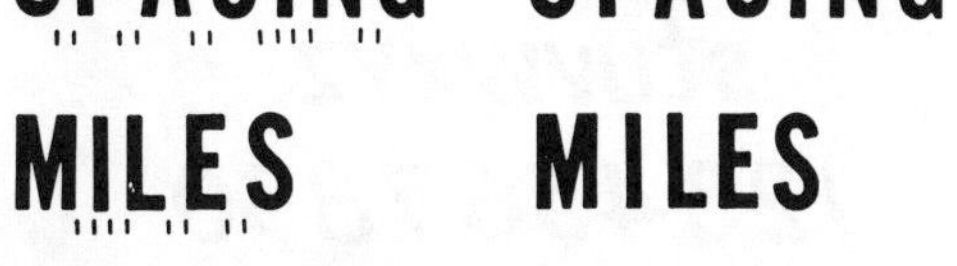

different classes of letters according to their regular or irregular appearance and according to whether they are narrow, average, or wide. They must be separated differently depending upon the combination in the word. Mechanical spacing should always be avoided.

Guidelines are simply what their name implies, lines to serve as guides for the lettering. They may be drawn with a straight-edge or a curve, but guidelines are better drawn with any of a number of patented devices designed for this purpose, such as the Ames lettering instrument or the Multi and Braddock-Rowe lettering angles. These devices have small holes in which a pencil point may be inserted. The device may then be moved along a drawing edge by moving the pencil. By placing the pencil in other holes, parallel lines may be drawn. Guide lines usually consist of three parallel lines, as shown in Figure 12.13. The bottom two determine the height of the lowercase letters, and the upper guideline indicates the height of the capitals and the ascenders of letters such as b, d, and f.

Stroking refers to the direction and order with which the various parts of letters are formed (Fig. 12.14). It is more or less important depending upon the kind of tools

SPACING   SPACING

MILES   MILES

Figure 12.12   Visual compared to mechanical spacing.

Figure 12.13   Guidelines.

we are using. For example, we cannot make an up stroke with a sharp pen or pencil without digging into the surface. Once proper stroking becomes a habit, letters will be better formed, regardless of tools.

If the cartographer does any freehand lettering in ink, he will do well to limit his efforts to simple designs. Such lettering is commonly confined to upright or slanted forms, often consisting of thick and thin strokes made with a square-tipped pen such as the Speedball Style C or the N or Z style of Graphos nibs. Simple lettering may be made with round or tubular nibs, so that whichever way the nib moves the line thickness remains the same. Figure 12.15 shows an example of each style.

Positioning the Lettering

Map reading is affected greatly by the positioning of the names.* When properly placed, the lettering clearly identifies the phenomenon to which it refers, without ambiguity. Equally important is the fact

*Eduard Imhof, "Die Anordnung der Namen in der Karte," *International Yearbook of Cartography*, **2**, 93–129 (1962).

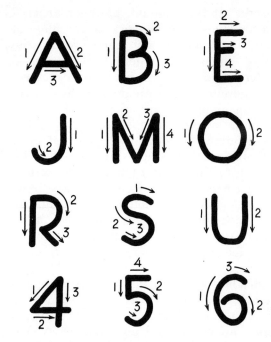

Figure 12.14   Examples of proper letter stroking in some of the letters and numbers of a simple Gothic alphabet.

that the positioning of the type has as much effect on the graphic quality of the map as do the selections of type styles, forms, and sizes. Incongruous, sloppy posi-

ABCDEFGHIJKL
MNOPQRSTUV
WXYZ
abcdefghijklmnop
qrstuvwxyz
1234567890

ABCDEFGHIJKLM
NOPQRSTUVW
XYZ
abcdefghijklmnopqr
stuvwxyz
1234567890

Figure 12.15   Two simple freehand alphabets. The left-hand is classed as upright Roman; the right-hand as slanted or inclined Gothic.

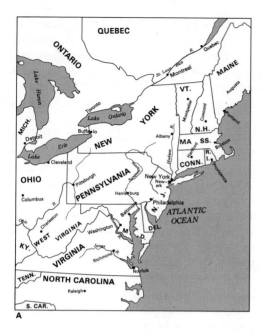

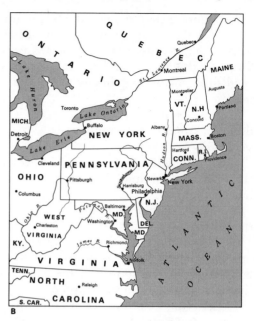

Figure 12.16   The general rules of type positioning have all been violated in map *A*. In contrast, map *B* attempts to set a better example.

tioning of lettering is as apparent to the reader as garish colors or poor line contrast (see Fig. 12.16).

Large map-making establishments tend to develop policies regarding the positioning of type partly to obtain uniformity, partly because it is cheaper, and partly because we can assign such activity to a machine. The excessive, systematic, overall result—*all* names parallel, for example —tends to set an unfortunate standard that should not be blindly followed. The cartographer, as in all other matters involving the structuring of his graphic composition, must be guided by principles and precepts based on the functioning of his map as a medium of communication. Generally speaking, the object to which a name applies should be easily recognized, the type should conflict with the other map material as little as possible, and the overall appearance should not be stiff and mechanical.

As was previously observed, one of the important functions of type on a map is to serve as a locative device. The lettering can do this in three ways: (1) by referring to point locations, such as cities; (2) by indicating the orientation and length of linear phenomena, such as mountain ranges; and (3) by designating the form and extent of areas, such as regions or states. The first rule of positioning type is to position the lettering so that it enhances the locative function as much as possible.

It will be convenient in the listing of the major principles of type positioning to organize them as they refer to point, line, or area phenomena. There are a few general rules, however, that are independent of such a locative organization. These are as follows.

1. Names should be either entirely on the land or on the water.

2. Lettering should generally be oriented to match the orientation structure of the map. In large-scale maps this means parallel with the upper and lower edges; in small-scale maps, parallel with the parallels.

3. Type should not be curved (i.e., different from 2 above) unless it is necessary to do so.

4. Disoriented lettering should never be set in a straight line, but should always have a slight curve.

5. Names should be letter spaced as little as necessary.

6. Where the continuity of names and other map data, such as lines and tones, conflicts with the lettering, the data, not the names, should be interrupted.

7. Lettering should never be upside down in any respect.

It is not uncommon that conflicts in precepts will occur because of particular combinations of requirements. There is no general rule for deciding such issues; the cartographer must make a decision in light of all the special factors.

Generally, lettering that refers to point locations should be placed above or below the point in question, preferably above and to the right.* The names of places located on one side of a river or boundary should be placed on that same side. Places on the shoreline of oceans and other large bodies of water should generally have their names entirely on the water. Where alternative names are shown (Köln, Cologne) the one should be symmetrically and indisputably arranged with the other so that no confusion can occur.

---

*The reader is referred to Imhof, *op. cit.,* for over 100 illustrations of good and poor cartographic practice in lettering maps. Although the text is in German, the graphic language of the illustrations is universal. They can be easily used because they are labeled "good" and "poor" in English.

Lettering intended to identify linear features always should be placed alongside and "parallel" to the river, boundary, road, etc., to which it refers, never separated from it by another symbol. Where there is curvature, the lettering should correspond. Ideally, we would place the type along an uncrowded extent where the lettering could be read horizontally, but this is usually not possible. Complicated curvatures should be avoided, and generally the names should be letter spaced somewhat, but not much. Such designations along rivers need to be repeated occasionally. If we must position lettering nearly vertically along a linear feature, if possible it should read upward on the left side of the map and downward on the right. Where it can be done, it is good practice to curve lettering so that the upper portions of the lowercase letters are closer together because there are more clues to letter form in the upper part. Names along linear items are better placed above than below the feature because there are fewer descenders than ascenders in lowercase lettering.

Normally, the lettering to identify areal phenomena will be placed within the boundaries of the region. The name should be letter spaced to extend across the area, but it should not crowd against the boundaries. As was observed in the general rules, where any tilting is necessary there should be added a clearly noticeable bit of curvature so that the name will not look like a printed label simply cut out and pasted on the map. Curvatures should be very simple and constant.

Geographical Names

Difficult as the solutions may be to the various problems concerning the positioning of the lettering, often an even more difficult question is the proper or appropriate spelling of the names we wish to use.

For example, do we name an important river in Europe *Donau* (German and Austrian), *Duna* (Hungarian), *Dunav* Yugoslavian and Bulgarian), *Dunarea* (Rumanian), or do we spell it *Danube,* a form not used by any country through which it flows! Is it *Florence* or *Firenze, Rome* or *Roma, Wien* or *Vienna, Thessalonkiē, Thessoloniki, Salonika* or *Saloniki,* or any of a number of other variants? The problem is made even more difficult by the fact that names change because official languages are altered or because internal administrative changes occur. The problem of spelling is difficult indeed.

The difficulties are of sufficient moment that governments that produce many maps have established agencies whose sole job it is to formulate policy and to specify the spelling to be used for names on maps and in official documents. Examples are the British Permanent Committee on Geographical Names (PCGN) and the United States Board on Geographical Names (BGN) of the Department of the Interior. The majority of such governmental agencies concern themselves only with domestic problems, but the two named above include the spelling of all geographical names as part of their function.

One of the major tasks of such an agency (and of every cartographer) is the determination of how a name that exists in its original form in a non-Latin alphabet shall be rendered in the Latin alphabet. Various systems of transliteration from one alphabet to another have been devised by experts, and the agencies have published the approved systems. The Board on Geographic Names has published numerous bulletins of place-name decisions and guides recommending treatment and sources of information for many foreign areas. These are available upon application. It is well for the cartographer to acquire, or at least to have available, such bulletins because he is frequently required to make decisions on matters of transliteration. Even more frequently, he will find himself using map sources that contain other alphabets, characters, and ideographs.

The general rule is to use the conventional English form whenever such exists. Thus, *Finland* (instead of *Suomi*) and *Danube River* would be preferred. Names of places and features in countries using the Latin alphabet may, of course, be used in their local official form if that is desirable. The problem is very complex, and it is difficult to be consistent. It is easy for English readers to accept *Napoli* and *Roma* for *Naples* and *Rome,* but it is more difficult for them to accept *Dilli, Mumbai,* and *Kalikata* for *Delhi, Bombay,* and *Calcutta.*

The problem is much too complex to be treated in any detail in this book, but it is well for the student to be aware of it. Above all, he must not let himself fall into toponymic blunders by placing on maps such names as *Rio Grande River, Lake Windermere,* or *Sierra Nevada Mountains.*

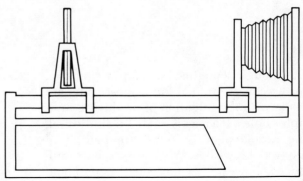

MAP 13 REPRODUCTION

Although sometimes a map is prepared for use in its original form, most maps are made with the intention of producing multiple copies. If a map is not to be duplicated, the cartographer is free to prepare it in any way he wishes. When it is to be reproduced, however, factors such as cost and requirements for the artwork place restrictions on the map maker. Too often, maps are prepared and then the search for an adequate duplicating process follows. In many cases it is discovered, too late, that there is no process that can economically reproduce the map.

The proper sequence in preparing a map is first to choose the process by which it will be duplicated, and then plan how the artwork can be prepared best to fit that method. In the selection of a process, we consider the kinds of copies we need, how many copies are required, the quality desired, and the size. To make intelligent choices and to be able to plan for the proper artwork, a knowledge of common duplicating processes is a necessity.

In this chapter an attempt will be made to acquaint the student with several processes to enable him to choose the best one for a particular situation. An understanding of the possibilities and limitations of the processes will be needed as a background for Chapter 14, which deals with the preparation of artwork.

Because there are a number of terms used in the reproduction trades, and in the cartographic construction process, that are likely to be unfamiliar to the student, the following glossary is provided.

Blacklight. Light rich in ultraviolet used for contact exposures for sensitized materials.

Continuous tone. Smooth and continuous transition of gray tones such as on an ordinary photograph.

Flap. Separate sheet on which a map or selected features of a map are drawn. (Other terms that are used to refer to these separate sheets are *overlay* and *separate.*) One map may require several flaps.

Opaque. Impervious to the rays of light. Opaque also refers to any of a variety of substances, whether white, black, or red, that prevent transmission of light.

Opaque, Actinically. Opaque to those wave lengths of light by which sensitized materials are affected.

Photoengraving. Photomechanical process for converting any object that can be photographed into a relief plate for letterpress printing.

Photolithography. Photomechanical process for converting an object that can be photographed into any of the various kinds of images on lithographic plates.

Pins, registry. Machined metal studs which are inserted into matched holes in flaps of a map to keep the flaps registered.

Proof. Any of various kinds of copies of a map that are used to check the accuracy or legibility of a map or to give an indication of the appearance of the final printed copies.

Register. When several pieces (flaps) of artwork are used to prepare a map, they must register or fit each other.

Reverse. Changing the relationship of the transparent areas and opaque areas on photographic film. If film is exposed to light passing through the transparent areas of a negative, for example, those areas will be opaque on the new film and the remainder of the new film will be transparent.

Scribecoat. An actinically opaque plastic coating (on a transparent base) that can be removed by scraping to produce open areas. The finished scribesheet can be used like a photographic negative in duplicating processes.

Screen, Contact. A variable-opacity screen that can be used in a camera or a vacuum frame to convert continuous tone to varying sized dots.

Screen, Halftone. A close network of perpendicular lines etched into glass and filled with an opaque pigment. When used at the photographic stage, it converts continuous tone to varying sized dots.

Screen, Tint. A film that is interposed between a negative and a sensitized surface to produce a pattern. Those with closely spaced small dots or narrow lines produce the illusion of solid gray.

Printing Process

Processes for reproducing text and illustrative material have been at hand since at least the fifth century when the Chinese carved block characters in wood, inked the raised portion, and transferred the impression to paper. It was not until about 1450, however, that movable type was invented and printing in its modern aspect was born. With movable type, letters of the alphabet were carved as single units and could then be assembled to produce text. By the process known as *letterpress,* a standard form of printing today, the paper receives the ink directly from surfaces standing in relief (Fig. 13.1). For drawings, the raised lines could be left standing by cutting in either metal or wood.

In the fifteenth century, another method of reproduction, *engraving,* also involving

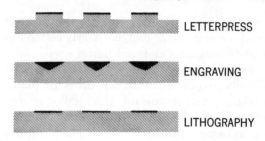

Figure 13.1   The basic processes of letterpress, engraving, and lithography operate in different ways to produce a surface from which ink may be transferred to paper.

ink and an uneven surface, was developed. Someone conceived the idea of cutting or engraving grooves in a flat metal plate and filling them with ink. The surface of the plate was then cleaned off and the plate with its ink-filled grooves was squeezed against a sheet of paper. The paper "took hold" of the ink and, when removed from the metal plate, the pattern of grooves appeared as ink lines. In a sense, this process of printing from an *intaglio* surface is just the opposite of letterpress printing, since the inking area is "down" instead of "up." Until well into the nineteenth century, most printed maps were reproduced by some form of the engraving process.

Through the intervening years, until the development of photography, the printing plates from which copies were to be made were prepared by hand. To produce the printing elements for a map, for example, an engraver had to cut away the relevant portions of wood or metal to produce the printing areas for the lines and lettering. Since the impression was to be transferred directly to paper, all work was done backward or wrong reading. Not only was this a laborious process, but the quality obviously depended greatly on the skill of the engraver and his interpretation of the original. Neither was it easy to produce the numerous fine lines necessary to give the illusion of changing values of tones; consequently, a very large proportion of printed matter consisted of text, with only relatively little illustrative material being used.

Progress in printing is probably most closely associated with advances in plate-making. An important development occurred in 1798 with the invention of lithography, a new printing process based on the incompatibility of grease and water. A drawing was made wrong reading directly on the smooth surface of a particular kind of limestone with greasy ink or crayon. The fats combined chemically with the elements in the stone to form a calcium oleate which had the property of repelling water. The unmarked portion of the surface could then be dampened, and when a greasy printing ink, which was repelled by the water, was rolled across the surface, the ink would adhere to the oily marked areas but not to the clean dampened ones. Paper pressed against the stone would pick up the ink. Because of the use of stone in the original form, this process was named *lithography*. Today, stone is not used, except in rare instances. Although thin metal plates have been substituted, it is still known as lithography. In contrast with printing from relief surfaces or grooves, it is a planographic process, that is, the surface of the lithographic printing plate is a plane, having no significant difference in elevation between the inked and noninked areas.

All three methods of printing involved preparation of the plates by hand. There was no mechanical way to accomplish this; and, as a consequence, arts such as engraving with a burin or graver directly in the metal, or drawing directly on the lithographic stone, were highly developed, and few possessed the necessary skills. The development of the specialized skills of the engraver or the lithographic artist had several effects on the development of cartography, the most important of which was that much of the detail of letter forms and other elements of the map design were delegated to the technician rather than being the responsibility of the cartographer.

Probably the most revolutionary development for the printing industry came during the nineteenth century when photography was incorporated into platemaking procedures. A process called photoengraving, by which printing plates could be engraved photomechanically, was developed almost concurrently with the photographic process.* Not only was the cost of preparing plates reduced, but the cartographer could now expect accurate images of his maps.

Still, however, it was possible only to print linework, that is, solid strokes of the pen. Tonal variation could only be accomplished by drawing closely spaced lines or by crosshatching of continuously changing tones. *Continuous tone*, such as the changing tones of an ordinary photograph produced with a hand camera, could not be printed until the development of the *halftone screen* during the 1880's. To produce a halftone screen, fine, closely spaced, parallel grooves are cut into two pieces of glass and filled with an opaque pigment. The two pieces are then cemented together with the lines at right angles to form a network of opaque lines and transparent squares. When the screen is placed between the film and the lens of the camera, light reflected from different tones on the artwork passes through the openings in the screen and produces dots on the emulsion of the film that vary in size with the amount of light transmitted. Because all the dots are small and closely spaced, they blend when viewed from normal reading distance and create the illusion of continuously changing tone (Fig. 13.2).

---

*Photolithography, the counterpart of photoengraving, is used to produce the lithographic plate.

Figure 13.2 How a halftone looks under magnification. The right-hand illustration shows clearly that the number of dots per unit area remains uniform; only their sizes vary. Compare with any photograph in this book (for example, Fig. 13.4) under a strong magnifying glass. (Courtesy of the Chicago Lithographic Institute, Inc.)

Classification of Duplicating Methods

Today there are a number of methods for the reproduction of maps. It is important for the cartographer to become acquainted with some of the methods and understand at least enough about them to make intelligent choices and to prepare his copy accordingly.

It is difficult to classify reproduction methods in a satisfactory manner because many processes require more than one technique, and the intermediate techniques in one may be an end in themselves in another, or they may be intermediate in several different processes. For example, photography is a step in the printing process but it can also be considered a separate reproduction process. Perhaps the most practical manner of classification is to group the reproduction methods on the basis of whether or not they involve a decreasing unit cost with increasing numbers of copies. It so happens that segregation on this basis also separates the common

processes according to whether or not they require printing plates and printing ink, that is, whether they are printing or nonprinting processes.

In the following descriptions only the widely used and generally available processes are considered. It should, however, be pointed out that there are a number of other processes that produce excellent results in specific "requirement situations," ranging from stencil reproduction and silk screen to gravure and collotype. These and others of the same category (not widely used in cartography) are not considered here, but the interested student can find abundant information about them in the graphic arts literature.

The nonprinting processes are distinguished by providing low-cost copies at a unit cost that ordinarily does not vary much with the "run," or number of copies produced. They are, therefore, ideal for short runs but are not appropriate for large runs. The printing processes, on the other hand, provide low-cost, single or multicolor copies at a unit cost that decreases with the number of copies run. The initial cost is high compared to the nonprinting methods. The printing processes are therefore not appropriate for very small runs.

To decide whether a printing or a nonprinting method will produce copies most economically, we must consider the size of the map and the number of copies required. As the size decreases and the copies required increase, the printing process tends to become the most economical. Nonprinting methods become more economical as the number of copies desired decreases and the size increases.

The widely used nonprinting processes are: (1) direct contact positive or diazo, for example, *Ozalid;* (2) direct contact negative, for example, blueprint; (3) photo copy, for example, *Photostat;* (4) film photograph; (5) xerography; and (6) various proofing materials. This is an unsystematic listing in several respects as there are many similarities and differences among the processes that would enable them to be grouped differently. On the other hand, these are the categories usually used in "the trade," so it is convenient not to deviate. Today the widely used printing processes are (1) letterpress and (2) lithography.

A source of considerable confusion for the beginner is the terminology employed in the printing and duplicating businesses. Over the years, cartographers have adopted some of this terminology to describe materials that they furnish for duplication. Because of the kinds of materials currently being prepared by cartographers, many of the terms that have been used in the past no longer seem appropriate. For example, the word copy can mean either text material (words) or artwork (drawings) or both. It also can be interpreted to mean one sheet of material, and also it seems to suggest that the material is in positive form. Referring to a duplicate of the copy as a copy or to duplicates as copies only adds to the confusion.

We will refer here to all map materials furnished for duplication as *artwork*, *art*, *original*, or *original drawing*. Negative materials will be referred to by the terms *negatives*, *negative scribecoat*, or *scribecoat*, and the term *negative* will be used to describe the film or other materials on which the light image is opposite from that desired as an end product, and which can be used in a duplicating process requiring transmitted light to make copies. A *positive* is a print made from a negative and a *film positive* is one made on transparent or translucent film.

NONPRINTING PROCESSES

The nonprinting processes chosen for discussion are those that are most likely to be available, and some others that the cartographer himself can utilize without specialized darkroom equipment. Ordinarily these processes are not used to produce great numbers of copies, and often only one color is possible. With some materials, however, color can be added to a copy for display, for teaching, or to facilitate research. Maps for which there are only sporadic requests or maps that must be revised often are usually most economically reproduced by these nonprinting methods.

In recent years, a number of new sensitized materials have been developed that can be used under ordinary lighting conditions. Some of these systems are invaluable for the cartographic process. Intermediate duplicating steps in the production of a map and the proofing process can be accomplished with a minimum of equipment. It is not uncommon for the cartographer to furnish the printer with materials that have been proofed and are ready for platemaking.

Direct Contact Positives

Direct contact positives are made by exposing a drawing on a translucent medium in contact with a printing paper sensitized with light-sensitive diazo compounds. The exposed paper is then ordinarily developed with ammonia fumes, but it may be treated with an aqueous solution. The resulting print is a positive reading print the same size as the original drawing. The trade name *Ozalid* applies to this type of process.

It is not possible when making direct contact diazo prints to enlarge or reduce. Consequently, the artwork must be designed for "same-size" reproduction. The process depends upon the translucence of the drawing surface; and, therefore, painting out imperfections is impossible, since paint is as opaque as the ink and would appear as a dark spot in the print. Creases on tracing paper or plastic and heavy erasures that affect the translucence are also frequently visible on the print. Prints are commonly obtained by feeding the original drawing into a machine against a roll of the diazo-sensitized paper; the exposure and dry developing take place rapidly within the machine. The drawing is returned and may be immediately reinserted for another exposure and copy. The image is not absolutely permanent, especially when exposed to sunlight, and may not be acceptable in those circumstances where permanence is required.

Ordinary diazo paper has a rather poor drafting surface, so this process is not too useful for obtaining copies of base maps on which other information is to be added. There are available, however, a great variety of sensitized diazo materials such as tracing cloth, cloth-backed paper, and various types of drafting films which do have adequate drafting surfaces. Cloth-backed paper, for example, is useful for construction of wall maps. We can add with ink to the black image of the base map, and color can be applied by the use of ordinary artists' oils.*

Zip-A-Tone, ordinary stick-up, and similar products on drawings do not produce very good results in the direct contact process unless they are placed on the back of the drawing to prevent "shadowing." An additional hazard with the standard forms of these materials is present in the *Ozalid* diazo process because the drawing must be fed around a roller, and such materials tend to curl off the paper if it is heated and

---

*For details see Randall D. Sale, "A Technique for Producing Colored Wall Maps," *The Professional Geographer*, **XIII** (**2**) (March 1961), pp. 19–21.

rolled. New types of preprinted sheets and stick-up lettering are available to obviate this difficulty. Another way of circumventing this problem is to have a film or paper "intermediate" made by a photographer. If the photographer prepares the intermediate on translucent paper or prints his ordinary film negative on another piece of film, the result is a direct reading film positive. It is called an intermediate because it is to be used to make the subsequent contact positives in place of the original drawing. If the photographer can arrange the printing of the film positive so that the emulsion side is on the bottom when the film positive "reads" correctly, then it may be fed onto the drum of an *Ozalid* machine with good results. It is not absolutely necessary to have the emulsion arranged as suggested, but the result will be more satisfactory.

Another use of intermediates is occasionally very helpful. It the cartographer or researcher has available a base map that he wishes to use as a compilation base for a series of maps but on which he does not wish to draw, he can proceed as follows. First, an intermediate may be prepared from the original; then he may add data to the intermediate; and finally, the corrected or modified intermediate can be used to produce subsequent direct contact positives.

Direct Contact Negatives

The direct contact negative process is here used as collective term to include several processes that produce results by similar techniques. A commonly employed variety is the standard blue-print process. In this process the artwork is laid next to iron-sensitized paper and is then exposed to special lights. The exposed print is wet developed and is, consequently, subject to some distortion. The print appears as a right-reading negative, that is, the darks and lights are the reverse of the original. If a positive copy is desired, a special "negative" can be made from the original, which is then used to produce prints that have dark lines on a white background. Another variety (B-W) using special papers produces positive copies directly. Whatever the variety, the prints will be the same size as the original.

The major use of these direct contact positive or negative processes in cartography is to obtain a few relatively inexpensive copies or copies on special materials for particular requirements. They are also useful for obtaining "same-size" copies of base maps, so that a variety of other maps requiring the same base data may be made, without the necessity of redrawing the base data for each map. Papers used for contact positives and negatives have a relatively poor drafting surface: and the combination of drawing ink lines and the blue, "black," red, or brown lines of the print are not particularly satisfactory for subsequent printing reproduction, although the results are not unusable.

Photocopy Process

This process provides prints in negative form on sensitized paper, without the necessity of any intermediate film step. It involves the exposure of the original drawing through a lens directly to the sensitized paper, which is then wet developed. The developing process and the subsequent drying frequently cause unequal shrinkage, so that some distortion of dimensions and directions is often present in a photocopy. *Photostat* is a term commonly used synonymously with photocopy, but it is a trade name referring to a specific machine and the paper used in it for photocopying.

The prints are reversed each time through the photocopy process. If a drawing of black lines on a white background is photocopied, the result will be a paper negative, that is, a reproduction with white lines on a black background. To gain positive copies (same as original) artwork, it is necessary to repeat the process using the negative as the artwork.

Maps may be enlarged or reduced in the photocopy process, and the only limitation is the size of the paper and the quality of the lens. Photocopy paper is usually limited to 18 × 24 in. sheets, but the edges are commonly rather badly distorted so that the effective size is somewhat smaller.

Film Photograph Process

Although photographic copies are usually more expensive than photocopies, for a variety of reasons they are often more desirable. Copies can be had in the form of paper prints, film prints, or film negatives. Prints can be made on any of a variety of drafting films that provide good drafting surfaces as well as being dimensionally stable.* Photography for cartographic purposes is usually accomplished with a camera that can make precise enlargements and reductions and can utilize large sheets of film. The camera consists of a movable copy frame that holds the artwork, a means for lighting the artwork, and a mount that permits positioning the lens at various distances from the film (Fig. 13.3). A vacuum is usually incorporated to hold the film flat, and either a vacuum or a glass cover is used to hold the artwork on the plane of the copy board. Since both the copy frame and the lens are movable, the distance between the lens and the film and the lens and the copy board can be varied to produce the desired reduction or enlargement.

*Cronaflex UC-4 Drafting Film, E.I. Dupont De Nemours and Company, Photo Products Department, Wilmington, Delaware; Photact Contact Polyester Film, Keuffel and Esser Co.; Kodagraph Ortho Matte Film, Eastman Kodak Company.

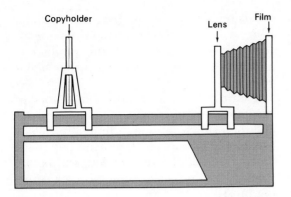

Figure 13.3 The basic components of a copy camera are a holder, which keeps the film flat, and a lens and copyholder, which can be moved perpendicular to the plane of the film. The positions of the lens and copyholder control the amount of enlargement or reduction.

The type of film used in the camera depends on the kind of original to be reproduced. If the artwork consists only of solid red or black, orthochromatic film (which is sensitive to light blue, green, and white) is used. The red and black portions do not affect the film, and those areas develop out, leaving them transparent on the negative. Corrections in the form of deletions are easily made by painting over transparent areas with a water soluble opaque. Additions are more difficult, since the rather brittle emulsion must be removed by cutting or scraping it away from the film base.

When the artwork has continuous tone, panchromatic film, sensitive to the visual spectrum, will record the changing values. A dark area on the artwork, which reflects little light, will appear more transparent than a lighter area that reflects more light. Making a print from the negative reverses the relationship and produces a replica of the original.

The continuous tone negative produced with panchromatic film is often not as useful in the cartographic process as is the halftone negative produced on orthochromatic film. Many of the sensitized materials used in the intermediate steps of map production and in proofing systems will not record tone changes unless they have been changed to varying sized dots.

The photographic process can produce quality copies at desired sizes on dimensionally stable materials; this makes it a useful process for compilation stages of map construction. Existing maps can be enlarged or reduced, and either negatives or positives can be used for tracing detail to a new map.

Xerography

Xerography occupies a unique position among photographic processes in that it is the only one that is completely dry, using no chemical solutions or chemical reactions. Producing the image depends on photoconductivity and surface electrification. A selenium coated aluminum drum is given a positive charge, which, upon exposure, disappears in those areas subjected to light. The remaining latent image is developed with a negatively charged powder that adheres to the image. Paper, which has been given a positive charge, is placed against the drum, receives the image by transfer of the powder and is fixed onto the paper with heat.

Because the *Xerox* process does not involve chemicals, time of immersion in solutions and subsequent drying are saved, and copies can be provided in just a few seconds. The image can be placed on a variety of kinds of paper or other surface materials since the transfer is physical.

Xerography provides excellent copies at a reasonable cost. With the use of a lens arrangement available reductions or enlargements can be accomplished; however, most of the *Xerox* copying machines that are available produce copies the same size as the original. Solid black areas or wide black lines do not reproduce faithfully since they tend to fade toward the centers.

In map construction, information from maps in books is sometimes difficult to trace because of difficulty in handling. Xerography provides a handy solution by quickly providing copies (including transparencies) that can be used flat on a drawing or light table.

Proofing Materials

Recently, a number of materials have been developed that are very useful in the cartographic process, especially for proofing. Most of the materials can be handled and processed under normal daylight conditions or under slightly subdued light. A cartographer who does not have darkroom facilities can make full color proofs of his maps, and he can even produce contact negatives and positives. Following will be a brief description of some available materials; however, specific uses will be discussed in Chapter 14.

3M Color-Key Proofing Film* is very thin clear acetate that has been coated with a colored emulsion that has been sensitized. A variety of colors are available, including the regular process colors: yellow, cyan, and magenta. The material can be used under daylight conditions, and the only equipment needed is a pressure or vacuum frame of some sort and a source of *blacklight,* rich in ultraviolet.

A negative of the map or a flap of the map is placed in contact with the proofing film (with emulsion of the proofing film away from the source of light), and an exposure of a minute or so, depending on the intensity of the light, is made. A special developer is then rubbed gently over the

*This film is available from distributors of the *Minnesota Mining and Manufacturing Company.*

emulsion until the image is developed. Rinsing under a water tap and drying completes the operation, and the product is a positive print in the particular color chosen.

A series of positives in different colors, of separate flaps of a map, can be prepared and then superimposed to form a full color proof of the map. Tint screens and halftones are faithfully reproduced, and colors combine in the same manner as on a printing press.

3M proofing sheets are also available for exposure with film positives. A special developer is used and a bleaching solution is applied after the film has been developed.

General Color-Guide,* like the *3M* proofing system, is a negative acting, ultraviolet sensitive film. It differs, however, in some important respects that are of special significance to cartographers. The film carrying the colored emulsion can easily be punched, whereas prepunched strips of some other material must be taped to the *3M* product for pin registration. *General Color-Guide* is developed by immersing it in a special salt bath and rinsing with warm water rather than by rubbing away the unexposed emulsion by hand. There is no shelf-life limitation, and if sheets are accidently exposed they may be returned to the box and reused in seventy-two hours.

Colored bichromate emulsions are available under several trade names,* and can be applied to special plastics by whirling or

*A product of *General Photo Products Co. Inc.,* 10 Paterson Avenue, Newton, New Jersey 07860. The material is available in various sheet sizes and in 42 inch rolls either 50 or 100 feet in length.

Brite-Line, Camden Products Co., Box 2233 Gardner Station, St. Louis, Missouri; *Kwik-Proof* and *Watercote, Direct Reproduction Corporation,* 835 Union St., Brooklyn, N.Y. 11215; *Proof-Kote, Litho Chemical and Supply Co.,* 46 Harriet Place, Lynbrook, N.Y. 11563.

Figure 13.4 Kwik-Proof emulsion is applied to the plastic sheet by using rather fast gentle strokes with the applicator. (Courtesy of Direct Reproduction Corporation.)

by hand application (Fig. 13.4). Again, several colors are available; however, any desired hue can easily be obtained by mixing proper amounts of the primary colors.

Smoother application of color is possible using a whirler, but with practice we can apply the sensitizer evenly over fairly large areas by hand. A small amount of the color is poured onto a plastic sheet that has been taped to a flat smooth surface. A small block applicator covered with a soft pad is then used with soft even strokes to spread the color evenly. Exposure is made, again with light in the ultraviolet end of the spectrum. Developing consists merely of rinsing under a water tap, and then a thorough drying. Once completely dry, another color can be added over the first image and the process repeated. As many colors as are desired can be applied and the result is comparable to a map which has been printed in several colors by separate press runs.

Kodagraph Wash-Off Contact Film produced by Kodak° provides the cartographer who lacks darkroom facilities with a capability for contacting film positives or negatives. An added advantage is that the material has a good drafting surface enabling additions and corrections.

Again, only a blacklight source and pressure or vacuum frame plus a flat tray for the developer is necessary. After a very short exposure, the film is developed for one minute and then rinsed under warm water until the unwanted emulsion has been washed away.

The film also has the ability to accept bichromated emulsions, so colors can easily be added to a film positive of a base map.

°Eastman Kodak Company, Rochester, New York 14650.

PRINTING PROCESSES

Most maps are printed by either letterpress or lithography. Books which have few illustrations have generally used the letterpress process, although there are recent trends toward greater use of lithography in book making. Large maps or books which contain many illustrations are usually more economically produced by the lithographic process. Under ideal conditions each method can produce excellent results, however, under conditions less than ideal, the quality can vary to a degree which is of concern to the cartographer.

Steps in the Printing Process

The basic operational steps in the printing process, from the time the printer receives the artwork until he delivers the printed maps, are much the same whether the process is letterpress or lithography, but the procedures and potentialities in each phase are different for the two kinds of printing. For a normal piece of art involving no complications the process consists, in the main, of the following operations:

(1). Photographing the original drawing.
(2). Processing the negative.
(3). Making the plate.
(4). Presswork.

The cartographer is concerned with one or all of the first three Beyond understanding the problems of the pressman, he is little concerned with the fourth stage.

Photographing the original drawing or copy is an exacting process requiring the use of what is called a copy camera, a large, rigidly mounted camera capable of making large or small exposures. The artwork is first placed in a vacuum frame with a glass cover, which holds it perfectly flat in front of the lens, and is then exposed for several seconds under illumination by arc lights. Relatively slow film is used in order to give the photographer greater control over the quality of the negative. For the printing processes (lithography and letterpress), the resulting negative must be composed of either opaque areas or transparent areas and nothing intermediate. Grays are not permitted on the negative as they are of course, in ordinary photography. The printing plate, to be made subsequently from the negative, must be entirely divided into two kinds of surfaces, one that takes ink and one that does not.

Any reduction that is to be made is done at this stage. The photographer can adjust the camera precisely, either by calculating the ratio of reduction or by actually measuring the image in the camera before the exposure.

It is important that the nature of the emulsion normally used be fully understood. This emulsion is orthochromatic, that is, sensitive to all colors except red. Consequently, red and black on the art work will appear clear or open on the negative, but light blue (and some similar colors) will not appear at all in the negative. For this reason, the cartographer should, as far as possible, draw any guidelines with a light blue pencil. Filters may be used in special circumstances.

Another process, scribing, which will be discussed in Chapter 14, permits the cartographer to draw his map in negative form and eliminate the photographic step in the printing process. Lines and symbols are hand engraved in a rather soft coating on a sheet of clear plastic. Where the coating is removed, light can be transmitted; however, the remaining coating is opaque to the wave lengths of light by which light-sensitive emulsions are affected. The scribed negative can be used in the same manner as an ordinary film negative.

Processing the Negative

Processing the negative is one of the more important steps in the cartographic technique, since it is in this stage that some results can be obtained better than in the drafting stage. No matter what processing or modification is to take place, however, the negative must be brought to perfection by removing all "pinholes" and blemishes in the emulsion so that the image is left sharp and clear. This is done by placing the negative on a light-table and opaquing with paint or scraping clean those spots and areas requiring repair or change (see Fig. 13.5).

Depending upon the circumstances, it is sometimes possible at this stage to add or subtract names and lines. If an important piece of information has been omitted or a word misspelled, a new piece of emulsion

Figure 13.5 **Before being stripped into position for platemaking, pinholes or unwanted words or symbols are removed by applying a water soluble opaque. (Courtesy Rand McNally and Company.)**

or emulsion-like material may be "stripped" into place on the negative. If material has been omitted, it can sometimes be added by "engraving" it in the emulsion. The emulsion, however, is quite brittle and tends to chip off, often resulting in ragged lines.

Screening of the negative involves interposing thin transparent sheets containing patterns of lines or dots (called screens) between the negative and the sensitized material. This accomplishes the same end result on the printing plate as applying *Zip-A-Tone* or similar shading film to, or drafting such shading on, the original drawing (Fig. 13.6). Better results are obtained with extra-fine patterns of lines or dots by applying them at this stage, after photography, than by making them "stand up" through the photographic process. Furthermore, opaquing can be done first and screening second, which is desirable because the

negative is more difficult to opaque if the emulsion already contains a fine pattern of lines or dots in addition to other data.

After a negative has been processed, it is marked for position on the plate and in general made ready for plate making.

Line and Halftone

All printed material belongs to either of two classes, line and halftone. The distinguishing characteristic that places a drawing in one or the other class is whether or not it contains any color shading or gray tones. If it does, it is halftone and must be dealt with in the reproduction process by a different (and more costly) procedure. The significance of this division results from the fact that the ordinary printing processes depend upon printing from a surface that is either inked or is not inked. There is no such thing in lithography and letterpress as "halfway inking."

It is possible, however, to make a printed area appear as if it were continuous tone, that is, shade from light to dark as if the surface had more or less ink on it. The process, halftoning, is accomplished by transforming the tone area into a large number of small dots of uniform spacing (tex-

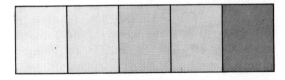

Figure 13.6 **Preprinted dot patterns can be applied to the artwork to give an illusion of gray; however, the effect can be destroyed if too much reduction is necessary. When tint screens are introduced at the platemaking stage, the amount of reduction need not be a consideration. The pre-printed patterns shown above were labelled as 10%, 20%, 30%, 40%, and 50%. It is evident that variations occur using these types of materials.**

ture) but of different sizes, depending upon whether it is to be dark or light. The dots print ink, and the spaces between do not. The dots usually are so close together that the eye is unable to distinguish them easily, so that the combination of inked spots and white spaces blends and appears as a tone.

The tone area is broken up by inserting a special screen directly in front of the emulsion. The light that passes through the screen is rendered as dots on the emulsion. All other things being equal, the closer the lines the smaller and closer together the dots will be. The closer together they are, the more difficult it is for the eye to see them individually and the smoother and more natural the result will appear. Screens range from 40 to 50 to more than 200 lines per inch. A screen of 120 lines to the inch provides more than 14,000 dots per square inch!

In recent years the contact screen, which is cheaper and easier to use, has tended to replace the halftone screen. It produces the varying sized dots through modulation of light by the optical action of a vignetted dot pattern of the screen acting on the film emulsion. The vignetted dots are produced by a dye that is thicker near the centers of the dots, and the size of the halftone dot depends on the amount of light that is able to pass through the dye.

The size of the printing dots relative to the white spaces between is dependent upon the darkness or lightness of the tones on the copy. It should be remembered, however, that unless special additional processing takes place, no part of a halftone will be without dots. All lines and lettering will therefore have fuzzy edges. Pure whites on the original drawing will, in ordinary halftoning, be printed with a covering of very small dots and therefore a light tone, whereas solid areas will reproduce with small white spaces rather than being completely solid. These effects can be removed by opaquing or scraping on the halftone negative, but this is difficult if the areas involved are complex.

Sometimes, for economic reasons, the cartographer is limited to line reproduction. However, in lieu of continuous tone, he may be able to achieve the desired effect by means of the following.

1. Shading either uniformly or for continuous-tone effect by hatching or stippling with pen and ink, "spatter painting" with an air brush, or shading on a coarse, rough surface.

2. Using preprinted symbols on the drawing.

3. Screening portions of the negative by covering them with shading film.

4. Using *Ross* or *Coquille* board for continuous-tone effect.

5. Using specially prepared drawing papers, such as *Craftint, Singletone,* or *Doubletone.*

Platemaking

Photolithography and photoengraving, although alike in principle, are each accomplished by ways whose differences are of significance for the cartographer. Both the letterpress and lithographic plates must of course be divided into image and non-image areas. On lithographic plates the image areas hold the greasy ink while the nonimage areas attract water and repel the ink. Modern lithography usually incorporates an offset arrangement on the printing press whereby the ink impression is transferred from the plate to a rubber cylinder and from the rubber to the paper. The image on the plate then must be right reading, whereas in plain lithography a wrong reading or mirror image is necessary to produce the proper impression on the paper. The *surface plate,* that is, one having the image on the same plane as the surface of the plate, is coated with substances such as bichromated albumen, casein, or soybean protein. When exposed to ultraviolet light, the emulsion hardens in the image areas and in the nonimage area is washed away. Another type of lithographic plate uses dissimilar metals to produce the image and nonimage areas. The printing surfaces, which are usually copper, are either etched slightly into the plate or are slightly raised above the surface. The advantage of these plates is that the images can produce longer press runs than the planographic image of the surface plate. One of these, the *deep etch*, is of special concern for the cartographer, since a positive rather than a negative is used to expose the plate.

The letterpress plate prints from a surface that stands in relief and is produced by etching away the metal from the nonimage areas. Since the impression will be transferred directly to the paper, the image on the plate must be in reverse or wrong reading.

An acquaintance with the process involved in producing plates will enable the cartographer to plan for reproduction more intelligently, and will serve to prevent a map from "turning out" unsatisfactorily in the reproduction process.

Surface Plates are exposed to an arc light with the negative placed so that the emulsion side is held tightly against the plate. Overexposing the emulsion of the plate does not affect its printing qualities: therefore, a single plate can be successively exposed to several negatives. The process, called double burning or double printing, permits two or more separate flaps to be combined on one plate. Line and halftone or line and screening can easily be combined. The platemaker, however, must be certain that for each exposure the negative is properly registered on the plate.

After proper exposures have been made, the plate is developed, placed on the press, and is ready to print. Most corrections on the plate are impractical if not impossible and are generally limited to deletions. If errors are found on the relatively inexpensive plate, it is usually discarded, the map artwork revised, a new negative made, and a new printing plate prepared.

Deep-Etch Plates are a form of lithographic printing in which the printing area is very slightly recessed into the plate. On an albumen plate, the image tends to wear away with long press runs. Dots in a halftone or in a screened area become ragged and smaller. To compensate for this and to provide a generally sharper image, the deep-etch plate was developed.

A film positive rather than a negative is used to expose the plate. Areas under the image are protected from the light, whereas the emulsion on the rest of the plate is exposed and hardens. When the plate is developed, the emulsion in the image area is washed off, leaving the bare metal which is subsequently etched to a depth of about .0002 to .0005 in.

Since positives rather than negatives are used to expose the plate, successive burnings are not possible. As an alternative, one or more positives can be sandwiched together and the exposure of all made simultaneously.

Letterpress Plates, after exposure to a negative, or to successive exposures to negatives, must be chemically etched to produce the raised or relief printing surface. Light transmitted through the transparent portion of the negative hardens the emulsion. This hardened emulsion protects the printing surfaces when the plate is exposed to an acid bath. The acid eats away the nonprinting areas, leaving a relief printing image on the surface of the plate.

The etching process is involved and requires several steps. Successive etchings (usually about four) are needed to reach the required depth for printing.* At this stage, a proof of a halftone from the plate would appear dull and lifeless. There would not be enough contrast because the dots in the highlight areas tend to be too large while those in the dark areas are likely to be too small. The engraver, therefore, must reduce chemically the size of the highlight dots and enlarge by polishing the dots in the darker areas. The engraver's work at this stage governs the quality of the halftone and, of course, adds to the cost of the plate.

In letterpress printing, elements of the composition (type and/or mounted engravings) are locked into a *form*, which is mounted in the press. Separate elements can quite easily be replaced, but errors on an engraving usually necessitate making a new plate. Because the letterpress plate is so expensive to produce, it is especially important for the cartographer to submit accurate artwork to the engraver.

Presses

No advantage would be gained by attempting to describe various types of printing presses. There is, however, a basic difference between letterpress and lithographic printing which is of concern to the cartographer. Most lithographic presses incorporate an offset arrangement whereby the impression from the plate is transferred to another cylinder with a rubber surface, called a blanket, and the image is then transferred from the blanket to the paper (Fig. 13.7). Actually, the offset arrangement is so standard that the terms offset and lithography have become almost synonymous. In letterpress printing, on the other

*With the use of zinc as the metal for the plate, required depth can be reached with one etching.

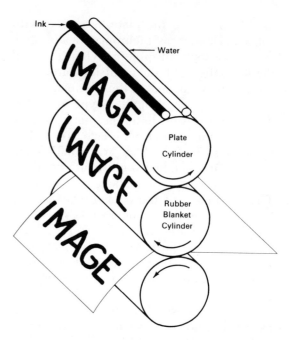

Figure 13.7   Since the image is passed from the plate to a rubber blanket and from the blanket to the paper, the plate must be right-reading.

hand, the paper receives the ink directly from the metal relief surface of the printing plate.

The soft rubber blanket that deposits the ink image on the paper in the offset process does not crush the paper fibers, and fine lines and dots can be faithfully reproduced even on soft absorbent paper. On the other hand, in the letterpress process, because pressure is necessary to cause the ink to adhere properly to the paper, the raised metal printing surface tends to squash the paper fiber. When soft unfinished paper is used in letterpress, fine lines and dots of halftones or screened areas tend to be enlarged and distorted. Very coarse halftone screens of 50 to 65 lines per inch must be used on newsprint paper, whereas 120 to 133 lines per inch can be used on finished paper.

If the cartographer can determine in advance the kind of printing and the grade of paper to be used, he can design his map for that particular combination. If he is preparing a map for a book that is to be produced by letterpress on unfinished stock, he should not choose fine patterns or expect halftone or fine screens to turn out well.

Color Reproduction

The reproduction of maps in color does not differ from black reproduction except that different colored inks are used. Each separate ink requires, of course, a separate printing plate, and thus a complete duplication of the steps in the whole process. Thus, generally speaking, the costs of color reproduction are many times that of the single color (usually black) reproduction. There are, however, two basically different color reproduction processes, and it is unwise to generalize further about relative costs. The two processes are called "flat color" and "process color." The major difference between them is that the color artwork for process color is prepared using color media in all hues on a single drawing, whereas that for flat color is prepared in black and white and usually requires at least one flap for each ink.

Flat color is the method most often used for maps and involves a straightforward procedure that varies little from the procedure described previously. For the flat color procedure the map is planned for a certain number of colored printing inks, and at least one separate drawing is prepared for each ink. Of course, many combinations of line and halftone effects are possible.

Process color, or, more properly, four-color process, is the name applied to an essentially different procedure. This method is based on the fact that almost all color combinations can be obtained by varying mixtures of magenta, yellow, cyan blue, and black. The black flap usually contains the border (if any), lettering, and graticule outlines. On a second flap all color work is done by painting, airbrush, etc. The color is photographed three times, each time through an appropriate color filter and a halftone screen, so that the three printing plates are halftones of the varying amounts of the primary colors. The black plate includes the line drawing and a halftone that is used as a toner to increase the shadow density and improve the overall contrast. When printed together again, the halftone dots and transparent inks merge and recreate the colors of the original drawing. The process is expensive because it is exacting. Much work by highly skilled persons is necessary on the halftone negatives (or on the plates in the case of letterpress), and the combination of halftone negatives, their modification, and careful processing throughout the printing process is time consuming.

Process color allows smooth gradations of hues and tints composed of mixtures of hues. The reproduction of color photographs and painted artwork in popular magazines is done by process color. Extremely careful manipulation of prepared halftone or contact screens can produce similar hue combinations by the flat color process, although the colors produced must necessarily be uniform over their map extent, that is, no gradations or continuous hue changes are ordinarily possible. Soil maps, altitude tint maps in atlases, and other maps containing many hue combinations are examples.

The preceding explanation of the principal color processes used in reproducing maps is greatly abbreviated and simplified, but it is included in order that the student may have an idea of the basic procedures. If the student is interested in color reproduction, he may investigate these interesting processes more carefully through reading (see bibliography) and especially by visits to printing plants where this type of work is done.

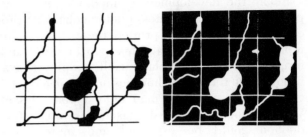

MAP **14** CONSTRUCTION

Before the development of methods for producing printing plates photomechanically, if a map was to be printed, the cartographer had either to make his map in the form of the printing plate itself, employ lithographic transfer techniques or rely on a craftsman to copy his map onto a plate. With photographic transfer, factual reproduction could be expected and more flexibility in the preparation of artwork was possible. Furthermore, photographic halftone reproduction allowed the printing of continuous tone variations so the map maker was no longer limited to using laboriously drawn substitutes.

The duplicating industries have made great strides during the past few years in perfecting new products and techniques. Some of these have solved problems cartographers have lived with for years. New sensitized materials are available for intermediate steps in the cartographic process that permit more efficient map construction and facilitate the preparation of trial runs or proofing. Compiling can be accomplished in a greater variety of ways, and the compilations can easily be transferred from one surface to another.

The cartographer is no longer forced to rely on the printer to process his artwork. Dimensionally stable materials are available that can be presented to the printer ready for platemaking. If desired, all the artwork necessary to produce multicolor maps can be produced by the cartographer without assistance from the duplicating trades.

To enable the cartographer to make use of new techniques and plan his work properly he must, of course, become acquainted with the new materials and processes. No one process is the best for all maps, and the production plan should be tailored for each situation. In a book of this size it is not possible to incorporate all the information available in the form of manuals, books, and catalogs. Also, because there are many combinations of materials and techniques that can be used, the treatment of the subject presents a problem in organization. The presentation is built on the background of the chapters on design, lettering, and reproduction, and with the exercise of imagination. The student should be able to devise workable plans for the construction of his maps that will result in well-designed maps that can be reproduced efficiently.

During the discussion of positive artwork, a variety of materials and equipment are mentioned along with information about possible uses and limitations. The list is not exhaustive but, on the other hand, those mentioned have been used and tested. The same holds true for the subsequent section, which deals with negative artwork. When reading both these parts, we should always keep in mind that there is a good deal of overlap, since negative and positive materials are often used in combination for one map. Finally, the section on complex artwork and proofing should provide the student with a basis for devising his own unique plans for solving complex construction and reproduction problems.

Map Artwork

Maps can be prepared in positive or negative form or in combinations of both. Preparing positive artwork involves placing symbols, lettering, and linework on a relatively white or translucent surface with ink or with a variety of preprinted materials. During the printing process, negatives of the artwork are prepared for use in making the printing plate, and the copies produced are then replicas of the original map.

Negative materials are prepared in an opposite manner, that is, producing the lines and symbols by cutting or scribing them from a coating that is actinically opaque.* The prepared negative can be used in the same manner as a film negative in duplicating processes.

Before beginning his map, the cartographer should decide which form of artwork would be most suitable, and select the one that will produce adequate copies at an economical cost. He must consider such things as the type of reproduction to be used, size of the copies, amount of detail, kinds of symbols, registration procedures, and the quality desired.

Whether artwork is prepared in negative or positive form or with a combination of the two, it often consists of more than one sheet of material. Each separate sheet is called a flap (Fig. 14.1). If a map is to be printed in one color, only one sheet of material may be necessary. More than one color or certain special effects such as areas of smooth gray tones usually require separate flaps that are combined at some stage in the reproduction process to produce the map on a single sheet.

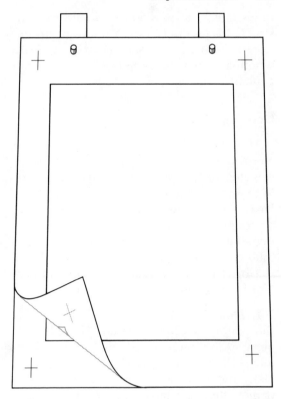

Figure 14.1 Each separate sheet of material used in the preparation of artwork for a map can be referred to as a flap. The 1/4 inch pins hold the flaps in register and the small crosses are used by the pressman when he positions the printing plates for successive colors.

*The negative need not be visually opaque, but opaque only to those wavelengths of light that affect light sensitive emulsions.

POSITIVE ARTWORK

Positive artwork destined for a printing process is usually prepared somewhat larger than the proposed reproduction size. Reduction sharpens the image somewhat and slight irregularities tend to disappear. At the larger size, symbols are easier to construct and more detail can be placed on the map. There is a temptation, however, to add so much detail that it is either lost or difficult to read at the reduced size. With a great amount of reduction, it is not easy to anticipate how certain symbols or patterns will appear. For these reasons, artwork is usually prepared no more than twice the reproduction size. If, in unusual situations, it is necessary to make the map several times larger, a sample area should be constructed according to the proposed design. A reduction of this sample will indicate whether or not the proposed design is acceptable.

Drawing Surfaces

The sheet materials to be used in the construction of the map may sometimes be selected on the basis of a particular attribute, but more often they are chosen on the basis of some overall qualities. Economy is sometimes a consideration, but the cost of materials is usually only a small part of the total cost of producing a map. The following brief listing includes the more important qualities for cartographic use.

1. Dimensional Stability. This refers to the ability of the material to withstand changes in temperature and humidity without shrinking or expanding. This is especially important when the artwork consists of more than one flap that must "fit" or register when reproduced. Because so much of modern map production requires several flaps to produce one map, this quality has assumed extreme importance in recent years, and a great amount of experimentation and research has been devoted to producing dimensionally stable drawing surfaces.

2. Ink Adherence. This refers to the ability of the surface to "hold on to" the ink. Some special inks have been developed for special surfaces; however, most of them tend to clog many kinds of pens. The material that will accept standard drafting inks is usually most desirable.

3. Translucence. This refers to the ease with which it is possible to see through the material, and the ability of the material to transmit light for the purpose of exposing sensitized materials. Translucence is of special concern in cartographic drafting, not only because a considerable amount of tracing is usually done, but also because much drafting for reproduction is done on separate flaps. Translucent material makes it possible to produce contact negatives and enables the cartographer to make proofs or produce other materials for the cartographic process without the use of a camera.

4. Erasing Quality. The ability to remove ink from a surface without damaging the drafting surface is necessary both for making corrections and for revising map artwork.

5. Strength. Some drawings must withstand repeated rolling and unrolling or even folding, and some receive wear from certain duplicating processes. For such drawings a strong material is required.

6. Reaction to Wetting. Many maps call for painting with various kinds of paints and inks. A material that curls excessively when wet is inappropriate for such a purpose.

There are many possible types of drawing surfaces, but only a few are useful for cartographic purposes. For maps that do not require great dimensional stability, a prepared tracing paper provides an excellent and economical drafting surface for ink.* Its translucence permits tracing even without the use of a light table, and good diazo or blue print copies can be made from it. It reacts to wetting, however, and it will curl if large areas are covered with ink or paint. Corrections can be made by gently scraping most of the ink away with a knife or razor blade and then rubbing gently with a medium hard eraser. If we are careful and do not excessively loosen the fibers

*Albanene Tracing Paper Number 107155 can be purchased from Kueffel and Esser Company.

of the paper, further drafting can be done in the corrected area.

Tracing paper is not particularly stable and tends to shrink or expand with changes in temperature and humidity. For small drawings, a foot or so square, it can be used successfully even if several flaps need to fit precisely. If, for some special reason, it is necessary to make separation drawings for large maps on tracing paper, we must take special precautions. Tracing paper does not change dimensions equally in both directions. The sheets, which should all be cut from the roll at one time, should have the grain of the paper running in the same direction. During the drafting, all the flaps should be kept at the same location. If some of the sheets are exposed to different temperature and humidity conditions, size changes will be unequal and registration will become a problem.

For preparing large maps where precise registry is required, a number of special plastic materials have been developed. Surfaces of these sheets have special coatings or have been roughened in some way to cause ink to adhere. Space does not permit a comparison of the various kinds; however, a material called *Cronaflex*, which has special and very desirable qualities, will be described.*

Cronaflex is available in rolls or in separate sheets and in varying thicknesses. Either side of the sheet provides an excellent drafting surface; however, lines produced with any given pen are just slightly wider than they are when drawn on prepared tracing paper. The surface can be cleaned of oil or grease with a soft cloth dampened with trichloroethylene. Cleaning can be accomplished before drafting begins or at any time during production

without affecting inked portions. Ink can be completely removed at any time with a cloth dampened with water without altering the drafting surface.

In addition to the ease of cleaning, correcting, and revising *Cronaflex,* its dimensional stability is a most important and useful quality. The material can be used in conjunction with very stable photographic films or scribe sheets and maintain precise registry. *Cronaflex* can also be provided with a photographic emulsion. Negatives of base maps can be photographically contacted and, when developed and dried, ink and preprinted materials can be added to complete the map.

For specialized jobs there are a variety of papers and other surfaces that are of use to the cartographer. *Ozalid* cloth-backed paper, for example, produces a good black image and has a surface that accepts ink well, and is therefore ideal for the preparation of wall maps. Another special purpose material useful to the cartographer is *Coquille* or *Ross* board. This material is prepared in a variety of rough surfaces so that when a carbon pencil or crayon is rubbed over it, the color remains only on the tops of the small bumps (Fig. 8.12). For this reason, varying shades of gray can be prepared with this material and can be reproduced without the necessity of being halftoned.

Drawing Pens

Although there are not as many kinds of pens as there are types of drawing surfaces, there are still a great number from which the cartographer must choose. Each pen has certain capabilities and limitations, and for an average job the cartographer may use two or three different kinds. Common to all pens, however, is the need to keep them clean for proper operation. Ragged lines, gray rather than black lines, and lines of inconsistent width can usually be accounted

---

°Cronaflex UC-4 Drafting Film is manufactured by E. I. Du Pont De Nemours and Company, Photo Products Department, Wilmington, Delaware.

for by dirty or clogged pens. We must clean our equipment even during use, since bits of lint or other debris are constantly being picked up from the drafting surface.

The standard ruling pen (Fig. 14.2) has probably been the most used of all drawing instruments in the past. Although new pens have been developed that are easier to use and produce more precise results, we still find occasions when we must resort to the ruling pen. The spacing of the blades is adjustable with the small screw on the side, so that lines of different thickness may be made with the same pen. The pen is filled by placing ink with a dropper be-

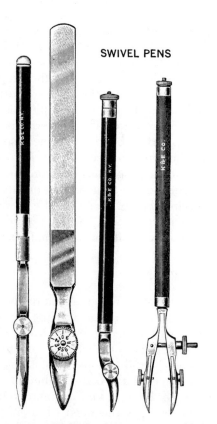

Figure 14.2 Kinds of ruling pens. The plain pen on the left is most frequently used. (Courtesy Keuffel and Esser Company.)

tween the adjustable blades. Ink should not be left long in the pen because it will dry a bit and cause an unequal flow. To clean the pen, a cloth can be inserted between the blades to wipe them dry.

It is often difficult for the beginner to produce consistent widths with the ruling pen. Careful adjustment of the blades is necessary, and a trained eye is required to judge minute differences in line weights.

The contour and railroad pens are modifications of the standard ruling pen (Fig. 14.2). In both cases, the blade assembly is on a swivel and it tends to "follow" the pen holder. The contour pen is useful for drawing curved lines freehand, and the railroad pen, with its two blades, is designed for parallel double lines. The latter can be used freehand or the blades can be locked to the pen holder for use with a straightedge.

A new type of pen, the *Pelican-Graphos,* which has come into wide use during the past twenty years, can in most cases be used in lieu of the standard ruling pen (Fig. 14.3). A set of nibs, which have been machined to produce lines of specific widths, can quickly be interchanged on a fountain-type pen. With these nibs we no longer need estimate line widths, but must learn only to handle the pen in a consistent manner. Most other nibs available for the *Pelican-Graphos* pen are designed for freehand work such as lettering or lines. Style "R" nibs are especially useful for producing round uniform dots.

Another pen that can produce excellent freehand lines was designed for *Leroy* lettering templates (Fig. 14.4). The *Leroy* pen has a cylinder for a point, and ink is fed through a small hole in the cylinder. When used freehand in its special holder, it is useful for those lines that should maintain a constant width no matter which direction they follow. Varying pressure makes little difference in the width of the line.

NIB: **KIND:** **WIDTHS SUPPLIED (in mm):**

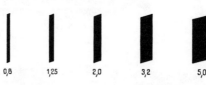

RULING NIBS
FOR
FINE LINES

0,1 0,12 0,16   0,2 0,25 0,3 0,4 0,5 0,6

RULING NIBS
for broad lines and for
writing posters

0,8 1,0 1,25 1,6 2,5 4,0 6,4 10,0

TUBULAR NIBS
FOR LETTERING
GUIDES

0,4 0,5 0,6 0,7 0,9 1,0 1,25 1,5 1,75 2,0 2,5 3,0

ROUND NIBS
for round
end lines

0,2 0,3 0,4 0,5 0,8 1,0 1,25 1,6 2,0 2,5 3,2 5,0

Right hand slant
nibs for
square end lines

0,8 1,25 2,0 2,5 3,2 4,0 5,0

Left hand slant
nibs for
square end lines

0,8 1,25 2,0 3,2 5,0

Drawing Nibs
FOR
free hand drawing

HB = medium hard

Figure 14.3 Illustration of the types of nibs and widths of lines that may be made by Pelican-Graphos nibs. All lines are made by individual nibs which can be inserted in a single fountain type pen. Special inserts are also available to provide the proper rate of flow from the reservoir depending upon the ink requirements of the nib. (Courtesy John Henschel and Company, New York.)

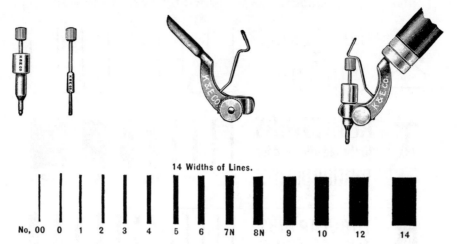

14 Widths of Lines.

No, 00 0 1 2 3 4 5 6 7N 8N 9 10 12 14

Figure 14.4   Leroy pens and penholder, and widths of lines made by various sizes of pens. (Courtesy Keuffel and Esser Company.)

The *Barch-Payzant* pen (Fig. 14.5) was also designed for lettering, but geographers find it a useful pen for making uniform lines or dots. It operates on a different principle from the *Leroy*, but the result is about the same. The flow of ink is adjustable with a *Barch-Payzant*, which makes it useful for drafting uniform dots.

Quill pens made of metal are needed to produce special kinds of lines and can be useful in drawing. A large variety is obtainable, and it is helpful to have a good selection on hand. Some are hard and stiff and make uniform lines; others are very flexible and are used for lines, such as rivers, that require a changing width on the drawing. A

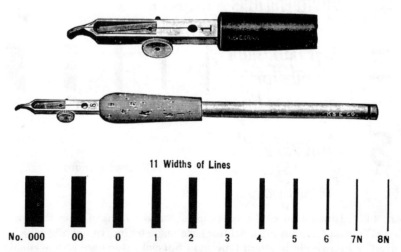

11 Widths of Lines

No. 000 00 0 1 2 3 4 5 6 7N 8N

Figure 14.5   Barch-Payzant pens and widths of lines made by various sizes of pens. (Courtesy Keuffel and Esser Company.)

favorite is the one called a "crow quill," a relatively stiff pen, which requires a special holder. Quill pens of any type may be dipped in the ink bottle, but a better practice is to use the ink dropper to apply a drop to the underside of the pen. This procedure helps to produce a finer line and allows frequent cleaning without excessive waste of ink.

Compasses, which are used for drawing arcs or circles, usually have blades comparable to the standard ruling pen; however, special attachments also permit the use of *Pelican-Graphos* pens. Usually compasses are designed so a pen or pencil can be interchanged (Fig. 14.6). For circles or arcs with long radii, a special beam compass, which can be extended by the insertion of extra linkage, is necessary. The drop compass, on the other hand, is used for making small circles. The pen is loose on the pointed shaft, and when the center has been located, the pen is dropped to the drafting surface and twirled. An alternative for constructing small circles is to use a *Leroy* pen inside the varying sized circles on any of several commercially available templates.

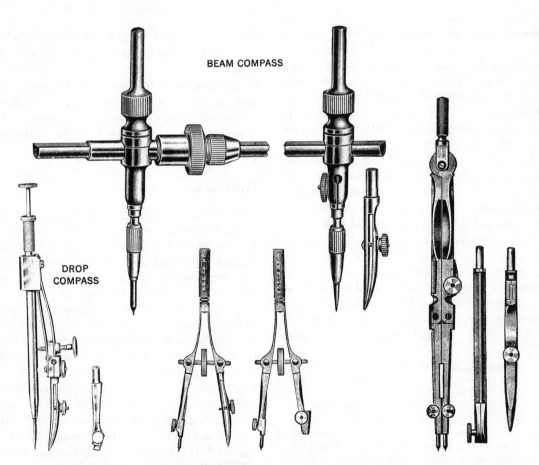

BEAM COMPASS

DROP COMPASS

Figure 14.6   Kinds of compasses. The beam of the beam compass may be several feet long. (Courtesy Keuffel and Esser Company.)

Positive Tints and Patterns

An indispensable part of many maps is the pattern of shading that must be applied to differentiate one area from another. This can be accomplished either by drafting them laboriously, as in the case of parallel lines or dotted "stippling," or by applying a commercially prepared pattern. Continuous tone shades of gray, as might be prepared by shading with a pencil or a brush, can only be employed when the halftone process is available. An appearance of shading can be created with *Coquille* or *Ross* board, but other linework is relatively difficult on these surfaces.

Preprinted materials, such as lines and dots printed on transparent film with an adhesive backing, are easy to use and save much of the time that was formerly necessary (Fig. 14.7). There is a considerable variety commercially available from most graphic arts supply houses. Catalogs may be obtained from the local outlet or from the manufacturer.* Several patterns are available in black, white, and red and, in some cases, in a variety of other colors. The colors (except red) are not frequently used in cartography. The patterns are available on thin transparent film coated with an adhesive, commonly wax, protected by a translucent backing sheet. Recently, some manufacturers have discontinued use of the wax backing, and have substituted heat-resisting adhesives. The material is placed over the area desired and is cut with a sharp needle or blade to fit. If cut without

*Para-Tone, Inc., 512 W. Burlington Avenue, La Grange, Illinois 60525, manufactures *Zip-A-Tone;* Craftint Mfg. Co., 1615 Collamer Avenue, Cleveland, Ohio 44110, manufactures *Craf-Tone;* Chart-Pak, Inc., Leeds, Massachusetts 01053, manufactures *Contak* as well as an adhesive tape useful as a substitute for drafted lines; *Artype,* Inc., 127 S. Northwest Highway, Barrington, Illinois 60010, manufactures dot and line screens with a regular progression of tonal values.

the backing sheet, the excess is stripped away and the pattern is burnished to the drawing. Extreme care must be exercised when cutting an adhesive pattern without the backing sheet over inked lines on a drawing. The stripping away of the excess will occasionally pull ink off the drawing.

For maps to be photographed for reproduction, patterns printed in translucent red are useful when the pattern to be used is dense; if it were black it would be impossible to see through it to know where to cut it. Red, like black, does not affect the ordinary film used in map reproduction. Large areas to appear ultimately as solid black may easily be constructed by using a solid red "pattern," especially on a paper that cannot be wetted. White patterns are useful for breaking up black areas or other patterns.

Knowledge of the relation of patterns to reduction and of the preparation of a graded series of values (darkness) requires considerable experimentation by the cartographer, as pointed out in Chapter 11.

Drafting the Map

Whether we make a map in positive or negative form, a worksheet of the compilation is prepared before any final drafting begins. The worksheet should be an accurate representation of everything to be shown on the map, and it is advantageous to prepare it at the scale planned for final drafting. Since it then can be merely traced, the cartographer can concentrate on drafting alone. When translucent material is used, tracing is easier and, if desired, the worksheet also can be used to make contact negatives or to contact images to other sensitized materials. For positive artwork the worksheet normally is done on one sheet of material. If categories of symbols are drawn with different colored pencils, this separation will assist at the drafting stage. One

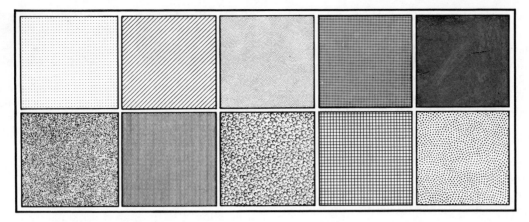

Figure 14.7   Examples of preprinted symbols, in this case Zip-A-Tone. Many of the maps and diagrams in this book have been prepared with the aid of these kinds of symbols.

size pen, then, can be used to trace the features represented by one color. For separation drawings, in which categories of symbols are drawn on separate flaps, the colors assist in the same way.

Registry. When the final art consists of more than one flap, some method for registering the flaps must be furnished the printer. A common method is to draft small crosses in each of the four margins outside the map border (Fig. 14.1). The register marks are retained on the negative and on the printing plate; as soon as the plate is properly positioned on the press, they are removed. They should be made with fine lines and placed far enough from the map proper to enable the printer to remove them at the printing stage without damaging the map. On the other hand, if they are placed far from the map, much more film is necessary to record their images.

Another, more precise method of registering the various flaps is to fasten them together with registry pins.° They consist of flat pieces of metal to which machined

quarter-inch studs have been attached. Studs of various lengths are available, some of which are short enough to allow their being used in vacuum frames. Precisely matched holes are punched into each proposed flap and into the worksheet before drafting begins. Special punches are manufactured for this purpose; however, they are quite expensive and we can easily make our own. With an ordinary two or three hole adjustable office punch, on which the punches have been *welded into a fixed position,* we can produce perfectly matching holes on a series of sheets. For greater distances between holes, we can fasten separate punches to lengths of metal or wood. Two pins, however, may not provide the rigidity needed for large sheets of material. An extra pin may be incorporated at the bottom of the sheet or a pin may be used at the midpoint on each of the four sides.

Another method of producing properly spaced holes is with the use of gummed tabs that have been prepunched.° Two tabs

°Available from the Chester F. Carlson Co., 2230 Edgewood Avenue, Minneapolis, Minnesota 55426.

°Various kinds of tabs and pins are available from Berkey Technical Corporation, 25-15 50th St., Woodside, N. Y. 11377.

are placed on the worksheet and on the pins placed in the holes. Each flap is then successively placed on the worksheet and another set of tabs is pressed on after carefully positioning the holes over the pins.

When all flaps are properly pinned before drafting begins, each sheet will fit precisely throughout the entire drafting operation, and interchanging flaps is accomplished quickly and accurately. Registry marks should be used in conjunction with the pinning system, since the printer will need their images on the negatives and plates.

Preparation of Drawings. Final preparation of the map should begin with the line work and other symbols that require the application of ink. When preprinted materials are burnished onto the map, the adhesive tends to be squeezed out along the edges and can become distributed over the drafting surface causing difficulty for further drafting. If ink must be added after this has happened, it is necessary to clean the area to be inked. Wax can be effectively removed by swabbing with trichloroethylene.

Artwork furnished the printer is converted to negative form, either by use of a camera or by direct contact in a vacuum frame. In either case good, sharp contrast between the symbols of the map and the drafting surface is important to produce quality negatives. The line artwork should be opaque black or red. Gray lines cause difficulty in determining exposure times, and the result is usually broken or weak linework on the printed copies. If the artwork is prepared specifically for direct contact negatives, white opaque cannot be used for deletions because it registers on the film the same as opaque black.

Figure 14.8   Lettering used in conjunction with a halftone should be placed on a separate flap and burned separately into the plate. When halftoned, the lettering (or any linework) tends to be fuzzy.

When the artwork for a map consists partly of line and partly of continuous tone, they should be prepared on separate flaps. In the halftone process even solid black areas are converted to dots and will, therefore, appear as dark grays. Lettering and other line symbols have a fuzzy rather than a sharp black appearance (Fig. 14.8). When separate flaps are prepared, the linework is photographed separately to produce a regular line negative and a halftone screen is used only to produce the negative of the continuous tone flap. The two negatives are then combined by exposing the one printing plate to each negative separately.

Lettering Flap. Making a separate flap or flaps for lettering is standard procedure when a map involves the use of color. On single color maps, however, the lettering is usually incorporated into a flap that contains other lines and symbols. In any case, there are usually situations when the lettering must fall over other map features. If colored areas or lines are not too dark, black lettering can usually be superimposed; however, dark or black lines should be interrupted to accomodate the names. A good practice is to leave spaces for the lettering when the linework is drafted.

NEGATIVE ARTWORK

Preparing map artwork in negative form is a relatively recent innovation in cartographic technique. The capability for the cartographer to produce lines and symbols in negative form came with the scribing process that has come into wide use since

1940. Since that time, many kinds of materials have appeared that can be used in conjunction with the scribecoat and permit the cartographer more easily to prepare his artwork in negative form. The artwork is normally made at reproduction size since the photographic step is not necessary, permitting the printer to go directly to the platemaking stage. Not only is the cost for reproduction less but, in general, cartographic quality is improved over that usually obtained by the use of positive network.

Scribe Sheets and Coatings

Scribing is a technique almost the opposite of drawing with pen and ink. In drawing, the desired lines and marks are applied by the draftsman; in scribing, the desired marks are obtained by the draftsman's removing material. He starts with a sheet of plastic film to which a translucent coating has been applied. Then, working over a light-table, the draftsman removes the coating by cutting and scraping to produce the lines and symbols. When he is finished, the sheet has the same general appearance as a negative made photographically. The scribe coating is compounded so that, among other properties, it is visually translucent but actinically opaque, that is, opaque to those wavelengths of light by which light-sensitive emulsions are especially affected.

During the development stages of scribing, glass was used as the base material for the plastic scribe coating. Glass, of course, is breakable, heavy, and difficult to store and has been replaced by various transparent plastics. The coating is available in several colors, most of which are translucent but actinically opaque. The translucent quality permits tracing on a light table, although this method is seldom used; instead, an image of the worksheet to guide the draftsman is normally made to appear in the emulsion on the scribing surface. The actual scribing operation is, however, still done on a light table, and the color of the coating seems to have a noticeable effect on the degree of eyestrain. Experiments have indicated that orange or yellow causes the least discomfort.

The scribe sheet coating, which is applied to the plastic base material in a whirler or by spraying, should be thin and plastic enough to permit easy removal. The cutting tool must remove the coating without cutting into the base sheet, which causes variations in line width. Any fragments of the coating left in an open area will, of course, block light and cause broken lines on the printing plate. If such fragments remain, they can be eliminated by pressing a piece of drafting tape onto the area. When the tape is lifted, the unwanted bits of material adhere to it and are removed from the base sheet.

Corrections or alterations on the scribecoat are made by painting over with a specially prepared opaque. New lines can then be cut; however, scribing in areas covered with the hand-applied opaque is somewhat less satisfactory than removing the original coating.

Scribing Tools

There are a number of kinds and designs of scribing tools, and scribing points are made from a variety of materials (Fig. 14.9). There are three basic kinds of instruments: the pen-type, the rigid graver, and the swivel graver. The pen-type is mainly for freehand work and is somewhat difficult to use in that it must be held at a correct and constant angle to produce consistent line widths. The rigid graver, on the other hand, with its tripod construction, assures a proper angle at all times. When points for different width lines are exchanged in the graver, adjustments must be made to produce again the correct angle. The most

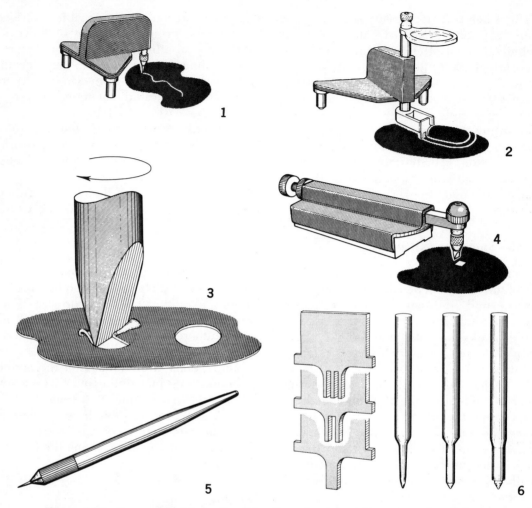

Figure 14.9 Some of the kinds of scribing instruments used to remove the scribe coating. At the top are shown a rigid graver (1) and a swivel graver (2), both of which are used for line work. The dot graver (3) and the "building" graver (4) are used to make small circular or square openings in the coating, such as for dots or buildings on a topographic map. The pentype graver (5) is used for freehand work, and at the right (6) are shown some of the types of graver points, such as the single or multiple chisel-edged points and needle points. (Courtesy the U.S. Geological Survey.)

convenient and efficient method is to provide each of the several gravers with different size points; then, to change line widths, a different graver with the desired point is selected. The swivel graver is used for points that produce two or more parallel lines or for chisel-type points used in making wide lines. The operating principle of the swivel graver is like that of the railroad pen, which is designed for producing parallel ink lines; however, the graver is much easier to use and produces much finer lines. Probably the best swivel graver is one that has enough weight to incorporate a spring mechanism to provide even pressure on the point. If the proper tension is set on the spring, the points will cut evenly and not scratch the surface of the base material.

Special kinds of gravers are designed for producing special symbols. A building graver, for example, can be adjusted so a chisel-type blade of a desired width is moved a specified distance. The adjustment, then, determines whether a square or a given size rectangle will be scribed. Another symbol that presents a special problem in scribing is the dot. The dot graver has a chisel point, which is spun into the coating and opens the dot. Dot gravers are available that are electrically operated.

The material from which the scribing points are made is of concern because it determines whether or not they will need sharpening. Because there is a certain amount of friction between the point and the base material, steel points tend to wear and, as a consequence, the line widths change. Jigs are commercially available with which they can be sharpened; however, this can be a tedious task. Jeweled points are advertised as never needing sharpening, and they do last for long periods of time. Although they are more expensive than steel points, time saved in sharpening may, in the long run, prove them to be more economical.

Scribing the Map

As in the case of positive artwork, a worksheet should be prepared from which the final artwork is prepared. In the scribing operation a worksheet is almost imperative, since scribing is most easily and accurately done if an image of the map is first placed on the scribecoat. Tracing an image through the scribecoat is difficult because of diffusion of the light.

The worksheet is usually produced in positive form on translucent material at the proposed scribing scale. This permits transfer of the image to the scribecoat base without the use of a camera or darkroom facilities. If for some reason it is necessary to prepare the worksheet at a larger scale, it can be reduced to the desired size by photography and then contacted to the scribe base.

Before beginning the worksheet, we should consider whether separation scribecoat negatives will be necessary. If linework is to appear on two or more scribecoats, it may be advisable to separate the information on separate flaps of the worksheet. For example, if drainage on the final map is to be printed in blue and the roads in red, each will be scribed on a separate sheet. If the information is separated on two flaps of a worksheet, a separate image can be placed on each scribe sheet or both images can be contacted to the scribesheets in different colors. Actually, contracting all the flaps in different colors to each scribecoat can reveal discrepancies in the worksheet flaps. A road, for example, may have been plotted in error so that it crosses a small lake. The multiple image would show this, and the proper position of the road would need to be established.

For very detailed and complicated maps, a modification of the multiple printing of the worksheet flaps might be used. Assume that the hyrographic features for the map were compiled from very accurate large-scale topographic sheets, roads were selected from a somewhat smaller-scale, more generalized map, and lines such as soil boundaries had been originally plotted on aerial photographs. Since the hydrographic features are likely to be in the most accurate planimetric position, that image is transferred to the scribecoat and the lines scribed. The soil boundaries are related to the hydrographic features and in some cases might end abruptly at lakes or streams. The second scribecoat will include the image of the worksheet of the soil boundaries and the scribed image of the hydrographic features. The soil boundaries are then scribed with the hydrographic detail as a guide. On the last scribecoat, the already scribed images of the hydrography and soil boundaries are combined with the worksheet flap of the roads. The delineation of the roads can then be adjusted to fit the previously scribed features.

Scribecoat is available in a sensitized form for diazo duplication. To produce an image we need only expose it to an arc light or blacklight while in contact with a positive of the worksheet and develop the image in ammonia fumes. If we prefer, we can purchase regular scribecoat and apply the diazo sensitizer ourselves.

Bichromate sensitizers are more versatile for contacting worksheet images, since successive applications of color enable the separation of categories of features. A disadvantage is that negatives rather than positives are needed. We can, however, produce our own negatives with either *Kodograph Contact Film* or *3m* orange proofing foils.

Letterpress—Lithography.

The image placed on the scribecoat negative must be "wrong reading" if the lithographic process is to be used. In the lithographic process the plate is prepared by placing the emulsion of the film or the coating of the scribesheet face down in contact with the plate. If the emulsion side is "up," light tends to creep because of the thickness of the film and the image on the plate will not be sharp.* A lithographic negative must,

therefore, have a wrong reading image when viewed from the emulsion side, so that when it is placed emulsion side against the plate, the resulting image on the plate will be right reading.

On the other hand, in photoengraving a letterpress plate, a very thin film is used so it can be "flopped" to produce a wrong reading plate. For letterpress printing, then, the image should be scribed right reading on the scribesheet so that the scribecoat emulsion can be in contact with the plate during exposure.

Lettering A separate positive flap is usually necessary when a map is scribed. If only a small number of names is required, it is possible to convert them to negative form, strip them to the scribecoat, and remove enough of the scribecoating to expose the name. This method is time consuming and is not often used.

To prepare the lettering flap the scribecoat is usually flipped over on a light table to produce a positive or right reading image. A sheet of transparent or translucent material is placed over the scribecoat, and the scribed symbols are used as a guide for placing the lettering. Upon completion, the lettering flap is placed beneath the scribecoat and the opaquing necessary to interrupt lines is done.

Very often, other symbols are placed on the lettering flap. Covering areas with prepared patterns can usually be done most conveniently on positive art, and some symbols that are difficult to scribe may be available in preprinted form. The lettering flap and the scribecoat are usually double burned at the platemaking stage; however, a negative of the lettering and the negative scribecoat can be combined in positive form on one piece of film. This film can then be treated as positive artwork for any of the duplicating processes.

---

*Using new improved films and a point source of light, the creeping of the light would be negligible; however, the best practice is to make exposures with the emulsion of the film in contact with the sensitized plate.

COMPLEX ARTWORK AND PROOFING

The cartographer with a good knowledge of duplicating and printing methods can take advantage of their capabilities to produce special effects for improving design and legibility. To use most efficiently the techniques of duplicating trades, the artwork must be carefully planned and the use of special materials is often necessary. The following section describes some special effects and provides some examples of their use. It is hoped that the student will be able to modify and expand these techniques and procedures to adapt them for solving his particular problems.

Open Window Negatives

For color reproduction or for laying tints in a given area, open window negatives are necessary. The negative must be actinically opaque except in that area in which the tint or color is desired. The open area, however, should fall exactly upon a bounding line. Preparing such a negative completely by hand can be a tedious job, especially if the area boundary is complicated. It can be done by cutting a thin film and peeling it away from a clear plastic base.*

Open window negatives can also be prepared photomechanically, and the only hand operation is peeling away the proper areas.†

The two kinds of material available are alike in appearance, since both have a thin red film adhered to a transparent sheet of plastic. Neither involves cutting since the lines bounding areas can be photochemically etched from the red film. By lifting the edge of an area with a knife, the whole area can be lifted away from the base sheet, leaving the transparent opening. Darkroom facilities are required for the use of *Peelcoat*, whereas *Striprite* may be processed under normal room lighting and can, therefore, be used in cartographic drafting rooms with limited equipment. Both materials produce good quality negatives that can be registered almost perfectly.

Since a positive is necessary to make *Striprite* negatives, we must convert the scribed negative of the lines to a film positive. A pin registry system, of course, is used throughout the process to insure registry of each open window negative with the lines on the scribecoat. The positive is placed in contact with the *Striprite* so that a negative image will result. Exposure is made with an arc lamp or mercury bulb, after which the sheet is developed and etched.

Tint Screens

Tonal values can also be added to maps at the platemaking stage by instructing the printer to insert a tint screen between the negative and the printing plate. These screens are composed of dots or closely spaced lines that block the light and produce the impression of gray on the printed copies. Since any reduction of the artwork has already been accomplished, there is no danger of the dots disappearing as there is when preprinted dots are placed on the original artwork. Either areas or symbols and lettering can be screened in this fashion, although the dots can give a fuzzy appearance if not bounded by a black line.

The value of the tone produced is designated in percentages of black; therefore, a

---

Rubilith and *Amberlith* are manufactured by Ulano Graphic Arts Supplies, Inc., 610 Dean St., Brooklyn, New York, 11238.

†*Striprite*, produced by Direct Reproduction Corporation, 811–13 Union Street, Brooklyn 15, N.Y. 11215, requires the use of a film positive for exposing. *Peelcoat Film*, which requires a negative for exposure, is available from the Keuffel and Esser Company.

10 percent screen will produce a very light tone and an 80 percent, a very dark tone. Note, however, that an 80 percent screen actually has very small opaque dots with most of the film transparent, whereas a 10 percent screen is mostly opaque with small transparent openings.

Screens produce dot patterns that are coarse or fine, depending on the number of dots per inch. A screen that has 65 dots per inch is very coarse and is called a 65 line screen. A 150 or 200 line screen, on the other hand, produces very fine dot patterns that can be detected only under magnification (Fig. 14.10). In letterpress printing, relatively coarse screens must be used except when printing on coated paper stock, whereas fine screens can be used in offset printing even on poor grades of paper.

An area on the map to be screened must be open on the negative. Positive artwork

Figure 14.10 Those with keen eyesight may be able to distinguish the dots on the 65 line tint screen (above) but are unable to do so on the 133 line screen (below).

for this open window negative is prepared by covering the area on the drawing with black ink or with a red material such as *Zip-a-Tone*. This, of course, is done on a separate flap and is carefully registered to the linework. When the negatives are arranged for platemaking, they are carefully placed, using the registry marks provided by the cartographer. If poor registry is discovered after the plate is made, adjustments can not be made and the plate must be discarded. This is especially significant in the case of expensive letterpress plates.

Up to three screens can be printed on one area, and in color printing a fourth can be added. Extreme care must be exercised when positioning the screens, however, to prevent the undesirable *moiré* effect (Fig. 14.11). Screens must be rotated so that their orientations are separated by angles of 30 degrees. It is customary to place one screen at 45 degrees, one at 75, and the third at 105. If a fourth screen must be used it should be assigned to the lightest color and rotated to an angle of 90 degrees.°

If tone alone is used to distinguish categories of information, that is, if there are no visible differences in texture, arrangement, or orientation, the tones on a map should generally be limited to three values. If more are incorporated many map readers have difficulty matching tones of the map with the legend.

Because screens can be superimposed, it may not be necessary to prepare a separate flap for each tone. If, for example, we wanted values of 20, 40, and 60 percent, only two flaps would be required. On flap number one, the areas of 20 and 60 percent would be opaqued and a 20 percent screen specified. A second flap would include the areas of 40 and 60 percent and would be

---

°Numerous graphic devices are used for positioning screens at the proper angle. Companies that manufacture tint screens usually have them available.

Figure 14.11 Tint screens should be aligned with a separation of 30 degrees. Other angles can produce various moiré patterns, such as the one illustrated.

printed in 40 percent tint. Since the two flaps overlap, the 60 percent area would receive 20 plus 40 percent of black. The cost would be reduced because of one less negative and one less plate exposure.

Most often when gray tints are abutted, a black line is needed between them because making them join exactly is not usually possible. If, however, the tones are a result of the superposition of screens, the black line may not be necessary (Fig. 14.12). When a screened area is adjacent to a solid area, the artwork should always carry the tint into the solid. If we attempt to make them just join, white spaces are almost sure to occur.

A very useful technique is to screen the areas of preprinted patterns that have been applied to the artwork, which often tends to look rather harsh when printed in solid black (Fig. 14.13). Also, lettering or symbols that might otherwise be lost can show clearly through the screened patterns. Again, a separate flap is prepared for screening. Selected patterns are substituted for the opaqued areas of regular screening. Unless very fine textured tint screens are to be used, rather coarse patterns should be chosen. It is also possible for a *moiré* to develop if fine dots or certain other preprinted patterns are used. As a precaution we should test the orientation of patterns and screens by superimposing

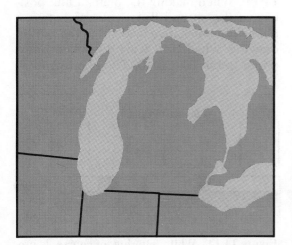
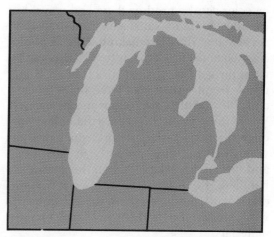

Figure 14.12 It is extremely difficult to make two tint screens match precisely (left). A tint superimposed on another (right), however, caused no difficulty.

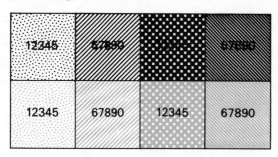

Figure 14.13 The patterns in the lower row have been screened to 30% of black. Not only does the pattern appear less harsh; lettering or symbols applied remain legible.

them on a light table. Lines, lettering, and symbols can also be screened, but very narrow lines and thin serifs on letters are likely to be lost. Lettering is probably most successfully screened with line screens, which do not produce the ragged edges caused by dot screens. Rather bold lines and sans-serif lettering, when screened, can add much to the utility and overall appearance of a map.

Reverse or Open Linework

White linework on a background of black or color can be produced several ways, but usually the most satisfactory method is to reverse it at the platemaking stage. A photocopy changes black to white, and this can be used directly on the artwork; but such materials frequently do not have sharp images, and the results are often unsatisfactory. White lines and lettering are difficult to produce directly on the artwork, since opaque white will not flow properly in most pens and it tends to flake off when the artwork is handled. Reversing the linework, on the other hand, produces sharp images from artwork prepared with the usual black opaque materials.

The linework that is to be open is produced on a separate flap and prepared in

the same way as any other flap. One of the other flaps will, of course, be an open window negative to produce either a tone or solid color for the background. Instructions to the printer merely indicate that a given flap is to be reversed from another. When the plate is exposed, a film positive of the flap to be reversed is placed, in registry, between the open window negative and the plate (Fig. 14.14). On the positive, the linework is opaque and blocks the light from those areas on the plate and produces a nonprinting area. The white paper then shows through the tint or solid area, and white linework is the result.

The positive used to reverse is sometimes referred to as a mask, and is indeed useful in many cases as a mask to block out unwanted portions of images. Suppose a shaded relief map is to be prepared for an area in which numerous lakes are to have a smooth gray tone. Shading the terrain can be done on the first flap without regard for the lake outlines and actually should overlap into the lake area. A second flap, on which all the lakes are opaqued, will provide an open window negative to be screened to produce an even tint (Fig. 14.15). When making the plate, a film positive of the second flap is placed between the halftone negative made from the terrain drawing, which will leave the plate unexposed in the lake areas. The open window

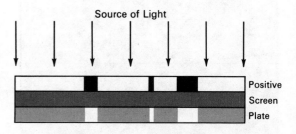

Figure 14.14 White lettering, of course, is produced in the same manner by inserting a positive at the plate-making stage.

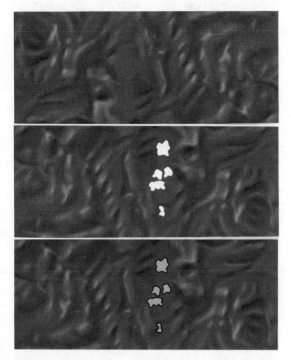

Figure 14.15 The shaded relief (above) has been accomplished without regard for the lakes. A film positive of the flap that is to lay the tint in the lake has been used to reverse the halftone dots out of the shaded relief (center). A negative of the lake flap then is screened and lays an even tint in the lakes (below).

negative of the lake flap with an attached screen is then used to lay the even tint in the masked areas.

Screened areas not only provide the necessary background for open or reverse line work but at the same time can permit overprinting of the solid color. If the screen is too light, white areas will not be distinguishable, and if the screen is too dark the solid color will not be discernible because the background will not provide enough contrast. A combination of a solid color and up to about 50 or 60% of the color is usually satisfactory for overprinting, while any tint down to about 30% is adequate for reversing (Fig. 14.16).

Artwork for Color Printing

Colored maps are usually produced from artwork that is prepared in black and white. The color results from separate press runs using different colored inks. At least one flap and often more than one is necessary for each color. The cartographer furnishes the printer a series of registered flaps with specific instructions for processing and printing.

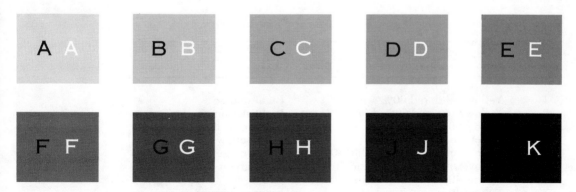

Figure 14.16 Lettering has been superimposed and reversed from tint screens from 10% in equal increments to solid black. We can determine from this illustration approximately the point where reversing or overprinting can be unsatisfactory.

A two-color map is usually one with black and one other color, although it can be two colors other than black. Three-color maps ordinarily would have black and any other two colors. A four-color map, however, most often uses the primary colors, red, yellow, and blue, along with black. Theoretically, any other hue can be produced by combinations of varying amounts of the primary colors. In color printing the combining is accomplished by successive press runs, and the amount of each color is controlled by tint screens. Charts showing the hues produced by superimposing different percentages of red, yellow, blue, and black are prepared by most printing firms. When choosing colors, the cartographer should use a chart prepared by the printer who will reproduce his map. If another chart must be used, a copy of it should be furnished the printer to indicate the precise color desired.

To demonstrate the procedure we might follow in planning and preparing the artwork for a colored map, consider an actual case of a map that requires separate colors for ten categories of information and light

	YELLOW	RED	BLUE
Solid	1 4 5 6 10	5 9	
80%			10
60%	2 3 9	1 2	
40%	7 8	7	
20%		4	6 7 Lakes
10%		3 6 10	2 5 9

Figure 14.17 **When complicated color combinations are necessary, a chart of this sort can be a helpful guide for the draftsman.**

blue for numerous small lakes.

From this listing we can prepare a chart that will reveal the number of flaps necessary to produce these colors and indicate

Table 14.1 gives a list of the colors selected from a chart.

Table 14.1

Area	Colors	Result
1	Yellow solid, red 60%	Light orange
2	Yellow 60%, red 60%, blue 10%	Reddish orange
3	Yellow 60%, red 10%	Medium yellow
4	Yellow solid, red 20%	Dark yellow
5	Yellow solid, red solid, blue 10%	Dark orange
6	Yellow solid, red 10%, blue 20%	Light green
7	Yellow 40%, red 40%, blue 20%	Beige
8	Yellow 40%	Light yellow
9	Yellow 60%, red solid, blue 10%	Red
10	Yellow solid, red 10%, blue 80%	Dark green
Lakes	Blue 20%	

which areas are to be opaqued on each flap (Fig. 14.17). Other flaps that are necessary will include a solid black flap with the base information, lettering, and area boundaries, a solid blue flap for the hydrographic features, and a 20% blue to lay a tint on the lakes. A separate flap for the lakes is also prepared to provide a film positive that can be used as a mask when exposing the plates for all other colors. This will prevent the other colors from falling within the lakes.

To produce the various flaps in positive form, it is customary to secure copies of the base map on a stable material with the image in light blue. Using the blue lines of the areas as a guide, black opaque is applied to the appropriate areas on each flap. When the open window negatives are prepared, the light blue images will disappear. A specified tint screen is then attached to the negatives and the plates are prepared.

To produce a map on which linework is to be in different colors, another method can be used in which all linework is placed on one flap. When the artwork is photographed, multiple negatives are made, one for each proposed color. Then each negative is opaqued to leave only that part of the image that is to be printed in a particular color. Since most film is dimensionally stable, excellent registry can be maintained.

Multiple Use of Negatives and Postives

With the capabilities of the printer to produce tints and reverse linework, the cartographer has many opportunities to enhance his map by various combinations of white, grays, and black. The decisions concerning the use of such combinations are made at the design stage, so that before drafting is begun, the artwork has been planned in a fashion that will produce the best results with a minimum of effort. A good knowledge of the possibilities and limitations of the printing processes is necessary for us to visualize how the various pieces of film are to be used; for example, we must keep in mind that a printing plate cannot be exposed to two negatives at the same time, that is, negatives require successive exposures. On the other hand, more than one film positive, or film positives and a negative, can be *sandwiched* together and used for a single exposure. On one printing plate, then, there can be an image that resulted from a series of exposures, any of which can incorporate more than one piece of film.

The following example will demonstrate that the cartographer is not as limited in one-color printing as he might believe.

Suppose a design program specifies that the hydrographic features and section lines of a map are to appear white, while the swamp areas, county boundaries, and township lines are to appear solid black on a background flat tone of 40% black (Fig. 14.18). Hydrographic features must be reversed from the background tint and from the swamp areas, but the black boundaries are to print over the white.

Four flaps are used to produce the map.

Flap A   All lowland areas.
Flap B   Entire area of map opaque to provide open window negative for overall tint screen.
Flap C   Hydrographic features and section lines.
Flap D   County boundaries and township lines.

All the flaps can be prepared in positive form with black ink and red *Zip-A-Tone*.

Following are the instructions given to the printer.

1. Lay a 60 percent tint screen on a negative of flap B, and using a positive of flap

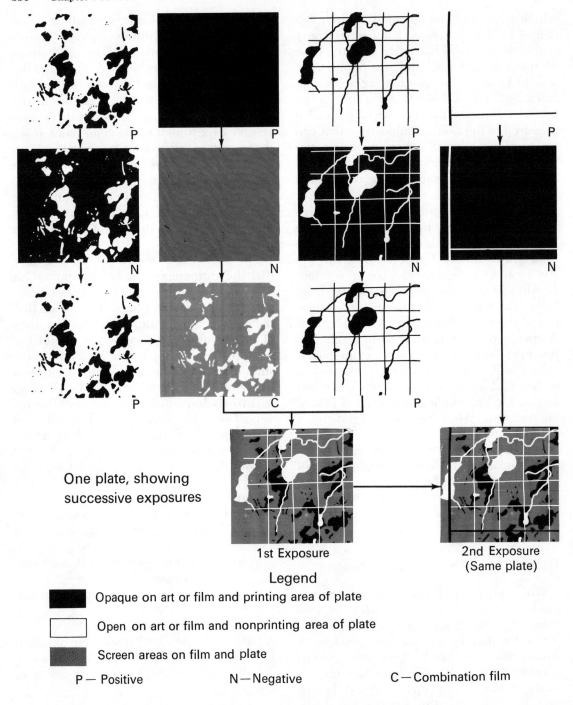

One plate, showing
successive exposures

1st Exposure

2nd Exposure
(Same plate)

Legend

Opaque on art or film and printing area of plate

Open on art or film and nonprinting area of plate

Screen areas on film and plate

P — Positive   N—Negative   C—Combination film

Figure 14.18 Possibilities for use of black and
white are unlimited. The student only need use
his imagination.

A as a mask, contact to a new film.

2. Double print on the plate.

(a) The new film with a positive of flap C as a mask.

(b) The negative of flap D.

When a 60 percent screen is used with the negative of flap B and contacted to the new film with a positive of flap A, the screen is converted to 40 percent. It will not cover the low wet areas, however, since the mask holds the light back and transparent areas result. If the plate were to be exposed to the new film, a 40 percent tint would result over the whole map except in the wet areas, which would be solid black. But, by combining a positive of flap C with the new film, the hydrographic features and the section lines block the light and produce a nonprinting area in the background tint and in the low wet areas.

Proofing

When a map is completed, it should be carefully and thoroughly checked before being sent for duplication. When facilities and materials are available, it is wise to proof a map if it consists of more than one flap. Even one-color maps can consist of as many as ten or more flaps, greatly increasing the chances for error and making editing difficult. A properly prepared proof will closely resemble the copies that will be printed from the artwork, and can, therefore, serve to evaluate the success of the design as well as reveal errors.

Artwork prepared on translucent material can be used to prepare proofs in the drafting room even though darkroom facilities are not available. The various proofing processes, such as *General Color Guide Film*, *3M* proofing materials, and the use of bichromate sensitizers, require the use of negatives for exposure. These direct contact negatives can be made using *Kodagraph Contact Film* or with *Orange 3M Color Key*, both of which are actinically opaque; either of these negatives can be used to make the proofs with any of the proofing systems. During each step of the proofing operation, the pin registration system must be employed. Because of the nature of the *3M* material, it cannot be easily punched and it is necessary, therefore, to use the gummed prepunched tabs or to punch strips of discarded film and tape them to the proofing sheets.

If facilities are not available for proofing, we can request that the printer furnish proofs after he has made the negatives but before he makes the printing plates. Errors found at this stage can often be corrected on the negative. If the artwork must be revised, we have lost only the cost of the negatives and not the cost of the plates, paper, and press time. For complicated colored maps we may wish to request press proofs, in which case the actual plates that will be used to print the final copies are used to print a small number. Errors found at this point save the cost of paper and press time. A final and important precaution is for the cartographer to be present while the map is actually being printed to inspect such things as registry and color quality.

APPENDIX

A COMMON LOGARITHMS

The common logarithm of a number is the power to which 10 must be raised to equal that number. For example, the number 356 $= 10^{2.55145}$; accordingly, the logarithm of 356 is 2.55145. The integer to the left of the decimal point in a logarithm is called the *characteristic*. The decimal fractional part of the logarithm (the numbers to the right of the decimal point) is called the *mantissa*. The logarithm of any number in which the same integers are in the same order, for example, 35,600, or 35.60, or 0.03560, has the same mantissa but a different characteristic. Characteristics are easy to determine; hence, tables of logarithms show only the mantissas.

When any number is equal to or greater than 1, then the characteristic of its logarithm is positive, and it is numerically *one less* than the number of places to the left of the decimal point in the original number. When any number is less than one, then the characteristic of its logarithm is negative, and it is numerically *one more* than the number of zeros immediately to the right of the decimal point. For example:

Number	Logarithm
	Etc.
35,600.0	4.55145
3,560.0	3.55145
356.0	2.55145
35.60	1.55145

3.560	0.55145
0.3560	−1.55145 or 9.55145 − 10
0.03560	−2.55145 or 8.55145 − 10
0.003560	−3.55145 or 7.55145 − 10
	Etc.

When we add or subtract logarithms that have negative characteristics, it is usually convenient to use the method of notation shown at the far right in the preceding table.

Logarithms are used for the multiplication and division of large or complex numbers when a calculator is not available and to determine powers and roots of numbers.

Multiplication is easily accomplished since when the logarithm of any number is *added* to the logarithm of any other number (for example, $\log x + \log y$), the sum of the two logarithms is the logarithm of the *product* of the two numbers, that is, $\log x + \log y = \log$ of xy. The numerical value of the product of xy is obtained by finding the mantissa of the sum in the table, noting the number (N) to which it refers (called the *antilogarithm*), and then placing the decimal point according to the value of the characteristic. Division is equally simple since $\log x - \log y = \log$ of $x \div y$. Consequently, multiplication and division of large or complex numbers are enormously simplified by using logarithms.

Powers and roots of numbers may be easily

obtained with the aid of logarithms as illustrated below. The logarithm of any number (*N*), when *directly multiplied* by a number, provides the logarithm of that power of the original number. For example, $356 \times 356 = 356^2 = 126,736$; the logarithm of 356 is 2.55145; $2.55145 \times 2 = 5.10290$, which is the logarithm of 126,736. Note that the logarithm is directly multiplied by the numerical value of the power, 2 in this example. Similarly, any root of any number (*N*) may be obtained by *directly dividing* the logarithm of the number by the numerical value of the root desired. For example, the logarithm 2.55145 divided by $2 = 1.275725$, which is the logarithm of 18.86, which is the square root of 356.

Interpolation. The mantissas in Table A.1 refer to values of *N* to four places. To find the logarithm of a number to five places, interpolation between successive mantissas is necessary. Likewise, when a mantissa value lies between two given in the table, interpolation is necessary to obtain the antilogarithm to five places.

Interpolation is facilitated by the tables of proportional parts shown alongside the mantissas, for example.

	18
1	1.8
2	3.6
3	5.4
4	7.2
5	9.0
6	10.8
7	12.6
8	14.4
9	16.2

The heading of the above table, 18, is the difference between the successive mantissas in a portion of the table near where it appears. The numbers on the left show the fifth place in *N*, and the numbers opposite show the corresponding decimal divisions of the mantissa interval.

TABLE A.1   FIVE-PLACE LOGARITHMS: 100—150

N	0	1	2	3	4	5	6	7	8	9
100	00 000	043	087	130	173	217	260	303	346	389
01	432	475	518	561	604	647	689	732	775	817
02	00 860	903	945	988	*030	*072	*115	*157	*199	*242
03	01 284	326	368	410	452	494	536	578	620	662
04	01 703	745	787	828	870	912	953	995	*036	*078
05	02 119	160	202	243	284	325	366	407	449	490
06	531	572	612	653	694	735	776	816	857	898
07	02 938	979	*019	*060	*100	*141	*181	*222	*262	*302
08	03 342	383	423	463	503	543	583	623	663	703
09	03 743	782	822	862	902	941	981	*021	*060	*100
110	04 139	179	218	258	297	336	376	415	454	493
11	532	571	610	650	689	727	766	805	844	883
12	04 922	961	999	*038	*077	*115	*154	*192	*231	*269
13	05 308	346	385	423	461	500	538	576	614	652
14	05 690	729	767	805	843	881	918	956	994	*032
15	06 070	108	145	183	221	258	296	333	371	408
16	446	483	521	558	595	633	670	707	744	781
17	06 819	856	893	930	967	*004	*041	*078	*115	*151
18	07 188	225	262	298	335	372	408	445	482	518
19	555	591	628	664	700	737	773	809	846	882
120	07 918	954	990	*027	*063	*099	*135	*171	*207	*243
21	08 279	314	350	386	422	458	493	529	565	600
22	636	672	707	743	778	814	849	884	920	955
23	08 991	*026	*061	*096	*132	*167	*202	*237	*272	*307
24	09 342	377	412	447	482	517	552	587	621	656
25	09 691	726	760	795	830	864	899	934	968	*003
26	10 037	072	106	140	175	209	243	278	312	346
27	380	415	449	483	517	551	585	619	653	687
28	10 721	755	789	823	857	890	924	958	992	*025
29	11 059	093	126	160	193	227	261	294	327	361
130	394	428	461	494	528	561	594	628	661	694
31	11 727	760	793	826	860	893	926	959	992	*024
32	12 057	090	123	156	189	222	254	287	320	352
33	385	418	450	483	516	548	581	613	646	678
34	12 710	743	775	808	840	872	905	937	969	*001
35	13 033	066	098	130	162	194	226	258	290	322
36	354	386	418	450	481	513	545	577	609	640
37	672	704	735	767	799	830	862	893	925	956
38	13 988	*019	*051	*082	*114	*145	*176	*208	*239	*270
39	14 301	333	364	395	426	457	489	520	551	582
140	613	644	675	706	737	768	799	829	860	891
41	14 922	953	983	*014	*045	*076	*106	*137	*168	*198
42	15 229	259	290	320	351	381	412	442	473	503
43	534	564	594	625	655	685	715	746	776	806
44	15 836	866	897	927	957	987	*017	*047	*077	*107
45	16 137	167	197	227	256	286	316	346	376	406
46	435	465	495	524	554	584	613	643	673	702
47	16 732	761	791	820	850	879	909	938	967	997
48	17 026	056	085	114	143	173	202	231	260	289
49	319	348	377	406	435	464	493	522	551	580
150	17 609	638	667	696	725	754	782	811	840	869
N	0	1	2	3	4	5	6	7	8	9

Prop. Parts

	44	43	42
1	4.4	4.3	4.2
2	8.8	8.6	8.4
3	13.2	12.9	12.6
4	17.6	17.2	16.8
5	22.0	21.5	21.0
6	26.4	25.8	25.2
7	30.8	30.1	29.4
8	35.2	34.4	33.6
9	39.6	38.7	37.8

	41	40	39
1	4.1	4	3.9
2	8.2	8	7.8
3	12.3	12	11.7
4	16.4	16	15.6
5	20.5	20	19.5
6	24.6	24	23.4
7	28.7	28	27.3
8	32.8	32	31.2
9	36.9	36	35.1

	38	37	36
1	3.8	3.7	3.6
2	7.6	7.4	7.2
3	11.4	11.1	10.8
4	15.2	14.8	14.4
5	19.0	18.5	18.0
6	22.8	22.2	21.6
7	26.6	25.9	25.2
8	30.4	29.6	28.8
9	34.2	33.3	32.4

	35	34	33
1	3.5	3.4	3.3
2	7.0	6.8	6.6
3	10.5	10.2	9.9
4	14.0	13.6	13.2
5	17.5	17.0	16.5
6	21.0	20.4	19.8
7	24.5	23.8	23.1
8	28.0	27.2	26.4
9	31.5	30.6	29.7

	32	31	30
1	3.2	3.1	3
2	6.4	6.2	6
3	9.6	9.3	9
4	12.8	12.4	12
5	16.0	15.5	15
6	19.2	18.6	18
7	22.4	21.7	21
8	25.6	24.8	24
9	28.8	27.9	27

TABLE A.1   FIVE-PLACE LOGARITHMS: 150—200

N	0	1	2	3	4	5	6	7	8	9
150	17 609	638	667	696	725	754	782	811	840	869
51	17 898	926	955	984	*013	*041	*070	*099	*127	*156
52	18 184	213	241	270	298	327	355	384	412	441
53	469	498	526	554	583	611	639	667	696	724
54	18 752	780	808	837	865	893	921	949	977	*005
55	19 033	061	089	117	145	173	201	229	257	285
56	312	340	368	396	424	451	479	507	535	562
57	590	618	645	673	700	728	756	783	811	838
58	19 866	893	921	948	976	*003	*030	*058	*085	*112
59	20 140	167	194	222	249	276	303	330	358	385
160	412	439	466	493	520	548	575	602	629	656
61	683	710	737	763	790	817	844	871	898	925
62	20 952	978	*005	*032	*059	*085	*112	*139	*165	*192
63	21 219	245	272	299	325	352	378	405	431	458
64	484	511	537	564	590	617	643	669	696	722
65	21 748	775	801	827	854	880	906	932	958	985
66	22 011	037	063	089	115	141	167	194	220	246
67	272	298	324	350	376	401	427	453	479	505
68	531	557	583	608	634	660	686	712	737	763
69	22 789	814	840	866	891	917	943	968	994	*019
170	23 045	070	096	121	147	172	198	223	249	274
71	300	325	350	376	401	426	452	477	502	528
72	553	578	603	629	654	679	704	729	754	779
73	23 805	830	855	880	905	930	955	980	*005	*030
74	24 055	080	105	130	155	180	204	229	254	279
75	304	329	353	378	403	428	452	477	502	527
76	551	576	601	625	650	674	699	724	748	773
77	24 797	822	846	871	895	920	944	969	993	*018
78	25 042	066	091	115	139	164	188	212	237	261
79	285	310	334	358	382	406	431	455	479	503
180	527	551	575	600	624	648	672	696	720	744
81	25 768	792	816	840	864	888	912	935	959	983
82	26 007	031	055	079	102	126	150	174	198	221
83	245	269	293	316	340	364	387	411	435	458
84	482	505	529	553	576	600	623	647	670	694
85	717	741	764	788	811	834	858	881	905	928
86	26 951	975	998	*021	*045	*068	*091	*114	*138	*161
87	27 184	207	231	254	277	300	323	346	370	393
88	416	439	462	485	508	531	554	577	600	623
89	646	669	692	715	738	761	784	807	830	852
190	27 875	898	921	944	967	989	*012	*035	*058	*081
91	28 103	126	149	171	194	217	240	262	285	307
92	330	353	375	398	421	443	466	488	511	533
93	556	578	601	623	646	668	691	713	735	758
94	28 780	803	825	847	870	892	914	937	959	981
95	29 003	026	048	070	092	115	137	159	181	203
96	226	248	270	292	314	336	358	380	403	425
97	447	469	491	513	535	557	579	601	623	645
98	667	688	710	732	754	776	798	820	842	863
99	29 885	907	929	951	973	994	*016	*038	*060	*081
200	30 103	125	146	168	190	211	233	255	276	298

Prop. Parts

	29	28
1	2.9	2.8
2	5.8	5.6
3	8.7	8.4
4	11.6	11.2
5	14.5	14.0
6	17.4	16.8
7	20.3	19.6
8	23.2	22.4
9	26.1	25.2

	27	26
1	2.7	2.6
2	5.4	5.2
3	8.1	7.8
4	10.8	10.4
5	13.5	13.0
6	16.2	15.6
7	18.9	18.2
8	21.6	20.8
9	24.3	23.4

	25
1	2.5
2	5.0
3	7.5
4	10.0
5	12.5
6	15.0
7	17.5
8	20.0
9	22.5

	24	23
1	2.4	2.3
2	4.8	4.6
3	7.2	6.9
4	9.6	9.2
5	12.0	11.5
6	14.4	13.8
7	16.8	16.1
8	19.2	18.4
9	21.6	20.7

	22	21
1	2.2	2.1
2	4.4	4.2
3	6.6	6.3
4	8.8	8.4
5	11.0	10.5
6	13.2	12.6
7	15.4	14.7
8	17.6	16.8
9	19.8	18.9

TABLE A.1   FIVE-PLACE LOGARITHMS: 200—250

N	0	1	2	3	4	5	6	7	8	9
200	30 103	125	146	168	190	211	233	255	276	298
01	320	341	363	384	406	428	449	471	492	514
02	535	557	578	600	621	643	664	685	707	728
03	750	771	792	814	835	856	878	899	920	942
04	30 963	984	*006	*027	*048	*069	*091	*112	*133	*154
05	31 175	197	218	239	260	281	302	323	345	366
06	387	408	429	450	471	492	513	534	555	576
07	597	618	639	660	681	702	723	744	765	785
08	31 806	827	848	869	890	911	931	952	973	994
09	32 015	035	056	077	098	118	139	160	181	201
210	222	243	263	284	305	325	346	366	387	408
11	428	449	469	490	510	531	552	572	593	613
12	634	654	675	695	715	736	756	777	797	818
13	32 838	858	879	899	919	940	960	980	*001	*021
14	33 041	062	082	102	122	143	163	183	203	224
15	244	264	284	304	325	345	365	385	405	425
16	445	465	486	506	526	546	566	586	606	626
17	646	666	686	706	726	746	766	786	806	826
18	33 846	866	885	905	925	945	965	985	*005	*025
19	34 044	064	084	104	124	143	163	183	203	223
220	242	262	282	301	321	341	361	380	400	420
21	439	459	479	498	518	537	557	577	596	616
22	635	655	674	694	713	733	753	772	792	811
23	34 830	850	869	889	908	928	947	967	986	*005
24	35 025	044	064	083	102	122	141	160	180	199
25	218	238	257	276	295	315	334	353	372	392
26	411	430	449	468	488	507	526	545	564	583
27	603	622	641	660	679	698	717	736	755	774
28	793	813	832	851	870	889	908	927	946	965
29	35 984	*003	*021	*040	*059	*078	*097	*116	*135	*154
230	36 173	192	211	229	248	267	286	305	324	342
31	361	380	399	418	436	455	474	493	511	530
32	549	568	586	605	624	642	661	680	698	717
33	736	754	773	791	810	829	847	866	884	903
34	36 922	940	959	977	996	*014	*033	*051	*070	*088
35	37 107	125	144	162	181	199	218	236	254	273
36	291	310	328	346	365	383	401	420	438	457
37	475	493	511	530	548	566	585	603	621	639
38	658	676	694	712	731	749	767	785	803	822
39	37 840	858	876	894	912	931	949	967	985	*003
240	38 021	039	057	075	093	112	130	148	166	184
41	202	220	238	256	274	292	310	328	346	364
42	382	399	417	435	453	471	489	507	525	543
43	561	578	596	614	632	650	668	686	703	721
44	739	757	775	792	810	828	846	863	881	899
45	38 917	934	952	970	987	*005	*023	*041	*058	*076
46	39 094	111	129	146	164	182	199	217	235	252
47	270	287	305	322	340	358	375	393	410	428
48	445	463	480	498	515	533	550	568	585	602
49	620	637	655	672	690	707	724	742	759	777
250	39 794	811	829	846	863	881	898	915	933	950
N	0	1	2	3	4	5	6	7	8	9

Prop. Parts

	22	21
1	2.2	2.1
2	4.4	4.2
3	6.6	6.3
4	8.8	8.4
5	11.0	10.5
6	13.2	12.6
7	15.4	14.7
8	17.6	16.8
9	19.8	18.9

	20
1	2
2	4
3	6
4	8
5	10
6	12
7	14
8	16
9	18

	19
1	1.9
2	3.8
3	5.7
4	7.6
5	9.5
6	11.4
7	13.3
8	15.2
9	17.1

	18
1	1.8
2	3.6
3	5.4
4	7.2
5	9.0
6	10.8
7	12.6
8	14.4
9	16.2

	17
1	1.7
2	3.4
3	5.1
4	6.8
5	8.5
6	10.2
7	11.9
8	13.6
9	15.3

TABLE A.1   FIVE-PLACE LOGARITHMS: 250—300

N	0	1	2	3	4	5	6	7	8	9
250	39 794	811	829	846	863	881	898	915	933	950
51	39 967	985	*002	*019	*037	*054	*071	*088	*106	*123
52	40 140	157	175	192	209	226	243	261	278	295
53	312	329	346	364	381	398	415	432	449	466
54	483	500	518	535	552	569	586	603	620	637
55	654	671	688	705	722	739	756	773	790	807
56	824	841	858	875	892	909	926	943	960	976
57	40 993	*010	*027	*044	*061	*078	*095	*111	*128	*145
58	41 162	179	196	212	229	246	263	280	296	313
59	330	347	363	380	397	414	430	447	464	481
260	497	514	531	547	564	581	597	614	631	647
61	664	681	697	714	731	747	764	780	797	814
62	830	847	863	880	896	913	929	946	963	979
63	41 996	*012	*029	*045	*062	*078	*095	*111	*127	*144
64	42 160	177	193	210	226	243	259	275	292	308
65	325	341	357	374	390	406	423	439	455	472
66	488	504	521	537	553	570	586	602	619	635
67	651	667	684	700	716	732	749	765	781	797
68	813	830	846	862	878	894	911	927	943	959
69	42 975	991	*008	*024	*040	*056	*072	*088	*104	*120
270	43 136	152	169	185	201	217	233	249	265	281
71	297	313	329	345	361	377	393	409	425	441
72	457	473	489	505	521	537	553	569	584	600
73	616	632	648	664	680	696	712	727	743	759
74	775	791	807	823	838	854	870	886	902	917
75	43 933	949	965	981	996	*012	*028	*044	*059	*075
76	44 091	107	122	138	154	170	185	201	217	232
77	248	264	279	295	311	326	342	358	373	389
78	404	420	436	451	467	483	498	514	529	545
79	560	576	592	607	623	638	654	669	685	700
280	716	731	747	762	778	793	809	824	840	855
81	44 871	886	902	917	932	948	963	979	994	*010
82	45 025	040	056	071	086	102	117	133	148	163
83	179	194	209	225	240	255	271	286	301	317
84	332	347	362	378	393	408	423	439	454	469
85	484	500	515	530	545	561	576	591	606	621
86	637	652	667	682	697	712	728	743	758	773
87	788	803	818	834	849	864	879	894	909	924
88	45 939	954	969	984	*000	*015	*030	*045	*060	*075
89	46 090	105	120	135	150	165	180	195	21C	225
290	240	255	270	285	300	315	330	345	359	374
91	389	404	419	434	449	464	479	494	509	523
92	538	553	568	583	598	613	627	642	657	672
93	687	702	716	731	746	761	776	790	805	820
94	835	850	864	879	894	909	923	938	953	967
95	46 982	997	*012	*026	*041	*056	*070	*085	*100	*114
96	47 129	144	159	173	188	202	217	232	246	261
97	276	290	305	319	334	349	363	378	392	407
98	422	436	451	465	480	494	509	524	538	553
99	567	582	596	611	625	640	654	669	683	698
300	47 712	727	741	756	770	784	799	813	828	842

Prop. Parts

	18	17	16	15	14
1	1.8	1.7	1.6	1.5	1.4
2	3.6	3.4	3.2	3.0	2.8
3	5.4	5.1	4.8	4.5	4.2
4	7.2	6.8	6.4	6.0	5.6
5	9.0	8.5	8.0	7.5	7.0
6	10.8	10.2	9.6	9.0	8.4
7	12.6	11.9	11.2	10.5	9.8
8	14.4	13.6	12.8	12.0	11.2
9	16.2	15.3	14.4	13.5	12.6

TABLE A.1   FIVE-PLACE LOGARITHMS: 300—350

N	0	1	2	3	4	5	6	7	8	9	Prop. Parts
300	47 712	727	741	756	770	784	799	813	828	842	
01	47 857	871	885	900	914	929	943	958	972	986	
02	48 001	015	029	044	058	073	087	101	116	130	
03	144	159	173	187	202	216	230	244	259	273	
04	287	302	316	330	344	359	373	387	401	416	
05	430	444	458	473	487	501	515	530	544	558	
06	572	586	601	615	629	643	657	671	686	700	
07	714	728	742	756	770	785	799	813	827	841	
08	855	869	883	897	911	926	940	954	968	982	
09	48 996	*010	*024	*038	*052	*066	*080	*094	*108	*122	
310	49 136	150	164	178	192	206	220	234	248	262	
11	276	290	304	318	332	346	360	374	388	402	
12	415	429	443	457	471	485	499	513	527	541	
13	554	568	582	596	610	624	638	651	665	679	
14	693	707	721	734	748	762	776	790	803	817	
15	831	845	859	872	886	900	914	927	941	955	
16	49 969	982	996	*010	*024	*037	*051	*065	*079	*092	
17	50 106	120	133	147	161	174	188	202	215	229	
18	243	256	270	284	297	311	325	338	352	365	
19	379	393	406	420	433	447	461	474	488	501	
320	515	529	542	556	569	583	596	610	623	637	
21	651	664	678	691	705	718	732	745	759	772	
22	786	799	813	826	840	853	866	880	893	907	
23	50 920	934	947	961	974	987	*001	*014	*028	*041	
24	51 055	068	081	095	108	121	135	148	162	175	
25	188	202	215	228	242	255	268	282	295	308	
26	322	335	348	362	375	388	402	415	428	441	
27	455	468	481	495	508	521	534	548	561	574	
28	587	601	614	627	640	654	667	680	693	706	
29	720	733	746	759	772	786	799	812	825	838	
330	851	865	878	891	904	917	930	943	957	970	
31	51 983	996	*009	*022	*035	*048	*061	*075	*088	*101	
32	52 114	127	140	153	166	179	192	205	218	231	
33	244	257	270	284	297	310	323	336	349	362	
34	375	388	401	414	427	440	453	466	479	492	
35	504	517	530	543	556	569	582	595	608	621	
36	634	647	660	673	686	699	711	724	737	750	
37	763	776	789	802	815	827	840	853	866	879	
38	52 892	905	917	930	943	956	969	982	994	*007	
39	53 020	033	046	058	071	084	097	110	122	135	
340	148	161	173	186	199	212	224	237	250	263	
41	275	288	301	314	326	339	352	364	377	390	
42	403	415	428	441	453	466	479	491	504	517	
43	529	542	555	567	580	593	605	618	631	643	
44	656	668	681	694	706	719	732	744	757	769	
45	782	794	807	820	832	845	857	870	882	895	
46	53 908	920	933	945	958	970	983	995	*008	*020	
47	54 033	045	058	070	083	095	108	120	133	145	
48	158	170	183	195	208	220	233	245	258	270	
49	283	295	307	320	332	345	357	370	382	394	
350	54 407	419	432	444	456	469	481	494	506	518	
N	**0**	**1**	**2**	**3**	**4**	**5**	**6**	**7**	**8**	**9**	Prop. Parts

Prop. Parts

	15
1	1.5
2	3.0
3	4.5
4	6.0
5	7.5
6	9.0
7	10.5
8	12.0
9	13.5

	14
1	1.4
2	2.8
3	4.2
4	5.6
5	7.0
6	8.4
7	9.8
8	11.2
9	12.6

	13
1	1.3
2	2.6
3	3.9
4	5.2
5	6.5
6	7.8
7	9.1
8	10.4
9	11.7

	12
1	1.2
2	2.4
3	3.6
4	4.8
5	6.0
6	7.2
7	8.4
8	9.6
9	10.8

TABLE A.1   FIVE-PLACE LOGARITHMS: 350—400

N	0	1	2	3	4	5	6	7	8	9
350	54 407	419	432	444	456	469	481	494	506	518
51	531	543	555	568	580	593	605	617	630	642
52	654	667	679	691	704	716	728	741	753	765
53	777	790	802	814	827	839	851	864	876	888
54	54 900	913	925	937	949	962	974	986	998	*011
55	55 023	035	047	060	072	084	096	108	121	133
56	145	157	169	182	194	206	218	230	242	255
57	267	279	291	303	315	328	340	352	364	376
58	388	400	413	425	437	449	461	473	485	497
59	509	522	534	546	558	570	582	594	606	618
360	630	642	654	666	678	691	703	715	727	739
61	751	763	775	787	799	811	823	835	847	859
62	871	883	895	907	919	931	943	955	967	979
63	55 991	*003	*015	*027	*038	*050	*062	*074	*086	*098
64	56 110	122	134	146	158	170	182	194	205	217
65	229	241	253	265	277	289	301	312	324	336
66	348	360	372	384	396	407	419	431	443	455
67	467	478	490	502	514	526	538	549	561	573
68	585	597	608	620	632	644	656	667	679	691
69	703	714	726	738	750	761	773	785	797	808
370	820	832	844	855	867	879	891	902	914	926
71	56 937	949	961	972	984	996	*008	*019	*031	*043
72	57 054	066	078	089	101	113	124	136	148	159
73	171	183	194	206	217	229	241	252	264	276
74	287	299	310	322	334	345	357	368	380	392
75	403	415	426	438	449	461	473	484	496	507
76	519	530	542	553	565	576	588	600	611	623
77	634	646	657	669	680	692	703	715	726	738
78	749	761	772	784	795	807	818	830	841	852
79	864	875	887	898	910	921	933	944	955	967
380	57 978	990	*001	*013	*024	*035	*047	*058	*070	*081
81	58 092	104	115	127	138	149	161	172	184	195
82	206	218	229	240	252	263	274	286	297	309
83	320	331	343	354	365	377	388	399	410	422
84	433	444	456	467	478	490	501	512	524	535
85	546	557	569	580	591	602	614	625	636	647
86	659	670	681	692	704	715	726	737	749	760
87	771	782	794	805	816	827	838	850	861	872
88	883	894	906	917	928	939	950	961	973	984
89	58 995	*006	*017	*028	*040	*051	*062	*073	*084	*095
390	59 106	118	129	140	151	162	173	184	195	207
91	218	229	240	251	262	273	284	295	306	318
92	329	340	351	362	373	384	395	406	417	428
93	439	450	461	472	483	494	506	517	528	539
94	550	561	572	583	594	605	616	627	638	649
95	660	671	682	693	704	715	726	737	748	759
96	770	780	791	802	813	824	835	846	857	868
97	879	890	901	912	923	934	945	956	966	977
98	59 988	999	*010	*021	*032	*043	*054	*065	*076	*086
99	60 097	108	119	130	141	152	163	173	184	195
400	60 206	217	228	239	249	260	271	282	293	304

Prop. Parts

13		12		11		10	
1	1.3	1	1.2	1	1.1	1	1.0
2	2.6	2	2.4	2	2.2	2	2.0
3	3.9	3	3.6	3	3.3	3	3.0
4	5.2	4	4.8	4	4.4	4	4.0
5	6.5	5	6.0	5	5.5	5	5.0
6	7.8	6	7.2	6	6.6	6	6.0
7	9.1	7	8.4	7	7.7	7	7.0
8	10.4	8	9.6	8	8.8	8	8.0
9	11.7	9	10.8	9	9.9	9	9.0

TABLE A.1   FIVE-PLACE LOGARITHMS: 400—450

N	0	1	2	3	4	5	6	7	8	9	Prop. Parts
400	60 206	217	228	239	249	260	271	282	293	304	
01	314	325	336	347	358	369	379	390	401	412	
02	423	433	444	455	466	477	487	498	509	520	
03	531	541	552	563	574	584	595	606	617	627	
04	638	649	660	670	681	692	703	713	724	735	
05	746	756	767	778	788	799	810	821	831	842	
06	853	863	874	885	895	906	917	927	938	949	**11**
07	60 959	970	981	991	*002	*013	*023	*034	*045	*055	1 1.1
08	61 066	077	087	098	109	119	130	140	151	162	2 2.2
09	172	183	194	204	215	225	236	247	257	268	3 3.3
410	278	289	300	310	321	331	342	352	363	374	4 4.4 / 5 5.5 / 6 6.6
11	384	395	405	416	426	437	448	458	469	479	7 7.7
12	490	500	511	521	532	542	553	563	574	584	8 8.8
13	595	606	616	627	637	648	658	669	679	690	9 9.9
14	700	711	721	731	742	752	763	773	784	794	
15	805	815	826	836	847	857	868	878	888	899	
16	61 909	920	930	941	951	962	972	982	993	*003	
17	62 014	024	034	045	055	066	076	086	097	107	
18	118	128	138	149	159	170	180	190	201	211	
19	221	232	242	252	263	273	284	294	304	315	
420	325	335	346	356	366	377	387	397	408	418	
21	428	439	449	459	469	480	490	500	511	521	**10**
22	531	542	552	562	572	583	593	603	613	624	1 1.0
23	634	644	655	665	675	685	696	706	716	726	2 2.0
24	737	747	757	767	778	788	798	808	818	829	3 3.0 / 4 4.0
25	839	849	859	870	880	890	900	910	921	931	5 5.0 / 6 6.0
26	62 941	951	961	972	982	992	*002	*012	*022	*033	7 7.0
27	63 043	053	063	073	083	094	104	114	124	134	8 8.0
28	144	155	165	175	185	195	205	215	225	236	9 9.0
29	246	256	266	276	286	296	306	317	327	337	
430	347	357	367	377	387	397	407	417	428	438	
31	448	458	468	478	488	498	508	518	528	538	
32	548	558	568	579	589	599	609	619	629	639	
33	649	659	669	679	689	699	709	719	729	739	
34	749	759	769	779	789	799	809	819	829	839	
35	849	859	869	879	889	899	909	919	929	939	
36	63 949	959	969	979	988	998	*008	*018	*028	*038	
37	64 048	058	068	078	088	098	108	118	128	137	**9**
38	147	157	167	177	187	197	207	217	227	237	1 0.9
39	246	256	266	276	286	296	306	316	326	335	2 1.8
440	345	355	365	375	385	395	404	414	424	434	3 2.7 / 4 3.6
41	444	454	464	473	483	493	503	513	523	532	5 4.5 / 6 5.4
42	542	552	562	572	582	591	601	611	621	631	7 6.3
43	640	650	660	670	680	689	699	709	719	729	8 7.2 / 9 8.1
44	738	748	758	768	777	787	797	807	816	826	
45	836	846	856	865	875	885	895	904	914	924	
46	64 933	943	953	963	972	982	992	*002	*011	*021	
47	65 031	040	050	060	070	079	089	099	108	118	
48	128	137	147	157	167	176	186	196	205	215	
49	225	234	244	254	263	273	283	292	302	312	
450	65 321	331	341	350	360	369	379	389	398	408	
N	0	1	2	3	4	5	6	7	8	9	Prop. Parts

TABLE A.1   FIVE-PLACE LOGARITHMS: 450—500

N	0	1	2	3	4	5	6	7	8	9
450	65 321	331	341	350	360	369	379	389	398	408
51	418	427	437	447	456	466	475	485	495	504
52	514	523	533	543	552	562	571	581	591	600
53	610	619	629	639	648	658	667	677	686	696
54	706	715	725	734	744	753	763	772	782	792
55	801	811	820	830	839	849	858	868	877	887
56	896	906	916	925	935	944	954	963	973	982
57	65 992	*001	*011	*020	*030	*039	*049	*058	*068	*077
58	66 087	096	106	115	124	134	143	153	162	172
59	181	191	200	210	219	229	238	247	257	266
460	276	285	295	304	314	323	332	342	351	361
61	370	380	389	398	408	417	427	436	445	455
62	464	474	483	492	502	511	521	530	539	549
63	558	567	577	586	596	605	614	624	633	642
64	652	661	671	680	689	699	708	717	727	736
65	745	755	764	773	783	792	801	811	820	829
66	839	848	857	867	876	885	894	904	913	922
67	66 932	941	950	960	969	978	987	997	*006	*015
68	67 025	034	043	052	062	071	080	089	099	108
69	117	127	136	145	154	164	173	182	191	201
470	210	219	228	237	247	256	265	274	284	293
71	302	311	321	330	339	348	357	367	376	385
72	394	403	413	422	431	440	449	459	468	477
73	486	495	504	514	523	532	541	550	560	569
74	578	587	596	605	614	624	633	642	651	660
75	669	679	688	697	706	715	724	733	742	752
76	761	770	779	788	797	806	815	825	834	843
77	852	861	870	879	888	897	906	916	925	934
78	67 943	952	961	970	979	988	997	*006	*015	*024
79	68 034	043	052	061	070	079	088	097	106	115
480	124	133	142	151	160	169	178	187	196	205
81	215	224	233	242	251	260	269	278	287	296
82	305	314	323	332	341	350	359	368	377	386
83	395	404	413	422	431	440	449	458	467	476
84	485	494	502	511	520	529	538	547	556	565
85	574	583	592	601	610	619	628	637	646	655
86	664	673	681	690	699	708	717	726	735	744
87	753	762	771	780	789	797	806	815	824	833
88	842	851	860	869	878	886	895	904	913	922
89	68 931	940	949	958	966	975	984	993	*002	*011
490	69 020	028	037	046	055	064	073	082	090	099
91	108	117	126	135	144	152	161	170	179	188
92	197	205	214	223	232	241	249	258	267	276
93	285	294	302	311	320	329	338	346	355	364
94	373	381	390	399	408	417	425	434	443	452
95	461	469	478	487	496	504	513	522	531	539
96	548	557	566	574	583	592	601	609	618	627
97	636	644	653	662	671	679	688	697	705	714
98	723	732	740	749	758	767	775	784	793	801
99	810	819	827	836	845	854	862	871	880	888
500	69 897	906	914	923	932	940	949	958	966	975

Prop. Parts

10	
1	1.0
2	2.0
3	3.0
4	4.0
5	5.0
6	6.0
7	7.0
8	8.0
9	9.0

9	
1	0.9
2	1.8
3	2.7
4	3.6
5	4.5
6	5.4
7	6.3
8	7.2
9	8.1

8	
1	0.8
2	1.6
3	2.4
4	3.2
5	4.0
6	4.8
7	5.6
8	6.4
9	7.2

TABLE A.1   FIVE-PLACE LOGARITHMS: 500—550

N	0	1	2	3	4	5	6	7	8	9	Prop. Parts
500	69 897	906	914	923	932	940	949	958	966	975	
01	69 984	992	*001	*010	*018	*027	*036	*044	*053	*062	
02	70 070	079	088	096	105	114	122	131	140	148	
03	157	165	174	183	191	200	209	217	226	234	
04	243	252	260	269	278	286	295	303	312	321	
05	329	338	346	355	364	372	381	389	398	406	
06	415	424	432	441	449	458	467	475	484	492	
07	501	509	518	526	535	544	552	561	569	578	**9**
08	586	595	603	612	621	629	638	646	655	663	1 0.9
09	672	680	689	697	706	714	723	731	740	749	2 1.8
510	757	766	774	783	791	800	808	817	825	834	3 2.7 4 3.6 5 4.5 6 5.4
11	842	851	859	868	876	885	893	902	910	919	7 6.3
12	70 927	935	944	952	961	969	978	986	995	*003	8 7.2
13	71 012	020	029	037	046	054	063	071	079	088	9 8.1
14	096	105	113	122	130	139	147	155	164	172	
15	181	189	198	206	214	223	231	240	248	257	
16	265	273	282	290	299	307	315	324	332	341	
17	349	357	366	374	383	391	399	408	416	425	
18	433	441	450	458	466	475	483	492	500	508	
19	517	525	533	542	550	559	567	575	584	592	
520	600	609	617	625	634	642	650	659	667	675	
21	684	692	700	709	717	725	734	742	750	759	
22	767	775	784	792	800	809	817	825	834	842	**8**
23	850	858	867	875	883	892	900	908	917	925	1 0.8
24	71 933	941	950	958	966	975	983	991	999	*008	2 1.6 3 2.4
25	72 016	024	032	041	049	057	066	074	082	090	4 3.2
26	099	107	115	123	132	140	148	156	165	173	5 4.0 6 4.8
27	181	189	198	206	214	222	230	239	247	255	7 5.6
28	263	272	280	288	296	304	313	321	329	337	8 6.4
29	346	354	362	370	378	387	395	403	411	419	9 7.2
530	428	436	444	452	460	469	477	485	493	501	
31	509	518	526	534	542	550	558	567	575	583	
32	591	599	607	616	624	632	640	648	656	665	
33	673	681	689	697	705	713	722	730	738	746	
34	754	762	770	779	787	795	803	811	819	827	
35	835	843	852	860	868	876	884	892	900	908	
36	916	925	933	941	949	957	965	973	981	989	
37	72 997	*006	*014	*022	*030	*038	*046	*054	*062	*070	**7**
38	73 078	086	094	102	111	119	127	135	143	151	1 0.7
39	159	167	175	183	191	199	207	215	223	231	2 1.4
540	239	247	255	263	272	280	288	296	304	312	3 2.1 4 2.8
41	320	328	336	344	352	360	368	376	384	392	5 3.5
42	400	408	416	424	432	440	448	456	464	472	6 4.2
43	480	488	496	504	512	520	528	536	544	552	7 4.9
44	560	568	576	584	592	600	608	616	624	632	8 5.6
45	640	648	656	664	672	679	687	695	703	711	9 6.3
46	719	727	735	743	751	759	767	775	783	791	
47	799	807	815	823	830	838	846	854	862	870	
48	878	886	894	902	910	918	926	933	941	949	
49	73 957	965	973	981	989	997	*005	*013	*020	*028	
550	74 036	044	052	060	068	076	084	092	099	107	
N	0	1	2	3	4	5	6	7	8	9	Prop. Parts

TABLE A.1   FIVE-PLACE LOGARITHMS: 550—600

N	0	1	2	3	4	5	6	7	8	9
550	74 036	044	052	060	068	076	084	092	099	107
51	115	123	131	139	147	155	162	170	178	186
52	194	202	210	218	225	233	241	249	257	265
53	273	280	288	296	304	312	320	327	335	343
54	351	359	367	374	382	390	398	406	414	421
55	429	437	445	453	461	468	476	484	492	500
56	507	515	523	531	539	547	554	562	570	578
57	586	593	601	609	617	624	632	640	648	656
58	663	671	679	687	695	702	710	718	726	733
59	741	749	757	764	772	780	788	796	803	811
560	819	827	834	842	850	858	865	873	881	889
61	896	904	912	920	927	935	943	950	958	966
62	74 974	981	989	997	*005	*012	*020	*028	*035	*043
63	75 051	059	066	074	082	089	097	105	113	120
64	128	136	143	151	159	166	174	182	189	197
65	205	213	220	228	236	243	251	259	266	274
66	282	289	297	305	312	320	328	335	343	351
67	358	366	374	381	389	397	404	412	420	427
68	435	442	450	458	465	473	481	488	496	504
69	511	519	526	534	542	549	557	565	572	580
570	587	595	603	610	618	626	633	641	648	656
71	664	671	679	686	694	702	709	717	724	732
72	740	747	755	762	770	778	785	793	800	808
73	815	823	831	838	846	853	861	868	876	884
74	891	899	906	914	921	929	937	944	952	959
75	75 967	974	982	989	997	*005	*012	*020	*027	*035
76	76 042	050	057	065	072	080	087	095	103	110
77	118	125	133	140	148	155	163	170	178	185
78	193	200	208	215	223	230	238	245	253	260
79	268	275	283	290	298	305	313	320	328	335
580	343	350	358	365	373	380	388	395	403	410
81	418	425	433	440	448	455	462	470	477	485
82	492	500	507	515	522	530	537	545	552	559
83	567	574	582	589	597	604	612	619	626	634
84	641	649	656	664	671	678	686	693	701	708
85	716	723	730	738	745	753	760	768	775	782
86	790	797	805	812	819	827	834	842	849	856
87	864	871	879	886	893	901	908	916	923	930
88	76 938	945	953	960	967	975	982	989	997	*004
89	77 012	019	026	034	041	048	056	063	070	078
590	085	093	100	107	115	122	129	137	144	151
91	159	166	173	181	188	195	203	210	217	225
92	232	240	247	254	262	269	276	283	291	298
93	305	313	320	327	335	342	349	357	364	371
94	379	386	393	401	408	415	422	430	437	444
95	452	459	466	474	481	488	495	503	510	517
96	525	532	539	546	554	561	568	576	583	590
97	597	605	612	619	627	634	641	648	656	663
98	670	677	685	692	699	706	714	721	728	735
99	743	750	757	764	772	779	786	793	801	808
600	77 815	822	830	837	844	851	859	866	873	880

Prop. Parts

8
1 | 0.8
2 | 1.6
3 | 2.4
4 | 3.2
5 | 4.0
6 | 4.8
7 | 5.6
8 | 6.4
9 | 7.2

7
1 | 0.7
2 | 1.4
3 | 2.1
4 | 2.8
5 | 3.5
6 | 4.2
7 | 4.9
8 | 5.6
9 | 6.3

TABLE A.1   FIVE-PLACE LOGARITHMS: 600—650

N	0	1	2	3	4	5	6	7	8	9	Prop. Parts
600	77 815	822	830	837	844	851	859	866	873	880	
01	887	895	902	909	916	924	931	938	945	952	
02	77 960	967	974	981	988	996	*003	*010	*017	*025	
03	78 032	039	046	053	061	068	075	082	089	097	
04	104	111	118	125	132	140	147	154	161	168	
05	176	183	190	197	204	211	219	226	233	240	
06	247	254	262	269	276	283	290	297	305	312	
07	319	326	333	340	347	355	362	369	376	383	
08	390	398	405	412	419	426	433	440	447	455	
09	462	469	476	483	490	497	504	512	519	526	
610	533	540	547	554	561	569	576	583	590	597	
11	604	611	618	625	633	640	647	654	661	668	
12	675	682	689	696	704	711	718	725	732	739	
13	746	753	760	767	774	781	789	796	803	810	
14	817	824	831	838	845	852	859	866	873	880	
15	888	895	902	909	916	923	930	937	944	951	
16	78 958	965	972	979	986	993	*000	*007	*014	*021	
17	79 029	036	043	050	057	064	071	078	085	092	
18	099	106	113	120	127	134	141	148	155	162	
19	169	176	183	190	197	204	211	218	225	232	
620	239	246	253	260	267	274	281	288	295	302	
21	309	316	323	330	337	344	351	358	365	372	
22	379	386	393	400	407	414	421	428	435	442	
23	449	456	463	470	477	484	491	498	505	511	
24	518	525	532	539	546	553	560	567	574	581	
25	588	595	602	609	616	623	630	637	644	650	
26	657	664	671	678	685	692	699	706	713	720	
27	727	734	741	748	754	761	768	775	782	789	
28	796	803	810	817	824	831	837	844	851	858	
29	865	872	879	886	893	900	906	913	920	927	
630	79 934	941	948	955	962	969	975	982	989	996	
31	80 003	010	017	024	030	037	044	051	058	065	
32	072	079	085	092	099	106	113	120	127	134	
33	140	147	154	161	168	175	182	188	195	202	
34	209	216	223	229	236	243	250	257	264	271	
35	277	284	291	298	305	312	318	325	332	339	
36	346	353	359	366	373	380	387	393	400	407	
37	414	421	428	434	441	448	455	462	468	475	
38	482	489	496	502	509	516	523	530	536	543	
39	550	557	564	570	577	584	591	598	604	611	
640	618	625	632	638	645	652	659	665	672	679	
41	686	693	699	706	713	720	726	733	740	747	
42	754	760	767	774	781	787	794	801	808	814	
43	821	828	835	841	848	855	862	868	875	882	
44	889	895	902	909	916	922	929	936	943	949	
45	80 956	963	969	976	983	990	996	*003	*010	*017	
46	81 023	030	037	043	050	057	064	070	077	084	
47	090	097	104	111	117	124	131	137	144	151	
48	158	164	171	178	184	191	198	204	211	218	
49	224	231	238	245	251	258	265	271	278	285	
650	81 291	298	305	311	318	325	331	338	345	351	
N	0	1	2	3	4	5	6	7	8	9	Prop. Parts

Prop. Parts

	8
1	0.8
2	1.6
3	2.4
4	3.2
5	4.0
6	4.8
7	5.6
8	6.4
9	7.2

	7
1	0.7
2	1.4
3	2.1
4	2.8
5	3.5
6	4.2
7	4.9
8	5.6
9	6.3

	6
1	0.6
2	1.2
3	1.8
4	2.4
5	3.0
6	3.6
7	4.2
8	4.8
9	5.4

TABLE A.1 FIVE-PLACE LOGARITHMS: 650—700

Prop. Parts	N	0	1	2	3	4	5	6	7	8	9
	650	81 291	298	305	311	318	325	331	338	345	351
	51	358	365	371	378	385	391	398	405	411	418
	52	425	431	438	445	451	458	465	471	478	485
	53	491	498	505	511	518	525	531	538	544	551
	54	558	564	571	578	584	591	598	604	611	617
	55	624	631	637	644	651	657	664	671	677	684
	56	690	697	704	710	717	723	730	737	743	750
	57	757	763	770	776	783	790	796	803	809	816
	58	823	829	836	842	849	856	862	869	875	882
	59	889	895	902	908	915	921	928	935	941	948
	660	81 954	961	968	974	981	987	994	*000	*007	*014
	61	82 020	027	033	040	046	053	060	066	073	079
	62	086	092	099	105	112	119	125	132	138	145
7	63	151	158	164	171	178	184	191	197	204	210
1 0.7	64	217	223	230	236	243	249	256	263	269	276
2 1.4	65	282	289	295	302	308	315	321	328	334	341
3 2.1	66	347	354	360	367	373	380	387	393	400	406
4 2.8	67	413	419	426	432	439	445	452	458	465	471
5 3.5	68	478	484	491	497	504	510	517	523	530	536
6 4.2	69	543	549	556	562	569	575	582	588	595	601
7 4.9	670	607	614	620	627	633	640	646	653	659	666
8 5.6	71	672	679	685	692	698	705	711	718	724	730
9 6.3	72	737	743	750	756	763	769	776	782	789	795
	73	802	808	814	821	827	834	840	847	853	860
	74	866	872	879	885	892	898	905	911	918	924
	75	930	937	943	950	956	963	969	975	982	988
	76	82 995	*001	*008	*014	*020	*027	*033	*040	*046	*052
	77	83 059	065	072	078	085	091	097	104	110	117
	78	123	129	136	142	149	155	161	168	174	181
	79	187	193	200	206	213	219	225	232	238	245
	680	251	257	264	270	276	283	289	296	302	308
	81	315	321	327	334	340	347	353	359	366	372
	82	378	385	391	398	404	410	417	423	429	436
6	83	442	448	455	461	467	474	480	487	493	499
1 0.6	84	506	512	518	525	531	537	544	550	556	563
2 1.2	85	569	575	582	588	594	601	607	613	620	626
3 1.8	86	632	639	645	651	658	664	670	677	683	689
4 2.4	87	696	702	708	715	721	727	734	740	746	753
5 3.0	88	759	765	771	778	784	790	797	803	809	816
6 3.6	89	822	828	835	841	847	853	860	866	872	879
7 4.2	690	885	891	897	904	910	916	923	929	935	942
8 4.8	91	83 948	954	960	967	973	979	985	992	998	*004
9 5.4	92	84 011	017	023	029	036	042	048	055	061	067
	93	073	080	086	092	098	105	111	117	123	130
	94	136	142	148	155	161	167	173	180	186	192
	95	198	205	211	217	223	230	236	242	248	255
	96	261	267	273	280	286	292	298	305	311	317
	97	323	330	336	342	348	354	361	367	373	379
	98	386	392	398	404	410	417	423	429	435	442
	99	448	454	460	466	473	479	485	491	497	504
	700	84 510	516	522	528	535	541	547	553	559	566
Prop. Parts	N	0	1	2	3	4	5	6	7	8	9

TABLE A.1   FIVE-PLACE LOGARITHMS: 700—750

N	0	1	2	3	4	5	6	7	8	9	Prop. Parts
700	84 510	516	522	528	535	541	547	553	559	566	
01	572	578	584	590	597	603	609	615	621	628	
02	634	640	646	652	658	665	671	677	683	689	
03	696	702	708	714	720	726	733	739	745	751	
04	757	763	770	776	782	788	794	800	807	813	
05	819	825	831	837	844	850	856	862	868	874	
06	880	887	893	899	905	911	917	924	930	936	
07	84 942	948	954	960	967	973	979	985	991	997	
08	85 003	009	016	022	028	034	040	046	052	058	
09	065	071	077	083	089	095	101	107	114	120	
710	126	132	138	144	150	156	163	169	175	181	
11	187	193	199	205	211	217	224	230	236	242	
12	248	254	260	266	272	278	285	291	297	303	
13	309	315	321	327	333	339	345	352	358	364	
14	370	376	382	388	394	400	406	412	418	425	
15	431	437	443	449	455	461	467	473	479	485	
16	491	497	503	509	516	522	528	534	540	546	
17	552	558	564	570	576	582	588	594	600	606	
18	612	618	625	631	637	643	649	655	661	667	
19	673	679	685	691	697	703	709	715	721	727	
720	733	739	745	751	757	763	769	775	781	788	
21	794	800	806	812	818	824	830	836	842	848	
22	854	860	866	872	878	884	890	896	902	908	
23	914	920	926	932	938	944	950	956	962	968	
24	85 974	980	986	992	998	*004	*010	*016	*022	*028	
25	86 034	040	046	052	058	064	070	076	082	088	
26	094	100	106	112	118	124	130	136	141	147	
27	153	159	165	171	177	183	189	195	201	207	
28	213	219	225	231	237	243	249	255	261	267	
29	273	279	285	291	297	303	308	314	320	326	
730	332	338	344	350	356	362	368	374	380	386	
31	392	398	404	410	415	421	427	433	439	445	
32	451	457	463	469	475	481	487	493	499	504	
33	510	516	522	528	534	540	546	552	558	564	
34	570	576	581	587	593	599	605	611	617	623	
35	629	635	641	646	652	658	664	670	676	682	
36	688	694	700	705	711	717	723	729	735	741	
37	747	753	759	764	770	776	782	788	794	800	
38	806	812	817	823	829	835	841	847	853	859	
39	864	870	876	882	888	894	900	906	911	917	
740	923	929	935	941	947	953	958	964	970	976	
41	86 982	988	994	999	*005	*011	*017	*023	*029	*035	
42	87 040	046	052	058	064	070	075	081	087	093	
43	099	105	111	116	122	128	134	140	146	151	
44	157	163	169	175	181	186	192	198	204	210	
45	216	221	227	233	239	245	251	256	262	268	
46	274	280	286	291	297	303	309	315	320	326	
47	332	338	344	349	355	361	367	373	379	384	
48	390	396	402	408	413	419	425	431	437	442	
49	448	454	460	466	471	477	483	489	495	500	
750	87 506	512	518	523	529	535	541	547	552	558	
N	0	1	2	3	4	5	6	7	8	9	Prop. Parts

Prop. Parts

7
1 0.7
2 1.4
3 2.1
4 2.8
5 3.5
6 4.2
7 4.9
8 5.6
9 6.3

6
1 0.6
2 1.2
3 1.8
4 2.4
5 3.0
6 3.6
7 4.2
8 4.8
9 5.4

5
1 0.5
2 1.0
3 1.5
4 2.0
5 2.5
6 3.0
7 3.5
8 4.0
9 4.5

TABLE A.1   FIVE-PLACE LOGARITHMS: 750—800

Prop. Parts		N	0	1	2	3	4	5	6	7	8	9
		750	87 506	512	518	523	529	535	541	547	552	558
		51	564	570	576	581	587	593	599	604	610	616
		52	622	628	633	639	645	651	656	662	668	674
		53	679	685	691	697	703	708	714	720	726	731
		54	737	743	749	754	760	766	772	777	783	789
		55	795	800	806	812	818	823	829	835	841	846
		56	852	858	864	869	875	881	887	892	898	904
		57	910	915	921	927	933	938	944	950	955	961
		58	87 967	973	978	984	990	996	*001	*007	*013	*018
		59	88 024	030	036	041	047	053	058	064	070	076
		760	081	087	093	098	104	110	116	121	127	133
		61	138	144	150	156	161	167	173	178	184	190
		62	195	201	207	213	218	224	230	235	241	247
	6	63	252	258	264	270	275	281	287	292	298	304
1	0.6											
2	1.2											
3	1.8	64	309	315	321	326	332	338	343	349	355	360
4	2.4	65	366	372	377	383	389	395	400	406	412	417
5	3.0	66	423	429	434	440	446	451	457	463	468	474
6	3.6											
7	4.2	67	480	485	491	497	502	508	513	519	525	530
8	4.8	68	536	542	547	553	559	564	570	576	581	587
9	5.4	69	593	598	604	610	615	621	627	632	638	643
		770	649	655	660	666	672	677	683	689	694	700
		71	705	711	717	722	728	734	739	745	750	756
		72	762	767	773	779	784	790	795	801	807	812
		73	818	824	829	835	840	846	852	857	863	868
		74	874	880	885	891	897	902	908	913	919	925
		75	930	936	941	947	953	958	964	969	975	981
		76	88 986	992	997	*003	*009	*014	*020	*025	*031	*037
		77	89 042	048	053	059	064	070	076	081	087	092
		78	098	104	109	115	120	126	131	137	143	148
		79	154	159	165	170	176	182	187	193	198	204
		780	209	215	221	226	232	237	243	248	254	260
		81	265	271	276	282	287	293	298	304	310	315
		82	321	326	332	337	343	348	354	360	365	371
	5	83	376	382	387	393	398	404	409	415	421	426
1	0.5											
2	1.0											
3	1.5	84	432	437	443	448	454	459	465	470	476	481
4	2.0	85	487	492	498	504	509	515	520	526	531	537
5	2.5	86	542	548	553	559	564	570	575	581	586	592
6	3.0											
7	3.5	87	597	603	609	614	620	625	631	636	642	647
8	4.0	88	653	658	664	669	675	680	686	691	697	702
9	4.5	89	708	713	719	724	730	735	741	746	752	757
		790	763	768	774	779	785	790	796	801	807	812
		91	818	823	829	834	840	845	851	856	862	867
		92	873	878	883	889	894	900	905	911	916	922
		93	927	933	938	944	949	955	960	966	971	977
		94	89 982	988	993	998	*004	*009	*015	*020	*026	*031
		95	90 037	042	048	053	059	064	069	075	080	086
		96	091	097	102	108	113	119	124	129	135	140
		97	146	151	157	162	168	173	179	184	189	195
		98	200	206	211	217	222	227	233	238	244	249
		99	255	260	266	271	276	282	287	293	298	304
		800	90 309	314	320	325	331	336	342	347	352	358
Prop. Parts		N	0	1	2	3	4	5	6	7	8	9

TABLE A.1   FIVE-PLACE LOGARITHMS: 800—850

N	0	1	2	3	4	5	6	7	8	9
800	90 309	314	320	325	331	336	342	347	352	358
01	363	369	374	380	385	390	396	401	407	412
02	417	423	428	434	439	445	450	455	461	466
03	472	477	482	488	493	499	504	509	515	520
04	526	531	536	542	547	553	558	563	569	574
05	580	585	590	596	601	607	612	617	623	628
06	634	639	644	650	655	660	666	671	677	682
07	687	693	698	703	709	714	720	725	730	736
08	741	747	752	757	763	768	773	779	784	789
09	795	800	806	811	816	822	827	832	838	843
810	849	854	859	865	870	875	881	886	891	897
11	902	907	913	918	924	929	934	940	945	950
12	90 956	961	966	972	977	982	988	993	998	*004
13	91 009	014	020	025	030	036	041	046	052	057
14	062	068	073	078	084	089	094	100	105	110
15	116	121	126	132	137	142	148	153	158	164
16	169	174	180	185	190	196	201	206	212	217
17	222	228	233	238	243	249	254	259	265	270
18	275	281	286	291	297	302	307	312	318	323
19	328	334	339	344	350	355	360	365	371	376
820	381	387	392	397	403	408	413	418	424	429
21	434	440	445	450	455	461	466	471	477	482
22	487	492	498	503	508	514	519	524	529	535
23	540	545	551	556	561	566	572	577	582	587
24	593	598	603	609	614	619	624	630	635	640
25	645	651	656	661	666	672	677	682	687	693
26	698	703	709	714	719	724	730	735	740	745
27	751	756	761	766	772	777	782	787	793	798
28	803	808	814	819	824	829	834	840	845	850
29	855	861	866	871	876	882	887	892	897	903
830	908	913	918	924	929	934	939	944	950	955
31	91 960	965	971	976	981	986	991	997	*002	*007
32	92 012	018	023	028	033	038	044	049	054	059
33	065	070	075	080	085	091	096	101	106	111
34	117	122	127	132	137	143	148	153	158	163
35	169	174	179	184	189	195	200	205	210	215
36	221	226	231	236	241	247	252	257	262	267
37	273	278	283	288	293	298	304	309	314	319
38	324	330	335	340	345	350	355	361	366	371
39	376	381	387	392	397	402	407	412	418	423
840	428	433	438	443	449	454	459	464	469	474
41	480	485	490	495	500	505	511	516	521	526
42	531	536	542	547	552	557	562	567	572	578
43	583	588	593	598	603	609	614	619	624	629
44	634	639	645	650	655	660	665	670	675	681
45	686	691	696	701	706	711	716	722	727	732
46	737	742	747	752	758	763	768	773	778	783
47	788	793	799	804	809	814	819	824	829	834
48	840	845	850	855	860	865	870	875	881	886
49	891	896	901	906	911	916	921	927	932	937
850	92 942	947	952	957	962	967	973	978	983	988
N	0	1	2	3	4	5	6	7	8	9

Prop. Parts

	6
1	0.6
2	1.2
3	1.8
4	2.4
5	3.0
6	3.6
7	4.2
8	4.8
9	5.4

	5
1	0.5
2	1.0
3	1.5
4	2.0
5	2.5
6	3.0
7	3.5
8	4.0
9	4.5

TABLE A.1   FIVE-PLACE LOGARITHMS: 850—900

N	0	1	2	3	4	5	6	7	8	9
850	92 942	947	952	957	962	967	973	978	983	988
51	92 993	998	*003	*008	*013	*018	*024	*029	*034	*039
52	93 044	049	054	059	064	069	075	080	085	090
53	095	100	105	110	115	120	125	131	136	141
54	146	151	156	161	166	171	176	181	186	192
55	197	202	207	212	217	222	227	232	237	242
56	247	252	258	263	268	273	278	283	288	293
57	298	303	308	313	318	323	328	334	339	344
58	349	354	359	364	369	374	379	384	389	394
59	399	404	409	414	420	425	430	435	440	445
860	450	455	460	465	470	475	480	485	490	495
61	500	505	510	515	520	526	531	536	541	546
62	551	556	561	566	571	576	581	586	591	596
63	601	606	611	616	621	626	631	636	641	646
64	651	656	661	666	671	676	682	687	692	697
65	702	707	712	717	722	727	732	737	742	747
66	752	757	762	767	772	777	782	787	792	797
67	802	807	812	817	822	827	832	837	842	847
68	852	857	862	867	872	877	882	887	892	897
69	902	907	912	917	922	927	932	937	942	947
870	93 952	957	962	967	972	977	982	987	992	997
71	94 002	007	012	017	022	027	032	037	042	047
72	052	057	062	067	072	077	082	086	091	096
73	101	106	111	116	121	126	131	136	141	146
74	151	156	161	166	171	176	181	186	191	196
75	201	206	211	216	221	226	231	236	240	245
76	250	255	260	265	270	275	280	285	290	295
77	300	305	310	315	320	325	330	335	340	345
78	349	354	359	364	369	374	379	384	389	394
79	399	404	409	414	419	424	429	433	438	443
880	448	453	458	463	468	473	478	483	488	493
81	498	503	507	512	517	522	527	532	537	542
82	547	552	557	562	567	571	576	581	586	591
83	596	601	606	611	616	621	626	630	635	640
84	645	650	655	660	665	670	675	680	685	689
85	694	699	704	709	714	719	724	729	734	738
86	743	748	753	758	763	768	773	778	783	787
87	792	797	802	807	812	817	822	827	832	836
88	841	846	851	856	861	866	871	876	880	885
89	890	895	900	905	910	915	919	924	929	934
890	939	944	949	954	959	963	968	973	978	983
91	94 988	993	998	*002	*007	*012	*017	*022	*027	*032
92	95 036	041	046	051	056	061	066	071	075	080
93	085	090	095	100	105	109	114	119	124	129
94	134	139	143	148	153	158	163	168	173	177
95	182	187	192	197	202	207	211	216	221	226
96	231	236	240	245	250	255	260	265	270	274
97	279	284	289	294	299	303	308	313	318	323
98	328	332	337	342	347	352	357	361	366	371
99	376	381	386	390	395	400	405	410	415	419
900	95 424	429	434	439	444	448	453	458	463	468

Prop. Parts

	6		5		4
1	0.6	1	0.5	1	0.4
2	1.2	2	1.0	2	0.8
3	1.8	3	1.5	3	1.2
4	2.4	4	2.0	4	1.6
5	3.0	5	2.5	5	2.0
6	3.6	6	3.0	6	2.4
7	4.2	7	3.5	7	2.8
8	4.8	8	4.0	8	3.2
9	5.4	9	4.5	9	3.6

TABLE A.1   FIVE-PLACE LOGARITHMS: 900—950

N	0	1	2	3	4	5	6	7	8	9	Prop. Parts
900	95 424	429	434	439	444	448	453	458	463	468	
01	472	477	482	487	492	497	501	506	511	516	
02	521	525	530	535	540	545	550	554	559	564	
03	569	574	578	583	588	593	598	602	607	612	
04	617	622	626	631	636	641	646	650	655	660	
05	665	670	674	679	684	689	694	698	703	708	
06	713	718	722	727	732	737	742	746	751	756	
07	761	766	770	775	780	785	789	794	799	804	
08	809	813	818	823	828	832	837	842	847	852	
09	856	861	866	871	875	880	885	890	895	899	
910	904	909	914	918	923	928	933	938	942	947	
11	952	957	961	966	971	976	980	985	990	995	
12	95 999	*004	*009	*014	*019	*023	*028	*033	*038	*042	**5**
13	96 047	052	057	061	066	071	076	080	085	090	**1** 0.5
14	095	099	104	109	114	118	123	128	133	137	**2** 1.0 / **3** 1.5
15	142	147	152	156	161	166	171	175	180	185	**4** 2.0 / **5** 2.5
16	190	194	199	204	209	213	218	223	227	232	**6** 3.0
17	237	242	246	251	256	261	265	270	275	280	**7** 3.5
18	284	289	294	298	303	308	313	317	322	327	**8** 4.0
19	332	336	341	346	350	355	360	365	369	374	**9** 4.5
920	379	384	388	393	398	402	407	412	417	421	
21	426	431	435	440	445	450	454	459	464	468	
22	473	478	483	487	492	497	501	506	511	515	
23	520	525	530	534	539	544	548	553	558	562	
24	567	572	577	581	586	591	595	600	605	609	
25	614	619	624	628	633	638	642	647	652	656	
26	661	666	670	675	680	685	689	694	699	703	
27	708	713	717	722	727	731	736	741	745	750	
28	755	759	764	769	774	778	783	788	792	797	
29	802	806	811	816	820	825	830	834	839	844	
930	848	853	858	862	867	872	876	881	886	890	
31	895	900	904	909	914	918	923	928	932	937	**4**
32	942	946	951	956	960	965	970	974	979	984	**1** 0.4
33	96 988	993	997	*002	*007	*011	*016	*021	*025	*030	**2** 0.8
34	97 035	039	044	049	053	058	063	067	072	077	**3** 1.2
35	081	086	090	095	100	104	109	114	118	123	**4** 1.6 / **5** 2.0
36	128	132	137	142	146	151	155	160	165	169	**6** 2.4
37	174	179	183	188	192	197	202	206	211	216	**7** 2.8
38	220	225	230	234	239	243	248	253	257	262	**8** 3.2
39	267	271	276	280	285	290	294	299	304	308	**9** 3.6
940	313	317	322	327	331	336	340	345	350	354	
41	359	364	368	373	377	382	387	391	396	400	
42	405	410	414	419	424	428	433	437	442	447	
43	451	456	460	465	470	474	479	483	488	493	
44	497	502	506	511	516	520	525	529	534	539	
45	543	548	552	557	562	566	571	575	580	585	
46	589	594	598	603	607	612	617	621	626	630	
47	635	640	644	649	653	658	663	667	672	676	
48	681	685	690	695	699	704	708	713	717	722	
49	727	731	736	740	745	749	754	759	763	768	
950	97 772	777	782	786	791	795	800	804	809	813	
N	0	1	2	3	4	5	6	7	8	9	Prop. Parts

TABLE A.1   FIVE-PLACE LOGARITHMS: 950—1000

Prop. Parts	N	0	1	2	3	4	5	6	7	8	9
	950	97 772	777	782	786	791	795	800	804	809	813
	51	818	823	827	832	836	841	845	850	855	859
	52	864	868	873	877	882	886	891	896	900	905
	53	909	914	918	923	928	932	937	941	946	950
	54	97 955	959	964	968	973	978	982	987	991	996
	55	98 000	005	009	014	019	023	028	032	037	041
	56	046	050	055	059	064	068	073	078	082	087
	57	091	096	100	105	109	114	118	123	127	132
	58	137	141	146	150	155	159	164	168	173	177
	59	182	186	191	195	200	204	209	214	218	223
	960	227	232	236	241	245	250	254	259	263	268
5	61	272	277	281	286	290	295	299	304	308	313
1 0.5	62	318	322	327	331	336	340	345	349	354	358
2 1.0	63	363	367	372	376	381	385	390	394	399	403
3 1.5	64	408	412	417	421	426	430	435	439	444	448
4 2.0	65	453	457	462	466	471	475	480	484	489	493
5 2.5	66	498	502	507	511	516	520	525	529	534	538
6 3.0	67	543	547	552	556	561	565	570	574	579	583
7 3.5	68	588	592	597	601	605	610	614	619	623	628
8 4.0	69	632	637	641	646	650	655	659	664	668	673
9 4.5	**970**	677	682	686	691	695	700	704	709	713	717
	71	722	726	731	735	740	744	749	753	758	762
	72	767	771	776	780	784	789	793	798	802	807
	73	811	816	820	825	829	834	838	843	847	851
	74	856	860	865	869	874	878	883	887	892	896
	75	900	905	909	914	918	923	927	932	936	941
	76	945	949	954	958	963	967	972	976	981	985
	77	98 989	994	998	*003	*007	*012	*016	*021	*025	*029
	78	99 034	038	043	047	052	056	061	065	069	074
	79	078	083	087	092	096	100	105	109	114	118
	980	123	127	131	136	140	145	149	154	158	162
4	81	167	171	176	180	185	189	193	198	202	207
1 0.4	82	211	216	220	224	229	233	238	242	247	251
2 0.8	83	255	260	264	269	273	277	282	286	291	295
3 1.2	84	300	304	308	313	317	322	326	330	335	339
4 1.6	85	344	348	352	357	361	366	370	374	379	383
5 2.0	86	388	392	396	401	405	410	414	419	423	427
6 2.4	87	432	436	441	445	449	454	458	463	467	471
7 2.8	88	476	480	484	489	493	498	502	506	511	515
8 3.2	89	520	524	528	533	537	542	546	550	555	559
9 3.6	**990**	564	568	572	577	581	585	590	594	599	603
	91	607	612	616	621	625	629	634	638	642	647
	92	651	656	660	664	669	673	677	682	686	691
	93	695	699	704	708	712	717	721	726	730	734
	94	739	743	747	752	756	760	765	769	774	778
	95	782	787	791	795	800	804	808	813	817	822
	96	826	830	835	839	843	848	852	856	861	865
	97	870	874	878	883	887	891	896	900	904	909
	98	913	917	922	926	930	935	939	944	948	952
	99	99 957	961	965	970	974	978	983	987	991	996
	1000	00 000	004	009	013	017	022	026	030	035	039
Prop. Parts	N	0	1	2	3	4	5	6	7	8	9

Useful Formulas and Relationships and Their Logarithms

Relationship	Number or Formula	Logarithm
Area of a circle	πr^2	
Area of a sphere	$4\pi r^2$	
The circumference to the diameter of a circle (π)	3.141593	0.49715
Statute miles in equatorial circumference	24 902	4.39623
Kilometers in equatorial circumference	40 075	4.60287
Miles in radius of sphere of equal area	3 958.7	3.59755
Kilometers in radius of sphere of equal area	6 370.9	3.80420
Square miles in area of earth	197 260 000 (approx.)	8.29504
Square kilometers in area of earth	510 900 000 (approx.)	8.70834
Inches in one foot	12.0	1.07918
Feet in one statute mile	5 280.0	3.72263
Inches in one statute mile	63 360.0	4.80182
Statute miles in one nautical mile (Int.)	1.1508	0.06099
Inches in one centimeter	0.3937	9.59517 − 10
Centimeters in one inch	2.540	0.40483
Statute miles in one kilometer	0.62137	9.79335 − 10
Kilometers in one statute mile	1.60935	0.20665
Feet in one kilometer	3 280.83	3.51598
Square statute miles in one square kilometer	0.3861	9.58670 − 10
Square kilometers in one square statute mile	2.590	0.41330

Note. Because of rounding of numbers and differences in definitions, some internal inconsistencies appear in the above table.

APPENDIX

B NATURAL TRIGONOMETRIC FUNCTIONS

Table B.1 gives the values of the *sine, cosine, tangent,* and *cotangent* of degrees from 0° to 90°: For degrees at the left use the column headings at the top; for degrees at the right use column headings at the bottom.

For fractions of degrees an appropriate amount of the difference between adjacent values may be used. For precise calculations, however, one should employ a more complete table and especially one showing the logarithms of the trigonometric functions.

The values of the *secant* and *cosecant* may be derived as follows: secant = 1 ÷ cos; cosecant = 1 ÷ sin.

TABLE B.1

NATURAL TRIGONOMETRIC FUNCTIONS

°	Sin	Tan	Cot	Cos	°	°	Sin	Tan	Cot	Cos	°
0	.0000	.0000	—	1.0000	90	23	.3907	.4245	2.356	.9205	67
1	.0174	.0175	57.290	.9998	89	24	.4067	.4452	2.246	.9135	66
2	.0349	.0349	28.636	.9994	88	25	.4226	.4663	2.144	.9063	65
3	.0523	.0524	19.081	.9986	87	26	.4384	.4877	2.050	.8988	64
4	.0698	.0699	14.301	.9976	86	27	.4540	.5095	1.963	.8910	63
5	.0872	.0875	11.430	.9962	85	28	.4695	.5317	1.881	.8829	62
6	.1045	.1051	9.514	.9945	84	29	.4848	.5543	1.804	.8746	61
7	.1219	.1228	8.144	.9925	83						
8	.1392	.1405	7.115	.9903	82	30	.5000	.5773	1.732	.8660	60
9	.1564	.1584	6.314	.9877	81	31	.5150	.6009	1.664	.8572	59
						32	.5299	.6249	1.600	.8480	58
10	.1736	.1763	5.671	.9848	80	33	.5446	.6494	1.540	.8387	57
11	.1908	.1944	5.145	.9816	79	34	.5592	.6745	1.483	.8290	56
12	.2079	.2126	4.705	.9781	78	35	.5736	.7002	1.428	.8191	55
13	.2249	.2309	4.331	.9744	77	36	.5878	.7265	1.376	.8090	54
14	.2419	.2493	4.011	.9703	76	37	.6018	.7535	1.327	.7986	53
15	.2588	.2679	3.732	.9659	75	38	.6157	.7813	1.280	.7880	52
16	.2756	.2867	3.487	.9613	74	39	.6293	.8098	1.235	.7771	51
17	.2924	.3057	3.271	.9563	73						
18	.3090	.3249	3.078	.9511	72	40	.6428	.8391	1.192	.7660	50
19	.3256	.3443	2.904	.9455	71	41	.6561	.8693	1.150	.7547	49
						42	.6691	.9004	1.111	.7431	48
20	.3420	.3640	2.747	.9397	70	43	.6820	.9325	1.072	.7313	47
21	.3584	.3839	2.605	.9336	69	44	.6947	.9657	1.035	.7193	46
22	.3746	.4040	2.475	.9272	68	45	.7071	1.0000	1.000	.7071	45
°	Cos	Cot	Tan	Sin	°	°	Cos	Cot	Tan	Sin	°

Trigonometric Functions of an Acute Angle in a Right Triangle

$$\text{sine} = \frac{\text{opposite side}}{\text{hypotenuse}}$$

$$\text{cosine} = \frac{\text{adjacent side}}{\text{hypotenuse}}$$

$$\text{tangent} = \frac{\text{opposite side}}{\text{adjacent side}}$$

$$\text{cotangent} = \frac{\text{adjacent side}}{\text{opposite side}}$$

$$\text{secant} = \frac{\text{hypotenuse}}{\text{adjacent side}}$$

$$\text{cosecant} = \frac{\text{hypotenuse}}{\text{opposite side}}$$

APPENDIX

C SQUARES, CUBES, AND ROOTS

TABLE C.1
SQUARES, CUBES, AND ROOTS

n	n^2	$\sqrt{n}$	$\sqrt{10n}$	n^3	$\sqrt[3]{n}$	$\sqrt[3]{10n}$
1	1	1.000	3.162	1	1.000	2.154
2	4	1.414	4.472	8	1.260	2.714
3	9	1.732	5.477	27	1.442	3.107
4	16	2.000	6.325	64	1.587	3.420
5	25	2.236	7.071	125	1.710	3.684
6	36	2.449	7.746	216	1.817	3.915
7	49	2.646	8.367	343	1.913	4.121
8	64	2.828	8.944	512	2.000	4.309
9	81	3.000	9.487	729	2.080	4.481
10	100	3.162	10.000	1 000	2.154	4.642
11	121	3.317	10.488	1 331	2.224	4.791
12	144	3.464	10.954	1 728	2.289	4.932
13	169	3.606	11.402	2 197	2.351	5.066
14	196	3.742	11.832	2 744	2.410	5.192
15	225	3.873	12.247	3 375	2.466	5.313
16	256	4.000	12.649	4 096	2.520	5.429
17	289	4.123	13.038	4 913	2.571	5.540
18	324	4.243	13.416	5 832	2.621	5.646
19	361	4.359	13.784	6 859	2.668	5.749
20	400	4.472	14.142	8 000	2.714	5.848
21	441	4.583	14.491	9 261	2.759	5.944
22	484	4.690	14.832	10 648	2.802	6.037
23	529	4.796	15.166	12 167	2.844	6.127
24	576	4.899	15.492	13 824	2.884	6.214
25	625	5.000	15.811	15 625	2.924	6.300

TABLE C.1

SQUARES, CUBES, AND ROOTS (Continued)

n	n^2	$\sqrt{n}$	$\sqrt{10n}$	n^3	$\sqrt[3]{n}$	$\sqrt[3]{10n}$
26	676	5.099	16.125	17 576	2.962	6.383
27	729	5.196	16.432	19 683	3.000	6.463
28	784	5.292	16.733	21 952	3.037	6.542
29	841	5.385	17.029	24 389	3.072	6.619
30	900	5.477	17.321	27 000	3.107	6.694
31	961	5.568	17.607	29 791	3.141	6.768
32	1 024	5.657	17.889	32 768	3.175	6.840
33	1 089	5.745	18.166	35 937	3.208	6.910
34	1 156	5.831	18.439	39 304	3.240	6.980
35	1 225	5.916	18.708	42 875	3.271	7.047
36	1 296	6.000	18.974	46 656	3.302	7.114
37	1 369	6.083	19.235	50 653	3.332	7.179
38	1 444	6.164	19.494	54 872	3.362	7.243
39	1 521	6.245	19.748	59 319	3.391	7.306
40	1 600	6.325	20.000	64 000	3.420	7.368
41	1 681	6.403	20.248	68 921	3.448	7.429
42	1 764	6.481	20.494	74 088	3.476	7.489
43	1 849	6.557	20.736	79 507	3.503	7.548
44	1 936	6.633	20.976	85 184	3.530	7.606
45	2 025	6.708	21.213	91 125	3.557	7.663
46	2 116	6.782	21.448	97 336	3.583	7.719
47	2 209	6.856	21.679	103 823	3.609	7.775
48	2 304	6.928	21.909	110 592	3.634	7.830
49	2 401	7.000	22.136	117 649	3.659	7.884
50	2 500	7.071	22.361	125 000	3.684	7.937
51	2 601	7.141	22.583	132 651	3.708	7.990
52	2 704	7.211	22.804	140 608	3.733	8.041
53	2 809	7.280	23.022	148 877	3.756	8.093
54	2 916	7.348	23.238	157 464	3.780	8.143
55	3 025	7.416	23.452	166 375	3.803	8.193
56	3 136	7.483	23.664	175 616	3.826	8.243
57	3 249	7.550	23.875	185 193	3.849	8.291
58	3 364	7.616	24.083	195 112	3.871	8.340
59	3 481	7.681	24.290	205 379	3.893	8.387
60	3 600	7.746	24.495	216 000	3.915	8.434
61	3 721	7.810	24.698	226 981	3.936	8.481
62	3 844	7.874	24.900	238 328	3.958	8.527
63	3 969	7.937	25.100	250 047	3.979	8.573
64	4 096	8.000	25.298	262 144	4.000	8.618
65	4 225	8.062	25.495	274 625	4.021	8.662
66	4 356	8.124	25.690	287 496	4.041	8.707
67	4 489	8.185	25.884	300 763	4.062	8.750
68	4 624	8.246	26.077	314 432	4.082	8.794
69	4 761	8.307	26.268	328 509	4.102	8.837

TABLE C.1
SQUARES, CUBES, AND ROOTS (*Continued*)

n	n^2	$\sqrt{n}$	$\sqrt{10n}$	n^3	$\sqrt[3]{n}$	$\sqrt[3]{10n}$
70	4 900	8.367	26.458	343 000	4.121	8.879
71	5 041	8.426	26.646	357 911	4.141	8.921
72	5 184	8.485	26.833	373 248	4.160	8.963
73	5 329	8.544	27.019	389 017	4.179	9.004
74	5 476	8.602	27.203	405 224	4.198	9.045
75	5 625	8.660	27.386	421 875	4.217	9.086
76	5 776	8.718	27.568	438 976	4.236	9.126
77	5 929	8.775	27.749	456 533	4.254	9.166
78	6 084	8.832	27.928	474 552	4.273	9.205
79	6 241	8.888	28.107	493 039	4.291	9.244
80	6 400	8.944	28.284	512 000	4.309	9.283
81	6 561	9.000	28.461	531 441	4.327	9.322
82	6 724	9.055	28.636	551 368	4.344	9.360
83	6 889	9.110	28.810	571 787	4.362	9.398
84	7 056	9.165	28.983	592 704	4.380	9.435
85	7 225	9.220	29.155	614 125	4.397	9.473
86	7 396	9.274	29.326	636 056	4.414	9.510
87	7 569	9.327	29.496	658 503	4.431	9.546
88	7 744	9.381	29.665	681 472	4.448	9.583
89	7 921	9.434	29.833	704 969	4.465	9.619
90	8 100	9.487	30.000	729 000	4.481	9.655
91	8 281	9.539	30.166	753 571	4.498	9.691
92	8 464	9.592	30.332	778 688	4.514	9.726
93	8 649	9.644	30.496	804 357	4.531	9.761
94	8 836	9.695	30.659	830 584	4.547	9.796
95	9 025	9.747	30.822	857 375	4.563	9.830
96	9 216	9.798	30.984	884 736	4.579	9.865
97	9 409	9.849	31.145	912 673	4.595	9.899
98	9 604	9.899	31.305	941 192	4.610	9.933
99	9 801	9.950	31.464	970 299	4.626	9.967
100	10 000	10.000	31.623	1 000 000	4.642	10.000

APPENDIX

 GEOGRAPHIC TABLES

TABLE D.1

LENGTHS OF DEGREES OF THE PARALLEL

Lat.	Meters	Statute miles	Lat.	Meters	Statute miles	Lat.	Meters	Statute miles
° ′			° ′			° ′		
0 00	111 321	69.172	30 00	96 488	59.956	60 00	55 802	34.674
1 00	111 304	69.162	31 00	95 506	59.345	61 00	54 110	33.623
2 00	111 253	69.130	32 00	94 495	58.716	62 00	52 400	32.560
3 00	111 169	69.078	33 00	93 455	58.071	63 00	50 675	31.488
4 00	111 051	69.005	34 00	92 387	57.407	64 00	48 934	30.406
5 00	110 900	68.911	35 00	91 290	56.725	65 00	47 177	29.315
6 00	110 715	68.795	36 00	90 166	56.027	66 00	45 407	28.215
7 00	110 497	68.660	37 00	89 014	55.311	67 00	43 622	27.106
8 00	110 245	68.504	38 00	87 835	54.579	68 00	41 823	25.988
9 00	109 959	68.326	39 00	86 629	53.829	69 00	40 012	24.862
10 00	109 641	68.129	40 00	85 396	53.063	70 00	38 188	23.729
11 00	109 289	67.910	41 00	84 137	52.281	71 00	36 353	22.589
12 00	108 904	67.670	42 00	82 853	51.483	72 00	34 506	21.441
13 00	108 486	67.410	43 00	81 543	50.669	73 00	32 648	20.287
14 00	108 036	67.131	44 00	80 208	49.840	74 00	30 781	19.127
15 00	107 553	66.830	45 00	78 849	48.995	75 00	28 903	17.960
16 00	107 036	66.510	46 00	77 466	48.136	76 00	27 017	16.788
17 00	106 487	66.169	47 00	76 058	47.261	77 00	25 123	15.611
18 00	105 906	65.808	48 00	74 628	46.372	78 00	23 220	14.428
19 00	105 294	65.427	49 00	73 174	45.469	79 00	21 311	13.242
20 00	104 649	65.026	50 00	71 698	44.552	80 00	19 394	12.051
21 00	103 972	64.606	51 00	70 200	43.621	81 00	17 472	10.857
22 00	103 264	64.166	52 00	68 680	42.676	82 00	15 545	9.659
23 00	102 524	63.706	53 00	67 140	41.719	83 00	13 612	8.458
24 00	101 754	63.228	54 00	65 578	40.749	84 00	11 675	7.255
25 00	100 952	62.729	55 00	63 996	39.766	85 00	9 735	6.049
26 00	100 119	62.212	56 00	62 395	38.771	86 00	7 792	4.842
27 00	99 257	61.676	57 00	60 774	37.764	87 00	5 846	3.632
28 00	98 364	61.122	58 00	59 135	36.745	88 00	3 898	2.422
29 00	97 441	60.548	59 00	57 478	35.716	89 00	1 949	1.211
						90 00	0	0

* Tables D.1 and D.2 are from U.S. Coast and Geodetic Survey; Table D.3 is from *Smithsonian Geographical Tables*.

TABLE D.2
LENGTHS OF DEGREES OF THE MERIDIAN

Lat.	Meters	Statute miles	Lat.	Meters	Statute miles	Lat.	Meters	Statute miles
°			°			°		
0–1	110 567.3	68.703	30–31	110 857.0	68.883	60–61	111 423.1	69.235
1–2	110 568.0	68.704	31–32	110 874.4	68.894	61–62	111 439.9	69.246
2–3	110 569.4	68.705	32–33	110 892.1	68.905	62–63	111 456.4	69.256
3–4	110 571.4	68.706	33–34	110 910.1	68.916	63–64	111 472.4	69.266
4–5	110 574.1	68.707	34–35	110 928.3	68.928	64–65	111 488.1	69.275
5–6	110 577.6	68.710	35–36	110 946.9	68.939	65–66	111 503.3	69.285
6–7	110 581.6	68.712	36–37	110 965.6	68.951	66–67	111 518.0	69.294
7–8	110 586.4	68.715	37–38	110 984.5	68.962	67–68	111 532.3	69.303
8–9	110 591.8	68.718	38–39	111 003.7	68.974	68–69	111 546.2	69.311
9–10	110 597.8	68.722	39–40	111 023.0	68.986	69–70	111 559.5	69.320
10–11	110 604.5	68.726	40–41	111 042.4	68.998	70–71	111 572.2	69.328
11–12	110 611.9	68.731	41–42	111 061.9	69.011	71–72	111 584.5	69.335
12–13	110 619.8	68.736	42–43	111 081.6	69.023	72–73	111 596.2	69.343
13–14	110 628.4	68.741	43–44	111 101.3	69.035	73–74	111 607.3	69.349
14–15	110 637.6	68.747	44–45	111 121.0	69.047	74–75	111 617.9	69.356
15–16	110 647.5	68.753	45–46	111 140.8	69.060	75–76	111 627.8	69.362
16–17	110 657.8	68.759	46–47	111 160.5	69.072	76–77	111 637.1	69.368
17–18	110 668.8	68.766	47–48	111 180.2	69.084	77–78	111 645.9	69.373
18–19	110 680.4	68.773	48–49	111 199.9	69.096	78–79	111 653.9	69.378
19–20	110 692.4	68.781	49–50	111 219.5	69.108	79–80	111 661.4	69.383
20–21	110 705.1	68.789	50–51	111 239.0	69.121	80–81	111 668.2	69.387
21–22	110 718.2	68.797	51–52	111 258.3	69.133	81–82	111 674.4	69.391
22–23	110 731.8	68.805	52–53	111 277.6	69.145	82–83	111 679.9	69.395
23–24	110 746.0	68.814	53–54	111 296.6	69.156	83–84	111 684.7	69.398
24–25	110 760.6	68.823	54–55	111 315.4	69.168	84–85	111 688.9	69.400
25–26	110 775.6	68.833	55–56	111 334.0	69.180	85–86	111 692.3	69.402
26–27	110 791.1	68.842	56–57	111 352.4	69.191	86–87	111 695.1	69.404
27–28	110 807.0	68.852	57–58	111 370.5	69.202	87–88	111 697.2	69.405
28–29	110 823.3	68.862	58–59	111 388.4	69.213	88–89	111 698.6	69.406
29–30	110 840.0	68.873	59–60	111 405.9	69.224	89–90	111 699.3	69.407

TABLE D.3

AREAS OF QUADRILATERALS OF EARTH'S SURFACE OF 1° EXTENT IN LATITUDE AND LONGITUDE

Lower latitude of quadrilateral °	Area in square miles		Lower latitude of quadrilateral °	Area in square miles
0	4 752.16		45	3 354.01
1	4 750.75		46	3 294.71
2	4 747.93		47	3 234.39
3	4 743.71		48	3 173.04
4	4 738.08		49	3 110.69
5	4 731.04			
6	4 722.61		50	3 047.37
7	4 712.76		51	2 983.08
8	4 701.52		52	2 917.85
9	4 688.89		53	2 851.68
			54	2 784.62
10	4 674.86		55	2 716.67
11	4 659.43		56	2 647.85
12	4 642.63		57	2 578.19
13	4 624.44		58	2 507.70
14	4 604.87		59	2 436.42
15	4 583.92			
16	4 561.61		60	2 364.34
17	4 537.93		61	2 291.51
18	4 512.90		62	2 217.94
19	4 486.51		63	2 143.66
			64	2 068.68
20	4 458.78		65	1 993.04
21	4 429.71		66	1 916.75
22	4 399.30		67	1 839.84
23	4 367.57		68	1 762.33
24	4 334.52		69	1 684.24
25	4 300.17			
26	4 264.51		70	1 605.62
27	4 227.56		71	1 526.46
28	4 189.33		72	1 446.81
29	4 149.83		73	1 366.69
			74	1 286.12
30	4 109.06		75	1 205.13
31	4 067.05		76	1 123.75
32	4 023.79		77	1 041.99
33	3 979.30		78	959.90
34	3 933.59		79	877.49
35	3 886.67			
36	3 838.56		80	794.79
37	3 789.26		81	711.83
38	3 738.80		82	628.64
39	3 687.18		83	545.24
			84	461.66
40	3 634.42		85	377.93
41	3 580.54		86	294.08
42	3 525.54		87	210.12
43	3 469.44		88	126.10
44	3 412.26		89	42.04

APPENDIX

 TABLE FOR ESTIMATING VALUES OF FRACTIONAL AREAS

Table E.1, devised by John K. Wright, enables us to solve without multiplication or division the fundamental equation required to estimate densities of fractional areas as diagrammed in Fig. E.1.*

The basic equation is

$$D_n = \frac{D}{1 - a_m} - \frac{D_m a_m}{1 - a_m}$$

in which D_n = the density in area n (see Fig. E.1)

D = average density of the area as a whole (number of units ÷ area)

D_m = estimated density in part m of area

a_m = the fraction (0.1 to 0.9) of the total area comprised in m

$1 - a_m$ = the fraction (0.1 to 0.9) of the total area comprised in n

Values of $\dfrac{D}{1 - a_m}$ and $\dfrac{D_m a_m}{1 - a_m}$ may be extracted from the table as follows. The table is entered at the top with a_m and entered at the side with D or D_m as arguments. When D is the argument, the right-hand column under the particular value of a_m gives values of $\dfrac{D_m a_m}{1 - a_m}$.

In order to obtain D_n, subtract the value ob-

The explanation and tables are here presented by permission of *Geographical Review*, published by the American Geographical Society of New York.

tained from entering the table with D_m as argument from the value obtained from entering the table with D as argument. (If this value is a minus quantity, then the value of D_m is too large to be consistent with the values of D and of a_m.)

For example: $D = 90$, $D_m = 15$, $a_m = 0.7$. From row 90 and the left-hand column under 0.7, extract value 300; from row 15 and the right-hand column under 0.7, extract value 35; $300 - 35 = 265 = D_n$.

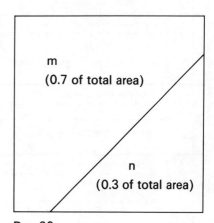

D = 90
Dm = 15
am = 0.7

Figure E.1 A hypothetical enumeration district with a density (D) of 90. The assumed density of area m (Dm) is 15, and the area of m (am) is estimated to be 0.7 of the total area of the district.

TABLE H.1

TABULAR AID TO CONSISTENCY IN ESTIMATING DENSITIES OF PARTS

a_m (column groups). Within each a_m group the two sub‑columns are $D_m a_m$ and D; the sub‑label beneath each is $1-a_m$.

D or D_m	0.9 $D_m a_m$	0.9 D	0.8 $D_m a_m$	0.8 D	0.7 $D_m a_m$	0.7 D	0.6 $D_m a_m$	0.6 D	0.5 $D_m a_m$	0.5 D	0.4 $D_m a_m$	0.4 D	0.3 $D_m a_m$	0.3 D	0.2 $D_m a_m$	0.2 D	0.1 $D_m a_m$	0.1 D	D or D_m
1	9	10	4	5	2.3	3	1.5	2	1	2	0.7	2	0.4	1	0.3	1	0.1	1	1
2	18	20	8	10	4.7	7	3.0	5	2	4	1.3	3	0.9	3	0.5	3	0.2	2	2
3	27	30	12	15	7.0	10	4.5	7	3	6	2.0	5	1.3	4	0.8	4	0.3	3	3
4	36	40	16	20	9.3	13	6.0	10	4	8	2.7	7	1.8	6	1.0	5	0.4	4	4
5	45	50	20	25	11.7	17	7.5	12	5	10	3.3	8	2.2	7	1.3	6	0.6	6	5
6	54	60	24	30	14.0	20	9.0	15	6	12	4.0	10	2.7	9	1.5	8	0.7	7	6
7	63	70	28	35	16.3	23	10.5	17	7	14	4.7	12	3.1	10	1.8	9	0.8	8	7
8	72	80	32	40	18.6	27	12.0	20	8	16	5.3	13	3.6	12	2.0	10	0.9	9	8
9	81	90	36	45	21.0	30	13.5	22	9	18	6.0	15	4.0	13	2.3	11	1.0	10	9
10	90	100	40	50	23.3	33	15.0	25	10	20	6.7	17	4.4	14	2.5	13	1.1	11	10
11	99	110	44	55	25.6	37	16.5	27	11	22	7.3	18	4.9	16	2.8	14	1.2	12	11
12	108	120	48	60	28.0	40	18.0	30	12	24	8.0	20	5.3	17	3.0	15	1.3	13	12
13	117	130	52	65	30.3	43	19.5	32	13	26	8.7	22	5.8	19	3.3	16	1.4	14	13
14	126	140	56	70	32.6	47	21.0	35	14	28	9.3	23	6.2	20	3.5	18	1.6	16	14
15	135	150	60	75	35.0	50	22.5	37	15	30	10.0	25	6.7	22	3.8	19	1.7	17	15
16	144	160	64	80	37.3	53	24.0	40	16	32	10.7	27	7.1	23	4.0	20	1.8	18	16
17	153	170	68	85	39.6	57	25.5	42	17	34	11.3	28	7.5	24	4.3	21	1.9	19	17
18	162	180	72	90	41.9	60	27.0	45	18	36	12.0	30	8.0	26	4.5	23	2.0	20	18
19	171	190	76	95	44.3	63	28.5	47	19	38	12.7	32	8.4	27	4.8	24	2.1	21	19
20	180	200	80	100	46.6	67	30.0	50	20	40	13.3	33	8.9	29	5.0	25	2.2	22	20
21	189	210	84	105	48.9	70	31.5	52	21	42	14.0	35	9.3	30	5.3	26	2.3	23	21
22	198	220	88	110	51.3	73	33.0	55	22	44	14.7	37	9.8	32	5.5	28	2.4	24	22
23	207	230	92	115	53.6	77	34.5	57	23	46	15.3	38	10.2	33	5.8	29	2.6	26	23
24	216	240	96	120	55.9	80	36.0	60	24	48	16.0	40	10.7	35	6.0	30	2.7	27	24
25	225	250	100	125	58.3	83	37.5	62	25	50	16.7	42	11.1	36	6.3	31	2.8	28	25
26	234	260	104	130	60.6	87	39.0	65	26	52	17.3	43	11.5	37	6.5	33	2.9	29	26
27	243	270	108	135	62.9	90	40.5	67	27	54	18.0	45	12.0	39	6.8	34	3.0	30	27
28	252	280	112	140	65.2	93	42.0	70	28	56	18.6	46	12.4	40	7.0	35	3.1	31	28
29	261	290	116	145	67.6	97	43.5	72	29	58	19.3	48	12.9	42	7.3	36	3.2	32	29
30	270	300	120	150	69.9	100	45.0	75	30	60	20.0	50	13.3	43	7.5	38	3.3	33	30
31	279	310	124	155	72.2	103	46.5	77	31	62	20.6	51	13.8	45	7.8	39	3.4	34	31
32	288	320	128	160	74.6	107	48.0	80	32	64	21.3	53	14.2	46	8.0	40	3.6	36	32
33	297	330	132	165	76.9	110	49.5	82	33	66	22.0	55	14.7	48	8.3	41	3.7	37	33
34	306	340	136	170	79.2	113	51.0	85	34	68	22.6	56	15.1	49	8.5	43	3.8	38	34
35	315	350	140	175	81.6	117	52.5	87	35	70	23.3	58	15.5	50	8.8	44	3.9	39	35
36	324	360	144	180	83.9	120	54.0	90	36	72	24.0	60	16.0	52	9.0	45	4.0	40	36
37	333	370	148	185	86.2	123	55.5	92	37	74	24.6	61	16.4	53	9.3	46	4.1	41	37
38	342	380	152	190	88.5	127	57.0	95	38	76	25.3	63	16.9	55	9.5	48	4.2	42	38
39	351	390	156	195	90.9	130	58.5	97	39	78	26.0	65	17.3	56	9.8	49	4.3	43	39
40	360	400	160	200	93.2	133	60.0	100	40	80	26.6	66	17.8	58	10.0	50	4.4	44	40
41	369	410	164	205	95.5	137	61.5	102	41	82	27.3	68	18.2	59	10.3	51	4.6	46	41
42	378	420	168	210	97.9	140	63.0	105	42	84	28.0	70	18.6	60	10.5	53	4.7	47	42
43	387	430	172	215	100.2	143	64.5	107	43	86	28.6	71	19.1	62	10.8	54	4.8	48	43
44	396	440	176	220	102.5	147	66.0	110	44	88	29.3	73	19.5	63	11.0	55	4.9	49	44
45	405	450	180	225	104.9	150	67.5	112	45	90	30.0	75	20.0	65	11.3	56	5.0	50	45
46	414	460	184	230	107.2	153	69.0	115	46	92	30.6	76	20.4	66	11.5	58	5.1	51	46
47	423	470	188	235	109.5	157	70.5	117	47	94	31.3	78	20.9	68	11.8	59	5.3	52	47
48	432	480	192	240	111.8	160	72.0	120	48	96	32.0	80	21.3	69	12.0	60	5.3	53	48
49	441	490	196	245	114.2	163	73.5	122	49	98	32.6	81	21.8	71	12.3	61	5.4	54	49

50	450	500	200	250	116.5	167	75.0	125	50	100	33.3	83	22.2	72	12.5	63	5.6	56	50
51	459	510	204	255	118.8	170	76.5	127	51	102	34.0	85	22.6	73	12.8	64	5.7	57	51
52	468	520	208	260	121.2	173	78.0	130	52	104	34.6	86	23.1	75	13.0	65	5.8	58	52
53	477	530	212	265	123.5	177	79.5	132	53	106	35.3	88	23.5	76	13.3	66	5.9	59	53
54	486	540	216	270	125.8	180	81.0	135	54	108	36.0	90	24.0	78	13.5	68	6.0	60	54
55	495	550	220	275	128.2	183	82.5	137	55	110	36.6	91	24.4	79	13.8	69	6.1	61	55
56	504	560	224	280	130.5	187	84.0	140	56	112	37.3	93	24.9	81	14.0	70	6.2	62	56
57	513	570	228	285	132.8	190	85.5	142	57	114	38.0	95	25.3	82	14.3	71	6.3	63	57
58	522	580	232	290	135.1	193	87.0	145	58	116	38.6	96	25.8	84	14.5	73	6.4	64	58
59	531	590	236	295	137.5	197	88.5	147	59	118	39.3	98	26.2	85	14.8	74	6.5	65	59
60	540	600	240	300	139.8	200	90.0	150	60	120	40.0	100	26.6	86	15.0	75	6.7	67	60
61	549	610	244	305	142.1	203	91.5	152	61	122	40.6	101	27.1	88	15.3	76	6.8	68	61
62	558	620	248	310	144.5	207	93.0	155	62	124	41.3	103	27.5	89	15.5	78	6.9	69	62
63	567	630	252	315	146.8	210	94.5	157	63	126	42.0	105	28.0	91	15.8	79	7.0	70	63
64	576	640	256	320	149.1	213	96.0	160	64	128	42.6	106	28.4	92	16.0	80	7.1	71	64
65	585	650	260	325	151.5	217	97.5	162	65	130	43.3	108	28.9	94	16.3	81	7.2	72	65
66	594	660	264	330	153.8	220	99.0	165	66	132	44.0	110	29.3	95	16.5	83	7.3	73	66
67	603	670	268	335	156.1	223	100.5	167	67	134	44.6	111	29.7	96	16.8	84	7.4	74	67
68	612	680	272	340	158.4	227	102.0	170	68	136	45.3	113	30.2	98	17.0	85	7.5	75	68
69	621	690	276	345	160.8	230	103.5	172	69	138	46.0	115	30.6	99	17.3	86	7.7	77	69
70	630	700	280	350	163.1	233	105.0	175	70	140	46.6	116	31.1	101	17.5	88	7.8	78	70
71	639	710	284	355	165.4	237	106.5	177	71	142	47.3	118	31.5	102	17.8	89	7.9	79	71
72	648	720	288	360	167.8	240	108.0	180	72	144	48.0	120	32.0	104	18.0	90	8.0	80	72
73	657	730	292	365	170.1	243	109.5	182	73	146	48.6	121	32.4	105	18.3	91	8.1	81	73
74	666	740	296	370	172.4	247	111.0	185	74	148	49.3	123	32.9	107	18.5	93	8.2	82	74
75	675	750	300	375	174.8	250	112.5	187	75	150	50.0	125	33.3	108	18.8	94	8.3	83	75
76	684	760	304	380	177.1	253	114.0	190	76	152	50.6	126	33.7	109	19.0	95	8.4	84	76
77	693	770	308	385	179.4	257	115.5	192	77	154	51.3	128	34.2	111	19.3	96	8.5	85	77
78	702	780	312	390	181.7	260	117.0	195	78	156	51.9	129	34.6	112	19.5	98	8.7	87	78
79	711	790	316	395	184.1	263	118.5	197	79	158	52.6	131	35.1	114	19.8	99	8.8	88	79
80	720	800	320	400	186.4	267	120.0	200	80	160	53.3	133	35.5	115	20.0	100	8.9	89	80
81	729	810	324	405	188.7	270	121.5	202	81	162	53.9	134	36.0	117	20.3	101	9.0	90	81
82	738	820	328	410	191.1	273	123.0	205	82	164	54.6	136	36.4	118	20.5	103	9.1	91	82
83	747	830	332	415	193.4	277	124.5	207	83	166	55.3	138	36.9	120	20.8	104	9.2	92	83
84	756	840	336	420	195.7	280	126.0	210	84	168	55.9	139	37.3	121	21.0	105	9.3	93	84
85	765	850	340	425	198.1	283	127.5	212	85	170	56.6	141	37.7	122	21.3	106	9.4	94	85
86	774	860	344	430	200.4	287	129.0	215	86	172	57.3	143	38.2	124	21.5	108	9.5	95	86
87	783	870	348	435	202.7	290	130.5	217	87	174	57.9	144	38.6	125	21.8	109	9.7	97	87
88	792	880	352	440	205.0	293	132.0	220	88	176	58.6	146	39.1	127	22.0	110	9.8	98	88
89	801	890	356	445	207.4	297	133.5	222	89	178	59.3	148	39.5	128	22.3	111	9.9	99	89
90	810	900	360	450	209.7	300	135.0	225	90	180	59.9	149	40.0	130	22.5	113	10.0	100	90
91	819	910	364	455	212.0	303	136.5	227	91	182	60.6	151	40.4	131	22.8	114	10.1	101	91
92	828	920	368	460	214.4	307	138.0	230	92	184	61.3	153	40.8	132	23.0	115	10.2	102	92
93	837	930	372	465	216.7	310	139.5	232	93	186	61.9	154	41.3	134	23.3	116	10.3	103	93
94	846	940	376	470	219.0	313	141.0	235	94	188	62.6	156	41.7	135	23.5	118	10.4	104	94
95	855	950	380	475	221.4	317	142.5	237	95	190	63.3	158	42.2	137	23.8	119	10.5	105	95
96	864	960	384	480	223.7	320	144.0	240	96	192	63.9	159	42.6	138	24.0	120	10.7	107	96
97	873	970	388	485	226.0	323	145.5	242	97	194	64.6	161	43.1	140	24.3	121	10.8	108	97
98	882	980	392	490	228.3	327	147.0	245	98	196	65.3	163	43.5	141	24.5	123	10.9	109	98
99	891	990	396	495	230.7	330	148.5	247	99	198	65.9	164	44.0	143	24.8	124	11.0	110	99
100	900	1 000	400	500	233.0	333	150.0	250	100	200	66.6	166	44.4	144	25.0	125	11.1	111	100

APPENDIX

F RADIUS INDEX VALUES FOR GRADUATED CIRCLES

The values in the body of Table F.1 provide directly the radius indices for scaling graduated circles. The values obtained from the table need only be divided by the chosen unit radius value to be used for scaling the circles. The basic procedure is described in Chapter 6.

The values in the table are the antilogarithms of the logarithms of $N \times 0.5718$. First suggested by John L. Thompson, they were machine computed according to a plan devised by George F. McCleary, Jr., Joel L. Morrison, and Morton W. Scripter. The constant 0.5718 is derived from the work of James J. Flannery.

Example 1. Suppose we wanted to draw graduated circles representing the populations of 3 cities with the following populations:

A – 435,210
B – 62,647
C – 126,538

The procedure would be as follows.

(a) By rounding, drop the last three digits.

(b) Enter the table for A horizontally with $N = 430$ and vertically in the column headed 5; find the index value 32.26.

(c) Enter the table with 60 and 3 as arguments for B; find the index value 10.69.

(d) Enter the table with 120 and 7 as arguments for C; find the index value 15.96.

(e) Select a unit quantity, for example, 0.5 = 4 mm. Divide each radius index by 0.5 and multiply by 0.4. The result may be directly plotted with a compass.

City	Index Value		Unit Quantity		Plotting Value
A	32.26	÷	0.5 × 4 mm	=	25.8 mm
B	10.69	÷	0.5 × 4 mm	=	8.6 mm
C	15.96	÷	0.5 × 4 mm	=	12.8 mm

Example 2. To find the radius index for N greater than 999, the largest argument in the table, for example, $N = 1,622$.

(a) Divide N into two whole numbers whose product equals 1622, for example, 811 and 2.

(b) Enter the table with these as arguments and obtain, respectively, 46.07 and 1.49.

(c) Multiply these two together to obtain the required radius index value, 68.64, ($46.07 \times 1.49 = 68.64$).

(d) Proceed as in *e* in Example 1.

N	0	1	2	3	4	5	6	7	8	9	N	0	1	2	3	4	5	6	7	8	9
0	0	1.00	1.49	1.87	2.21	2.51	2.79	3.04	3.28	3.51	500	34.94	34.98	35.02	35.06	35.10	35.14	35.17	35.21	35.25	35.29
10	3.73	3.94	4.14	4.33	4.52	4.70	4.88	5.05	5.22	5.39	510	35.33	35.37	35.41	35.45	35.49	35.53	35.57	35.61	35.65	35.69
20	5.55	5.70	5.86	6.01	6.15	6.30	6.44	6.58	6.72	6.86	520	35.73	35.77	35.81	35.85	35.89	35.92	35.96	36.00	36.04	36.08
30	6.99	7.12	7.26	7.38	7.51	7.64	7.76	7.88	8.00	8.12	530	36.12	36.16	36.20	36.24	36.28	36.31	36.35	36.39	36.43	36.47
40	8.24	8.36	8.48	8.59	8.70	8.82	8.93	9.04	9.15	9.26	540	36.51	36.55	36.58	36.62	36.66	36.70	36.74	36.78	36.82	36.85
50	9.36	9.47	9.58	9.68	9.79	9.89	9.99	10.09	10.19	10.29	550	36.89	36.93	36.97	37.01	37.05	37.08	37.12	37.16	37.20	37.24
60	10.39	10.49	10.59	10.69	10.78	10.88	10.98	11.07	11.16	11.26	560	37.27	37.31	37.35	37.39	37.43	37.46	37.50	37.54	37.58	37.62
70	11.35	11.44	11.54	11.63	11.72	11.81	11.90	11.99	12.08	12.16	570	37.65	37.69	37.73	37.77	37.80	37.84	37.88	37.92	37.96	37.99
80	12.25	12.34	12.43	12.51	12.60	12.68	12.77	12.85	12.94	13.02	580	38.03	38.07	38.11	38.14	38.18	38.22	38.25	38.29	38.33	38.37
90	13.10	13.19	13.27	13.35	13.43	13.52	13.60	13.68	13.76	13.84	590	38.40	38.44	38.48	38.52	38.55	38.59	38.63	38.66	38.70	38.74
100	13.92	14.00	14.08	14.16	14.23	14.31	14.39	14.47	14.54	14.62	600	38.77	38.81	38.85	38.89	38.92	38.96	39.00	39.03	39.07	39.11
110	14.70	14.77	14.85	14.93	15.00	15.08	15.15	15.23	15.30	15.37	610	39.14	39.18	39.22	39.25	39.29	39.33	39.36	39.40	39.44	39.47
120	15.45	15.52	15.59	15.67	15.74	15.81	15.89	15.96	16.03	16.10	620	39.51	39.54	39.58	39.62	39.65	39.69	39.73	39.76	39.80	39.84
130	16.17	16.24	16.31	16.38	16.45	16.52	16.59	16.66	16.73	16.80	630	39.87	39.91	39.94	39.98	40.02	40.05	40.09	40.12	40.16	40.20
140	16.87	16.94	17.01	17.08	17.15	17.21	17.28	17.35	17.42	17.48	640	40.23	40.27	40.30	40.34	40.38	40.41	40.45	40.48	40.52	40.55
150	17.55	17.62	17.68	17.75	17.82	17.88	17.95	18.01	18.08	18.15	650	40.59	40.63	40.66	40.70	40.73	40.77	40.80	40.84	40.88	40.91
160	18.21	18.28	18.34	18.40	18.47	18.53	18.60	18.66	18.73	18.79	660	40.95	40.98	41.02	41.05	41.09	41.12	41.16	41.19	41.23	41.26
170	18.85	18.92	18.98	19.04	19.10	19.17	19.23	19.29	19.35	19.42	670	41.30	41.34	41.37	41.41	41.44	41.48	41.51	41.55	41.58	41.62
180	19.48	19.54	19.60	19.66	19.73	19.79	19.85	19.91	19.97	20.03	680	41.65	41.69	41.72	41.76	41.79	41.83	41.86	41.90	41.93	41.97
190	20.09	20.15	20.21	20.27	20.33	20.39	20.45	20.51	20.57	20.63	690	42.00	42.04	42.07	42.10	42.14	42.17	42.21	42.24	42.28	42.31
200	20.69	20.75	20.81	20.87	20.92	20.98	21.04	21.10	21.16	21.22	700	42.35	42.38	42.42	42.45	42.49	42.52	42.55	42.59	42.62	42.66
210	21.27	21.33	21.39	21.45	21.50	21.56	21.62	21.68	21.73	21.79	710	42.69	42.73	42.76	42.80	42.83	42.86	42.90	42.93	42.97	43.00
220	21.85	21.90	21.96	22.02	22.07	22.13	22.19	22.24	22.30	22.35	720	43.04	43.07	43.10	43.14	43.17	43.21	43.24	43.27	43.31	43.34
230	22.41	22.47	22.52	22.58	22.63	22.69	22.74	22.80	22.85	22.91	730	43.38	43.41	43.44	43.48	43.51	43.55	43.58	43.61	43.65	43.68
240	22.96	23.02	23.07	23.13	23.18	23.23	23.29	23.34	23.40	23.45	740	43.71	43.75	43.78	43.82	43.85	43.88	43.92	43.95	43.98	44.02
250	23.50	23.56	23.61	23.66	23.72	23.77	23.82	23.88	23.93	23.98	750	44.05	44.09	44.12	44.15	44.19	44.22	44.25	44.29	44.32	44.35
260	24.04	24.09	24.14	24.20	24.25	24.30	24.35	24.41	24.46	24.51	760	44.39	44.42	44.45	44.49	44.52	44.55	44.59	44.62	44.65	44.69
270	24.56	24.61	24.67	24.72	24.77	24.82	24.87	24.92	24.97	25.03	770	44.72	44.75	44.79	44.82	44.85	44.89	44.92	44.95	44.98	45.02
280	25.08	25.13	25.18	25.23	25.28	25.33	25.38	25.43	25.48	25.54	780	45.05	45.08	45.12	45.15	45.18	45.22	45.25	45.28	45.31	45.35
290	25.59	25.64	25.69	25.74	25.79	25.84	25.89	25.94	25.99	26.04	790	45.38	45.41	45.45	45.48	45.51	45.54	45.58	45.61	45.64	45.67
300	26.09	26.14	26.19	26.24	26.28	26.33	26.38	26.43	26.48	26.53	800	45.71	45.74	45.77	45.81	45.84	45.87	45.90	45.94	45.97	46.00
310	26.58	26.63	26.68	26.73	26.78	26.82	26.87	26.92	26.97	27.02	810	46.03	46.07	46.10	46.13	46.16	46.20	46.23	46.26	46.29	46.33
320	27.07	27.12	27.16	27.21	27.26	27.31	27.36	27.40	27.45	27.50	820	46.36	46.39	46.42	46.45	46.49	46.52	46.55	46.58	46.62	46.65
330	27.55	27.60	27.64	27.69	27.74	27.79	27.83	27.88	27.93	27.97	830	46.68	46.71	46.74	46.78	46.81	46.84	46.87	46.90	46.94	46.97
340	28.02	28.07	28.12	28.16	28.21	28.26	28.30	28.35	28.40	28.44	840	47.00	47.03	47.06	47.10	47.13	47.16	47.19	47.22	47.26	47.29
350	28.49	28.54	28.58	28.63	28.68	28.72	28.77	28.81	28.86	28.91	850	47.32	47.35	47.38	47.42	47.45	47.48	47.51	47.54	47.57	47.61
360	28.95	29.00	29.04	29.09	29.14	29.18	29.23	29.27	29.32	29.36	860	47.64	47.67	47.70	47.73	47.76	47.80	47.83	47.86	47.89	47.92
370	29.41	29.46	29.50	29.55	29.59	29.64	29.68	29.73	29.77	29.82	870	47.95	47.98	48.02	48.05	48.08	48.11	48.14	48.17	48.20	48.24
380	29.86	29.91	29.95	30.00	30.04	30.09	30.13	30.18	30.22	30.26	880	48.27	48.30	48.33	48.36	48.39	48.42	48.46	48.49	48.52	48.55
390	30.31	30.35	30.40	30.44	30.49	30.53	30.57	30.62	30.66	30.71	890	48.58	48.61	48.64	48.67	48.71	48.74	48.77	48.80	48.83	48.86
400	30.75	30.79	30.84	30.88	30.93	30.97	31.01	31.06	31.10	31.14	900	48.89	48.92	48.95	48.98	49.02	49.05	49.08	49.11	49.14	49.17
410	31.19	31.23	31.28	31.32	31.36	31.41	31.45	31.49	31.53	31.58	910	49.20	49.23	49.26	49.29	49.33	49.36	49.39	49.42	49.45	49.48
420	31.62	31.66	31.71	31.75	31.79	31.84	31.88	31.92	31.96	32.01	920	49.51	49.54	49.57	49.60	49.63	49.66	49.69	49.73	49.76	49.79
430	32.05	32.09	32.13	32.18	32.22	32.26	32.30	32.35	32.39	32.43	930	49.82	49.85	49.88	49.91	49.94	49.97	50.00	50.03	50.06	50.09
440	32.47	32.52	32.56	32.60	32.64	32.68	32.73	32.77	32.81	32.85	940	50.12	50.15	50.18	50.21	50.24	50.27	50.31	50.34	50.37	50.40
450	32.89	32.94	32.98	33.02	33.06	33.10	33.14	33.18	33.23	33.27	950	50.43	50.46	50.49	50.52	50.55	50.58	50.61	50.64	50.67	50.70
460	33.31	33.35	33.39	33.43	33.47	33.52	33.56	33.60	33.64	33.68	960	50.73	50.76	50.79	50.82	50.85	50.88	50.91	50.94	50.97	51.00
470	33.72	33.76	33.80	33.84	33.89		33.93	34.01	34.05	34.09	970	51.03	51.06	51.09	51.12	51.15	51.18	51.21	51.24	51.27	51.30
480	34.13	34.17	34.21	34.25	34.29	34.33	34.37	34.41	34.45	34.49	980	51.33	51.36	51.39	51.42	51.45	51.48	51.51	51.54	51.57	51.60
490	34.53	34.57	34.62	34.66	34.70	34.74	34.78	34.82	34.86	34.90	990	51.63	51.66	51.69	51.72	51.75	51.78	51.81	51.84	51.87	51.90
N	0	1	2	3	4	5	6	7	8	9	N	0	1	2	3	4	5	6	7	8	9

APPENDIX

G DIRECTIONS FOR THE CONSTRUCTION OF PROJECTIONS

The directions and tabular data in this Appendix are arranged in four sections dealing with the construction of (1) cylindrical projections, (2) conic projections, (3) pseudocylindrical projections, and (4) azimuthal projections. Only projection systems that are commonly employed for smaller-scale maps for appropriate areas are given here. Directions for the construction of others may be found in technical treatises and manuals. If a projection is desired for use at large scales, the fuller tables and more complete data of such sources should be consulted.

Directions for the construction of the following projections are given as numbered.

A. Cylindrical Projections
    1. Mercator's (conformal)
    2. Plane chart
    3. Cylindrical equal-area
B. Conic Projections
    4. Conic with two standard parallels
    5. Lambert's conformal
    6. Albers' (equal-area)
    7. Bonne's (equal-area)
C. Pseudocylindrical Projections
    8. Sinusoidal (equal-area)
    9. Mollweide's (equal-area)
    10. Eckert's IV (equal-area)
    11. Flat polar quartic (equal-area)

D. Azimuthal Projections
    12. Stereographic (conformal)
    13. Orthographic
    14. Lambert's equal-area
    15. Azimuthal equidistant
    16. Shifting the center of the azimuthal projection

A. Cylindrical Projections

All cylindrical projections in their conventional form may be constructed with a straightedge, dividers, a scale, and a triangle. In these projections all meridians are the same length and all parallels are the same length; merely their spacing varies. In practice, the length of the equator or standard parallel and the length of a meridian are determined. These are drawn at right angles to one another. The standard parallel is then subdivided for the longitudinal interval desired. The meridians are drawn through these points as parallel lines. The spacing of the parallels is then plotted along a meridian, and the parallels are drawn.

1. *Mercator's.* Numerous tables of the spacing of the parallels on Mercator's projection are available. The values in Table G.1 are taken from *Special Publication No. 68* of the United States

Table G.1 *Distances of the Parallels from the Equator on Mercator's Projection in Minutes of Longitude at the Equator (From Deetz and Adams,* Elements of Map Projection)

0°	000.000	50°	3 456.581
5°	298.348	55°	3 948.830
10°	599.019	60°	4 507.133
15°	904.422	65°	5 157.629
20°	1 217.159	70°	5 943.955
25°	1 540.134	75°	6947.761
30°	1 876.706	80°	8 352.176
35°	2 230.898		(The pole is infi-
40°	2 607.683		nitely distant.)
45°	3013.427		

Coast and Geodetic Survey, which shows values of the distance of each minute of latitude from the equator. A 5° graticule interval is presented here. For any smaller interval it is necessary to consult the original table.

The values of Table G.1 are given in minutes of longitude on the equator, which is simply a convenient unit of distance. The scale of the table is, of course, 1:1, making it necessary to reduce the values to the scale desired. In addition to the reduction in scale, it is also necessary to convert the values to inches or centimeters for plotting purposes. How this is done will be illustrated in the following discussion. A similar procedure can be followed for values in other tables.

Note. Values for each minute of latitude are given in the reference from which the above abbreviated table is taken.

Suppose it were desired to construct the conventional form of Mercator's projection from the values in Table G.1 with a linear scale along the equator of 1:50,000,000. To divide each value in the table by 50,00,000 would give the map distances in "minutes of longitude at the equator," which would be useless for plotting purposes. Suppose further, then, that it were desired also to have the values converted to inches at the chosen scale. The procedure is as follows:

(a) The circumference of the nominal globe at a scale of 1:50,000,000 is calculated by dividing the actual circumference by 50,000,000. The result is about 801.5 mm.

(b) Since there are 21,600 minutes of longitude in a circle (360° x 60'), dividing 801.5 mm by 21,600 gives the length of 1 minute of longitude in millimeters on a globe at a scale of 1:50,000,000, about 0.0371 mm.

(c) Each value in the table, therefore, need only be multiplied by 0.0371 mm, and the result can be plotted directly in millimeters on the map.

2. Plane Chart. The conventional form of this projection is more precisely called the equirectangular projection, the name plane chart being reserved for the phase wherein the equator is standard. In any case, its construction is relatively simple.

A standard parallel is chosen, and its length is calculated or determined from tables. The length of any parallel (ϕ) may be calculated by multiplying the circumference of the earth by the cosine of the latitude. It may also be obtained by referring to the table of lengths of the degrees of the parallel (Table D.1, Appendix D) and multiplying the value given for the latitude by the number of degrees the map is to extend. It must then, of course, be reduced to scale. The results, however determined, are then marked on a straight horizontal line. This is the standard parallel. Vertical lines are drawn through the points to establish the meridians. The other parallels are determined by pointing off the actual distance between the parallels (reduced to scale) as determined from the table of meridional parts (Table D.2, Appendix D).

If the equator is made the standard parallel, the projection will be made up of squares; any other standard parallels will make rectangles whose north-south dimension is the long one.

3. Cylindrical (equal-area). The cylindrical equal-area projection is, like all cylindrical projections, relatively easy to construct, requiring only a straightedge, dividers, scale, and a triangle. In its conventional form the equator alone or any pair of parallels spaced equally from the equator may be chosen as standard. Since the angular deformation is zero along the great circle or small circles chosen as standard, these may be selected so that they pass through the areas of significance. If the projection is desired for only a portion of the earth rather than the whole, the projection is designed merely as a part of a larger, incomplete world projection.

This projection requires some calculation, but the formulas are elementary and are accomplished merely through the use of arithmetic. The general formulas for any form of the projection are as follows.

(a) Length of all parallels is $2R\pi \times \cos\theta$.
(b) Length of all meridians is $2R \div \cos\theta$.
(c) Distance of each parallel from equator is $R \sin\phi \div \cos\theta$.

R is the radius of the generating globe of chosen *area scale*; θ is the standard parallel; ϕ is the latitude.

The procedure for construction is similar to that for other rectangular projections. Perpendicular lines are drawn to represent the equator and a meridian (Fig. G.1). The length of the parallel chosen as standard (*ab*) is marked off on the equator. The length of a meridian (*cd*) is determined. These dimensions define a rectangle forming the poles and the bounding meridian of the projection. The distances of the parallels from the equator are then laid off on a meridian, and the parallels are drawn parallel to the equator as in quadrant C, Figure G.1. The parallels are, of course, equally subdivided by the meridians as in quadrant D.

B. Conic Projections

Conic projections may be constructed either from tables that provide X and Y plane coordinate values with which to locate the intersections of the graticule, or by the use of a straightedge and a beam compass capable of drawing large arcs. In the latter procedure it is necessary to determine the radii of the arcs representing the parallels and the spacing of the meridians on the parallels. In most conventional conic projections the meridians are equally spaced along each parallel, and the parallels are arcs of circles that may or may not be concentric; the meridians are usually straight lines.

In conventional form all conic projections are symmetrical around the central meridian. Thus, it is only necessary to draw one side of the projection; the other side may be copied by either folding the paper or copying it onto another paper with the aid of a light-table. Another easy procedure is to place a sheet of paper over the projection and prick the intersections of the graticule with a needle. The paper may then be "flopped over" and the points pricked through in their proper place on the other side of the central meridian.

4. Conic with Two Standard Parallels. The

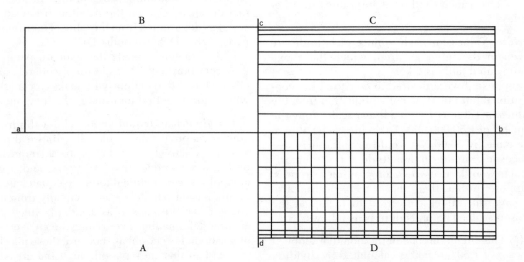

Figure G.1   Construction of the cylindrical equal-area projection.

earth's surface may be projected to a cone assumed to be set upon the sphere and, of course, tangent at one small circle; conventionally, this is made a parallel. If a limited latitudinal segment is being mapped, the deformation away from the standard small circle does not reach serious proportions. Nevertheless, a projection better in every way can be made by assuming the cone to intersect the earth's surface so that two small circles or parallels are standard. This is sometimes called a secant conic projection. As can be seen from Figure G.2, the scale would not be correct along the meridians if the surface were actually projected from some one point. If the point were the center of the sphere, the parallels would be too close together between the standard parallels and too far apart outside them. It is better to construct the projection so that the scale is more nearly correct along all the meridians. There are several ways of constructing this sort of projection, but only one is illustrated here—the conic with two standard parallels in which all the meridians are standard.*

*The interested reader is referred to an analysis of similar projections, John Leighly, "Extended Uses of Polyconic Projection Tables," *Annals of the Association of American Geographers,* **46,** 150–173 (1956).

To construct the conic projection with two standard parallels, first draw a vertical line, *ab* in Figure G.3, in the center of the paper. Select the two parallels, ϕ_1 and ϕ_2, to be made standard, and determine their actual distance apart on the generating globe of the desired linear scale. If *c* and *d* are then positions on *ab*, and if the radius *R* of the globe to be projected at scale has been determined, distance *cd* may be determined by

$$cd = 2\pi R \frac{\phi_1 - \phi_2}{360}$$

This distance may also be determined by reducing to scale the actual spacing on the earth. Points *c* and *d* should be placed in a convenient location on the paper so that the developed projection will be centered properly. Points *c* and *d* being the intersections of the standard parallels with the central meridian, it is next necessary to determine the position on *ab* of the center *x*, from which the two concentric arcs may be drawn. This may be done by

$$\frac{xc}{cd} = \frac{\cos \phi_1}{\cos \phi_2 - \cos \phi_1}$$

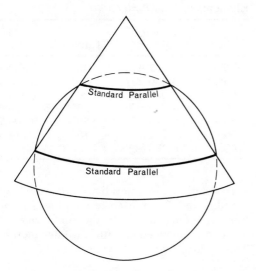

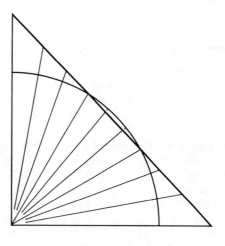

Figure G.2   The basis of a geometrically projected secant conic with two standard parallels.

or

$$xc = \frac{\cos \phi_1 \times cd}{\cos \phi_2 - \cos \phi_1}$$

Draw arcs with x as the center through c and d. Extend these arcs (*ecg* and *fdh*) a distance equal to the length of the standard parallels, which is to be included between the extreme meridians of the map. Space the arcs of the other parallels equally on *ab*.

The meridians are equally spaced on the parallels and are correctly spaced to scale on the two standard parallels. The most nearly accurate means of doing this is by first determining the chord distances of the bounding meridian, chord *cg* on ϕ_1 and chord *dh* on ϕ_2.

To determine the chord distance *cg* on ϕ_1:

(a) Determine longitude on ϕ_1 to be represented by *cg*, for example, 25°.

(b) The length of ϕ_1 on the sphere of the chosen scale $= 2 \pi R \cos \phi_1$.

(c) Let *xc* be radius r. The circumference of a circle with radius $r = 2 \pi r$.

(d) Therefore, angle λ_1 (angle *cxg* on Fig. G.3) may be determined by

$$\lambda_1 = \frac{2\pi R \cos \phi_1}{2\pi R} \times 25° = \frac{R \cos \phi_1}{r} \times 25°$$

(e) Chord distance $cg = 2r \sin \dfrac{\lambda_1}{2}$

Lay off the chord distance *cg*. Determine in similar fashion chord distance *dh*. Subdivide the parallels in question (with dividers) into the desired number of equal parts for the meridians. Join homologous points with straight lines.

5. Lambert's Conformal Conic. This projection is similar in appearance to Albers' and the simple conic. It, too, has straight-line meridians that meet at a common center: the parallels are arcs of circles, two of which are standard; and the parallels and meridians meet at right angles. The only difference is in the spacings of the parallels and meridians. In the Lambert Conic they are so spaced as to satisfy the condition of conformality, that is, $a = b$ at every point. It is advisable, if a satisfactory distribution of scale error is desired, to space the

standard parallels so that they include between them about two-thirds of the meridional section to be mapped.

The calculation of this projection requires considerable mathematical computations. On account of its relatively wide use for air-navigation maps, many tables for its construction with various standard parallels have been published. For example, a table for the construction of a map of the United States with standard parallels at 29° and 45° is given in the United States Coast and Geodetic Survey, *Special Publication No. 52*.

A Lambert conic with standard parallels at 36° and 54° is useful for middle-latitude areas. Table G.2 gives the radii of the parallels in meters for a map at a scale of 1:1. It is, of course, necessary to reduce each value to the scale desired.

To construct the projection draw a line, *cd* in Figure G.4, which will be the central meridian. The line must be sufficiently long so that it will include the center of the arcs of latitude. With a beam compass, describe arcs with radii, reduced to scale, taken from Table G.2.

Table G.2 Table for the Construction of a Lambert Conformal Conic Projection. Standard Parallels 36° and 54° (from Deetz and Adams, Elements of Map Projection)

Latitude	Radii, Meters	Latitude	Radii, Meters
75°	2 787 926	40°	6 833 183
70°	3 430 294	35°	7 386 250
65°	4 035 253	30°	7 946 911
60°	4 615 579	25°	8 519 065
55°	5 179 774	20°	9 106 796
50°	5 734 157	15°	9 714 516
45°	6 283 826		

To determine the meridians, it is necessary to calculate the chord distance on a lower parallel from its intersection with the central meridian (0°) to its intersection with an outer meridian. This is done by the following formula:

$$\text{chord} = 2r \sin \frac{n\lambda}{2}$$

where $n = 0.7101$; λ = longitude out from central meridian; and r = radius of the parallel in question.

This procedure is similar to the general process for obtaining the length of a chord that subtends any central angle. The formula introduces a factor n to take into account the special meridian spacing of the projection.

Example. On parallel 30° the chord of 45° (see Fig. G.4) out from the central meridian =

(a) $0.7101 \times 45° = 31° 57' 14'' = n\lambda$
(b) $n\lambda/2 = 15° 58' 37''$

(c) Sin 15° 58′ 37″ = 0.27534
(d) $2r$ = 15,893,822 meters
(e) 15,893,822 × 0.27534 = 4,376,200 meters

The value thus determined, *ab* in Figure G.4, is reduced by the desired scale ratio and measured out from the intersection of the parallel and the central meridian to the bounding meridian. If point *b*, thus located, is connected by means of a straightedge with the same center used in describing the parallels, this will determine the outer meridian. If a long straightedge

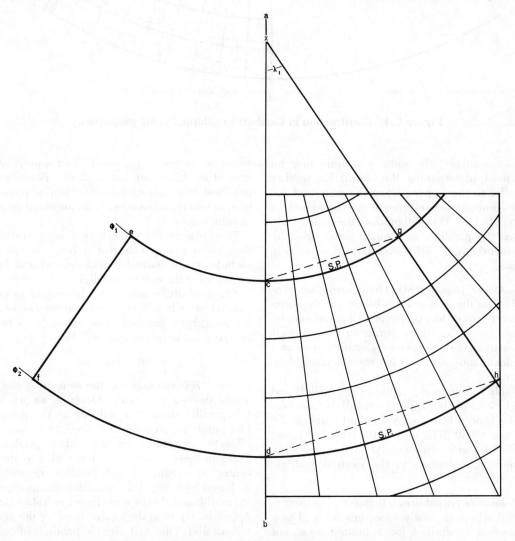

Figure G.3 Consturction of the conic with two standard parallels.

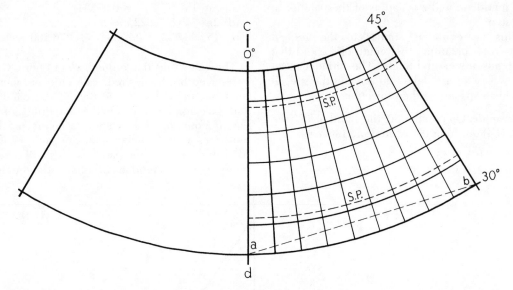

Figure G.4   Construction of Lambert's conformal conic projection.

is not available, the same procedure may be followed (determining the chord) for another parallel in the upper part of the map, and the two points, thus determined, may be joined by a straight line. This will produce the same result. Since the parallels are equally subdivided by the meridians, the other meridians may be easily located.

6. Albers' (equal-area). The construction procedure for this projection is essentially the same as for the preceding projection, Lambert's conic. Like Lambert's, Albers' conic, in conventional orientation, is suited to representation of an area predominantly east-west in extent in the middle latitudes.

Table G.3 gives the radii of the parallels and the lengths of chords on two parallels for a map of the United States with standard parallels 29° 30′ and 45° 30′. The scale of the table is 1:1, and the values are in meters. As in the preceding projection, reduction to the desired scale is necessary.

7. Bonne's (equal-area). Bonne's projection is useful when an easily constructed equal-area projection is desired for a limited area, but

when the tables for constructing an appropriate case of an Albers are not available. Bonne's is modified, like most useful conic forms of projection, in that it cannot actually be projected on an enveloping cone.

To construct the projection draw a vertical line, *cb* in Figure G.5, that is long enough to include the latitudinal extent desired and the center *(c)* of the central parallel.

Theoretically, a cone is made tangent to the parallel (ϕ), which is selected near the center of the area to be mapped. The radius *(r)* of this central parallel on the map will be

$$r = R \cot \phi = oc$$

in which R is the radius of the generating globe of the desired area scale. Describe an arc for this parallel through o with oc as the radius. Plot points on cb north and south of o, spaced correctly at scale for the other parallels. Through these points draw arcs with c as their center. The radius of each parallel may easily be found by subtracting or adding the appropriate meridional distance (as found in Table D.2, Appendix D), from the r value used for the central parallel. This will provide parallels of con-

Table G.3 Table for Construction of Albers Equal-Area Projection with Standard Parallels 29°30' and 45°30' (from Deetz and Adams, Elements of Map Projection)

Latitude	Radius of Parallel, Meters
20°	10 253 177
21°	10 145 579
22°	10 037 540
23°	9 929 080
24°	9 820 218
25°	9 710 969
26°	9 601 361
27°	9 491 409
28°	9 381 139
29°	9 270 576
29° 30'	9 215 188
30°	9 159 738
31°	9 048 648
32°	8 937 337
33°	8 825 827
34°	8 714 150
35°	8 602 328
36°	8 490 392
37°	8 378 377
38°	8 266 312
39°	8 154 228
40°	8 042 163
41°	7 930 152
42°	7 818 231
43°	7 706 444
44°	7 594 828
45°	7 483 426
45° 30'	7 427 822
46°	7 372 288
47°	7 261 459
48°	7 150 987
49°	7 040 925
50°	6 931 333
51°	6 822 264
52°	6 713 780

Long. from Central Meridian	Chord Distances in Meters On Latitude 25°	On Latitude 45°
1°	102 185	78 745
5°	510 867	393 682
25°	2 547 270	1 962 966
30°		2 352 568

centric arcs properly spaced on the central meridian.

Each of the parallels of Bonne's projection is standard, that is, the linear scale along it is correct and, therefore, the meridians are equally spaced on any given parallel. The positions of the meridians must be separately determined on *each* parallel; then smooth curves are drawn through homologous points. It is easiest to determine the chord distance for *each* parallel from the central meridian (*cb* in Fig. G.5) to one meridian near the edge of the map and then divide the intervening space along the parallel equally with dividers. The chord distances may be determined in the manner outlined in the description of the construction of Lambert's projection, *but* the central angle λ to be used in the computation must first be calculated. It will not be the actual angle on the earth between the meridians; instead it will be λ_1, which is the actual angle on the projection that corresponds

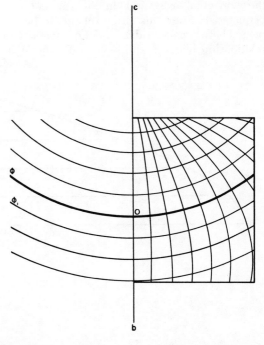

Figure G.5 Construction of Bonne's projection.

to λ on the parallel circle of the earth. For example, if the position of the meridian 30° from cb in Figure G.5 is desired on parallel ϕ_1, then λ_1 (which represents 30° in the general formula for a chord distance) is obtained by

$$\lambda_1 = \frac{R\cos\phi_1}{r} \times 30°$$

in which ϕ is the latitude, and $r = co$ plus the distance on the projection between ϕ and ϕ_1.

C. Pseudocylindrical Projections

Most useful projections of this class are equivalent, and they are constructed to an area scale. The linear dimensions, for construction purposes, of a pseudocylindrical projection depend upon the shape of the bounding meridian that encloses the projection. It is obvious that the axes of two dissimilar shapes would be different if both shapes enclose the same area, that is, if they were the same area scale. Most such projections have a vertical axis half the length of the horizontal axis; and, in the conventional "equatorial" form, this is the relation to be expected between a meridian and the equator. The relationship between the scale of the nominal globe and the particular projection being constructed is merely one that states the length of the central meridian on the projection compared to the radius (R) of the nominal globe of the same area scale. The equator being twice the length of the central meridian, no further calculation is necessary.

In the majority of the pseudocylindrical projections, the meridians are equally spaced along the parallels. The spacing of the parallels along the central meridian varies from projection to projection. These values are available in tubular form.

8. Sinusoidal. The sinusoidal projection is particularly simple to construct, since the spacings of the parallels on the central meridian and meridians on the parallels are the same (to scale) as they are on the earth. The length of the central meridian is 3.1416 times the radius (R) of a nominal globe of equal area. The equator is twice the length of the central meridian.

To construct the projection a horizontal line representing the equator, ab in Figure G.6 is drawn twice the length of the central meridian. The equator is bisected, and at point o a perpendicular central meridian (cd) is constructed.

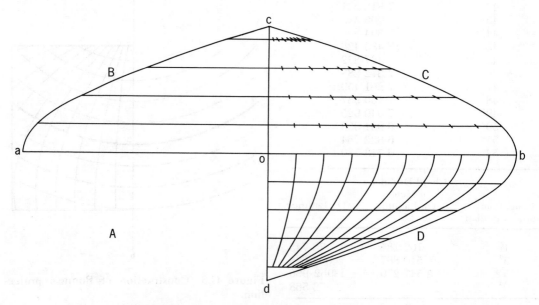

Figure G.6   Construction of the sinusoidal projection.

The positions of the parallels on *cd* are determined by spacing them as they are on the globe. For a small-scale projection this means equally; for a large-scale projection the exact spacings may be taken from Table D.2, Appendix D. Through the points thus established the parallels are drawn parallel to the equator as in quadrant *B*. The lengths of the various parallels are their true lengths (to scale) as they are on the earth and may be determined by multiplying the length of the equator (*ab*) by the cosine of the latitude. One-half of this value is plotted on each side of the central meridian. The meridians are drawn by subdividing, as in quadrant *C*, each parallel equally, with dividers, and drawing smooth curves (with a French curve) through homologous points as in quadrant *D*.

The linear scale along the parallels and the central meridian is correct (same as that of the nominal globe), and is the square root of the area scale; that is, if the area scale is $1:50,000,000^2$ the linear scale along the parallels and the central meridian will be $1:50,000,000$. This is the only equivalent pseudocylindrical projection in which this relationship exists, although some others come close to it.

9. Mollweide's. Mollweide's projection does not have the simple relationship to the sphere that characterizes the sinusoidal. The meridians of the sinusoidal are sine curves, which produce a pointed appearance, whereas in Mollweide's the meridians are ellipses, which provide a projection shape that is somewhat less of a radical departure from the globe impression. The length of the central meridian is 2.8284 times the radius (*R*) of the nominal globe of equal area. The equator is twice the length of the central meridian.

To construct the projection, a horizontal line representing the equator, *ab* in Figure G.7, is drawn twice the length of the central meridian. The equator is bisected, and at point *o* a perpendicular central meridian (*cd*) is constructed. A circle whose radius is *oc* is constructed around point *o*. This contains a hemisphere. The spacing of the parallels on the central meridian is given in Table G.4, in which *oc* equals 1.

These positions are plotted on the central meridian, and the parallels are drawn through the points parallel to the equator. Each parallel is extended outside the hemisphere circle a distance equal to its length inside the circle. Thus, in quadrant *B* of Figure G.7, *ef = fg, hi = ij, kl = lm*, etc. Each parallel is subdivided equally with dividers, as in quadrant *C*, to establish the position of the meridians. The meridians are drawn through homologous points, as in quadrant *D*, with the aid of a curve.

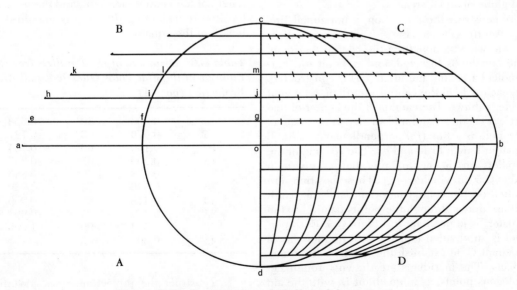

Figure G.7   Construction of Mollweide's projection.

Table G.4 Distances of the Parallels from the Equator in Mollweide's Projection (oc = 1) (from Deetz and Adams, *Elements of Map Projection*)

0°	0.000	50°	0.651
5°	0.069	55°	0.708
10°	0.137	60°	0.762
15°	0.205	65°	0.814
20°	0.272	70°	0.862
25°	0.339	75°	0.906
30°	0.404	80°	0.945
35°	0.468	85°	0.978
40°	0.531	90°	1.000
45°	0.592		

Table G.5 Distances of the Parallels from the Equator in Eckert's IV Projection (oc = 1)

0°	0.000	50°	0.718
5°	0.078	55°	0.775
10°	0.155	60°	0.827
15°	0.232	65°	0.874
20°	0.308	70°	0.915
25°	0.382	75°	0.950
30°	0.454	80°	0.976
35°	0.525	85°	0.994
40°	0.592	90°	1.000
45°	0.657		

10. Eckert's IV.

Eckert's IV projection is representative of a large group of projections in which, in conventional form, the pole is represented by a line rather than by a point, as is the case of Mollweide's and the sinusoidal. The rather excessive shearing of the higher latitudes is somewhat lessened by this device, at the expense, however, of increased angular deformation in the lower latitudes. As in the other pseudocylindrical projections, the length of the central meridian is half the length of the equator. In Eckert's IV projection the length of the central meridian is 2.6530 times the radius (R) of a nominal globe of equal area.

To construct the projection, a horizontal line representing the equator, *ab* in Figure G.8, is drawn twice the length of the central meridian. The equator is bisected, and at point *o* a perpendicular central meridian (*cd*) is constructed. On each side of the central meridian, a tangent circle is drawn. In quadrant *B* the center of one circle (*e*) is on *ab* and midway between *a* and *o*. The pole is a line (*fg*) perpendicular to *cd* and equal in length to *ao* and *cd*. The spacing of the parallels on the central meridian is given in Table G.5, in which *oc* = 1. These positions are plotted on the central meridian, and the parallels are drawn through the points parallel to the equator, as in quadrant *B*, Figure G.8. Each parallel is subdivided equally with dividers, as in quadrant *C*, to establish the position of the meridians. The meridians are drawn through homologous points, as in quadrant *D*, with the aid of a curve.

11. Flat Polar Quartic.

This projection is representative of a group in which a line represents the pole in conventional form, like Eckert's IV, but in which that line is less than half the length of the equator and the bounding meridian is a complex curve. On this account, neither the spacing of the parallels nor their lengths can be easily derived by construction. A table of lengths of X and Y coordinates is necessary. Tables G.6 and G.7 list the necessary information to construct the projection with a 5° grid interval. In the flat polar quartic equal-area projection, the length of the central meridian is 2.6513 times the radius (R) of a nominal globe of equal area. The equator is 2.2214 times the length of the central meridian, and the length of the line representing the pole is one-third the length of the equator.

Table G.6 Distances of the Parallels from the Equator in the Flat Polar Quartic Equal-Area Projection (oc = 1)

0°	0.000	50°	0.668
5°	0.070	55°	0.727
10°	0.140	60°	0.784
15°	0.209	65°	0.837
20°	0.278	70°	0.886
25°	0.346	75°	0.930
30°	0.413	80°	0.966
35°	0.479	85°	0.991
40°	0.544	90°	1.000
45°	0.607		

To construct the projection, draw a vertical line, *cd* in Figure G.9, representing the central

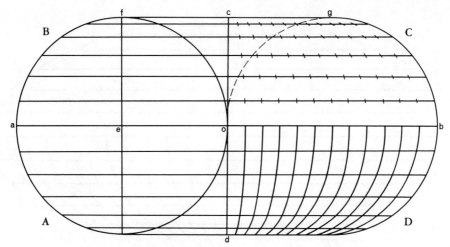

Figure G.8 Construction of Eckert's IV projection.

meridian to scale. Draw a perpendicular *ab* through *o*, midway between *c* and *d*. Lay off distance *ao* equal to 1.1107 *cd*. Length *ob* = *ao*. Draw *fg* parallel to *ab* at *c*. Distances *fc* and *cg* = 1/3 *ao*. From Table G.6 determine the distance of each parallel from the equator on the central meridian. In the table, length *oc* = 1. These positions are plotted on the central meridian, and through these points parallels to *ab* are drawn as in quadrant *B*. The length of each parallel is obtained from Table G.7 in which it is shown as a proportion of length *ob*. Subdivide each parallel equally with dividers as in quadrant *C* of Figure G.9 to establish the positions of the meridians. The meridians are drawn through homologous points, as in quadrant *D*, with the aid of a curve.

Table G.7 Lengths of the Parallels in the Flat Polar Quartic Equal-Area Projection (ob = 1)

0°	1.000	50°	0.752
5°	0.998	55°	0.700
10°	0.990	60°	0.643
15°	0.979	65°	0.581
20°	0.961	70°	0.517
25°	0.939	75°	0.453
30°	0.911	80°	0.394
35°	0.879	85°	0.350
40°	0.842	90°	0.333
45°	0.800		

A complete table for the construction of the projection with a 1° graticule by means of X and Y coordinates is given in Coast and Geodetic Survey, *Special Publication No. 245*, by F. Webster McBryde and Paul D. Thomas, who devised the projection. The values in the tables shown are based upon those in this reference.

D. Azimuthal Projections

Although azimuthal projections seem to have more in common with each other than with any other class of projections, the uses and common methods of construction are quite varied. Some can be easily constructed geometrically; some cannot in any way be constructed geometrically. Some are most expeditiously put together by using X and Y coordinates to locate intersections in the graticule; others by transforming one projection into another. It is this last method that is the key to understanding these projections. Any azimuthal projection can be transformed into any other by merely relocating the graticule intersections along their azimuths from the center of the projection, since the projections vary only as to the radical scale from the center of the projection.

There are a great many azimuthal projections (theoretically, an infinite number are possible), but only a few have desirable properties. Of

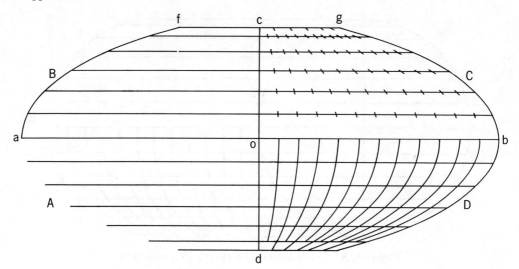

Figure G.9   Construction of the flat polar quartic equal-area projection.

these few, one, the gnomonic, is used primarily as a planning map in connection with navigation, and a cartographer is rarely called upon to construct it. The remaining common azimuthal projections, the Lambert's equal-area, the orthographic, the stereographic, and the azimuthal equidistant, are much in demand, and methods for their construction are suggested here. In the interests of brevity, not all the possibilities will be detailed, and the reader is referred to any of the standard works on map projections for a fuller account of possible procedures.

12. Stereographic. All that is necessary to construct this projection is a set of tables, a straight-edge, and a beam compass. It is relatively easy to calculate the values necessary to center the projection anywhere. The method will be described below, but Table G.8, necessary to calculate the projection centered on 40°, is included here to illustrate the construction procedure and to illustrate the construction and calculation terminology. The values in Table G.8 refer to a nominal globe with a diameter of unity. This means that the values in the table need only to be multiplied by the number of centimeters, or other units, contained in the diameter of the nominal globe of chosen scale.

To construct the projection, first draw a verti-

cal line, *ab* in Figure G.10. Locate on *ab* the center of the projection *(o)*, in this case 40°, and with *o* as the center describe a circle the diameter of which is twice the diameter of the nominal globe. The parallels are drawn by locating on *ab* the upper *(U)* and lower *(L)* points of each parallel. The center of the circle representing the parallel is midway between the points *U* and *L* for each parallel. For example, from Table G.8, the upper intersection with *ab* for the parallel of 20° is 1.73205 above *o* and is at *U* in Figure G.10. The lower intersection is 0.17633 below *o* (shown by the minus sign) and is located at *L*. Midway between these points is the center of the circle representing the parallel of 20°. The other parallels are similarly located and drawn with a compass. The centers of the arcs representing the meridians are all located along a straight line, *cd* in Figure G.10, perpendicular to *ab*, which is the homolatitude of the center of the projection. (The homolatitude of any point is the same latitude in the opposite hemisphere.) In Figure G.10 the homolatitude is located 0.83910 below *o*. The *bow* distance is the distance from the central meridian *(ab)* along the homolatitude *(cd)* to the intersection of the meridian with the homolatitude. The center distance is the distance along the homolatitude, on the opposite side of *ab*, to the center

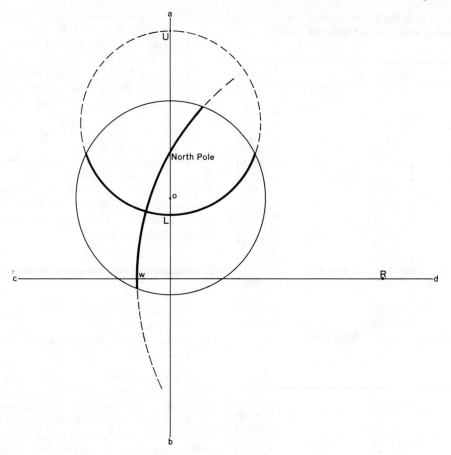

Figure G.10   Construction of the stereographic projection.

of the arc that represents the meridian. In Figure G.10, the *bow* distance for the meridian of 30° is 0.34986 and the intersection is at *w* on *cd*. The center distance is 2.26104 on the other side of *ab* and is located at *R*. The arc drawn through *w* must pass through the pole. The other meridians are drawn in similar fashion, first for one side and then repeated for the other side. They may, of course, be numbered in any desired sequence, depending upon what part of the earth is being mapped.

It will be observed that the spacing of the parallels on the central meridian increases away from the center. Since the parallels are evenly spaced on the earth, this establishes the radial scale for the projection. The scale is the same

from the center to the periphery in any direction.

A simplified method of calculating the kind of values exemplified by Table G.8 has been worked out by Professor James A. Barnes. It requires little more than reading the values from a table of natural trigonometric functions. The resulting values apply to a projection based on a nominal globe with a diameter of unity (D = 1.0). To convert the values to scale, multiply each by the diameter of the nominal globe.

The formulas and symbols are:

α = angle of tilt (position of center)
ϕ = latitude
λ = longitude

Table G.8   Table for Constructing the Stereographic Projection Centered on 40° (D = 1)

Parallels	Upper		Lower
North Pole		0.46631	
80°	0.57735		0.36397
70°	0.70021		0.26795
60°	0.83910		0.17633
50°	1.00000		0.08749
40°	1.19175		0.00000
30°	1.42815		−0.08749
20°	1.73205		−0.17633
10°	2.14451		−0.26795
0°	2.74748		−0.36397
10°	3.73205		−0.46631
20°	5.67128		−0.57735
30°	11.43005		−0.70021
40°	−0.83910	−0.83910	−0.83910
50°	−1.00000		−11.43005
60°	−1.19175		−5.67128
70°	−1.42815		−3.73205
80°	−1.73205		−2.74748
South Pole		−2.14451	

Homolatitude = −0.83910

Meridians	Bow	Center
10°	0.11421	7.40335
20°	0.23018	3.58658
30°	0.34986	2.26104
40°	0.47513	1.55573
50°	0.60872	1.09537
60°	0.75368	0.75368
70°	0.91406	0.47513
80°	1.09537	0.23018
90°	1.30541	0.00000

U = upper intersection of parallel with central meridian

L = lower intersection of parallel with central meridian

Q = intersection of homolatitude of center point and central meridian

N = intersection of meridian and homolatitude

M = center of meridian arc on homolatitude

Calculation of the Position of the Parallels

For north latitudes and equator: $U = + \tan 1/2 (180° - \phi - \alpha)$.

For south latitudes less than α: $U = + \tan 1/2 (180° + \phi - \alpha)$.

For south latitudes greater than α: $U = - \tan 1/2 (\phi + \alpha)$.

For north latitudes greater than α: $L = + \tan 1/2 (\phi - \alpha)$.

For north latitudes less than α and equator: $L = - \tan 1/2 (\alpha - \phi)$.

For south latitudes less than α: $L = - \tan 1/2 (\phi + \alpha)$.

For south latitudes greater than α: $L = - \tan 1/2 (180° + \alpha - \phi)$.

(Note. Plus and minus signs indicate whether the value is above or below the center of the projection. The center of any parallel is midway between U and L.)

Calculation of the Position of the Meridians

$Q = - \tan \alpha$

$N = \sec \alpha \times \tan 1/2 \lambda$

$M = \sec \alpha \times \cot \lambda$

(Note. N and M correspond to *Bow* and *Center* values in Table G.8.)

13. Orthographic. The orthographic projection is rather like an architect's elevation. The principle of its construction can be seen in Figure G.11 (showing the projection centered on the pole), where the latitudinal spacing on the globe is projected by parallel lines to the central meridian of the projection. Being an azimuthal projection, all great circles through the center are straight lines and azimuths from the center are correct. Since in this form the pole is the center, all great circles that pass through it are meridians; hence, all meridians on the projection are straight lines and are correctly arranged around the pole.

The construction of the projection centered on the equator is no more involved. The procedure is illustrated in Figure G.12. The parallel spacing on the central meridian is the same as in the polar case, but the parallels are horizontal lines. The positions of the meridians on the parallels are carried over from the polar case as illustrated. Since all four quadrants are images of one another, only one need be drawn. The others may be traced.

The orthographic projection centered on the pole or the equator is seldom used. The projec-

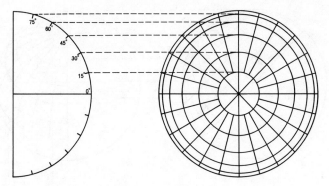

Figure G.11   Construction of the orthographic projection centered on the pole.

tion is more often centered on some point of interest between the pole and the equator. This may be accomplished by employing a polar and equatorial case, as in Figure G.13. It may also be drawn by using the equatorial case as a nomograph, as described below in Section 16. An appropriate method is to trace a photograph of the globe centered at the desired spot. A photograph of the globe is not a true orthographic, since some perspective convergence is bound to occur in the photographing, even when the camera is at a considerable distance. Nevertheless, the result is a true azimuthal projection and is very nearly the same as the orthographic. Since the only useful precise property is its azimuthality, nothing is lost by using a photograph.

14. Lambert's (equal-area). Like the orthographic, Lambert's azimuthal equal-area projection is most useful when centered in the area of interest, although the projection is frequently seen in the polar case to accompany other projections that distort areas considerably. The polar case is easily constructed as illustrated in Figure G.14. A segment of the globe is drawn with R as the radius of the nominal globe of chosen area scale. The chord distances from the pole to the parallels are carried up to a tangent with a compass and establish the positions of the parallels on the projection. The meridians, as in other azimuthal projections, are straight lines through the poles.

The equatorial case of the projection is somewhat more difficult to construct graphically and is more easily accomplished by plotting the X

and Y coordinate positions of the grid intersections from tables. Table G.9 gives values for every 10°.

As with most of the azimuthal projections, the oblique case centered on some area of interest is the most useful. The oblique case may be derived from the equatorial by using the latter as a nomograph in the manner outlined in Sec-

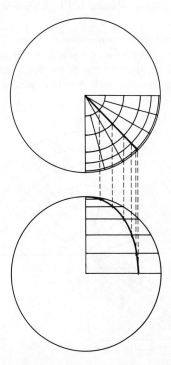

Figure G.12   Construction of the orthographic projection centered on the equator.

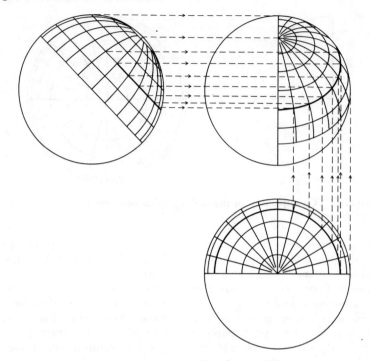

Figure G.13   Construction of the oblique orthographic projection.

tion 16 below. Coordinates are given in Table G.10 for a graticule centered at latitude 40°, which is an appropriate place for maps of the United States or North America, among others, to be centered.

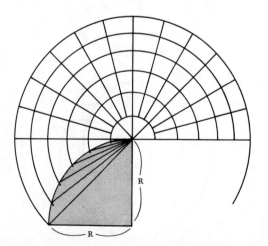

Figue G.14   Construction of Lambert's equal-area projection centered on the pole.

Since this projection is equal-area, it should be constructed to an area scale. The tables just referred to have been prepared on the basis of $R = 1$, so that each value in the tables needs only to be multiplied by the length of the radius of the nominal globe of chosen area scale.

15. Azimuthal Equidistant. It is not difficult to construct the azimuthal equidistant projection centered on any spot. The polar case is constructed by first drawing an appropriate set of meridians and then constructing equally spaced circles concentric around the pole. This may be extended to include the whole earth, in which case the bounding circle is the opposite pole, as in Figure G.15. As may be expected, the area and linear distortions become large as periphery is approached. A world map centered on a pole is not a very appropriate use of the azimuthal equidistant projection.

For an oblique case the simplest procedure is to prepare first a stereographic projection centered at the desired latitude. This may then be transformed (to any other azimuthal projection) by merely relocating the positions of the in-

TABLE G.9

TABLE FOR THE CONSTRUCTION OF A LAMBERT AZIMUTHAL EQUAL-AREA PROJECTION CENTERED ON THE EQUATOR.   COORDINATES IN UNITS OF THE EARTH'S RADIUS ($R = 1$) (From Deetz and Adams)

Lat.	Long. 0°		Long. 10°		Long. 20°		Long. 30°		Long. 40°	
	x	y	x	y	x	y	x	y	x	y
0°	0	0.000	0.174	0.000	0.347	0.000	0.518	0.000	0.684	0.000
10°	0	0.174	0.172	0.175	0.343	0.177	0.512	0.180	0.676	0.185
20°	0	0.347	0.166	0.349	0.331	0.352	0.493	0.359	0.651	0.369
30°	0	0.518	0.156	0.519	0.311	0.525	0.463	0.535	0.610	0.548
40°	0	0.684	0.142	0.686	0.283	0.693	0.420	0.705	0.553	0.722
50°	0	0.845	0.124	0.848	0.245	0.855	0.364	0.868	0.478	0.887
60°	0	1.000	0.101	1.003	0.199	1.010	0.295	1.023	0.386	1.041
70°	0	1.147	0.073	1.149	0.144	1.156	0.212	1.167	0.277	1.183
80°	0	1.286	0.039	1.287	0.078	1.291	0.114	1.299	0.148	1.308
90°	0	1.414	0.000	1.414	0.000	1.414	0.000	1.414	0.000	1.414

Lat.	Long. 50°		Long. 60°		Long. 70°		Long. 80°		Long. 90°	
	x	y	x	y	x	y	x	y	x	y
0°	0.845	0.000	1.000	0.000	1.147	0.000	1.286	0.000	1.414	0.000
10°	0.835	0.192	0.987	0.201	1.132	0.212	1.267	0.227	1.393	0.246
20°	0.804	0.382	0.949	0.399	1.086	0.421	1.213	0.448	1.329	0.484
30°	0.752	0.567	0.886	0.591	1.011	0.621	1.125	0.659	1.225	0.707
40°	0.679	0.744	0.798	0.773	0.906	0.809	1.002	0.854	1.083	0.909
50°	0.586	0.911	0.685	0.942	0.773	0.981	0.849	1.028	0.909	1.083
60°	0.471	1.065	0.548	1.095	0.614	1.132	0.668	1.175	0.707	1.225
70°	0.335	1.203	0.387	1.228	0.430	1.257	0.463	1.291	0.484	1.329
80°	0.178	1.321	0.204	1.336	0.224	1.353	0.238	1.372	0.246	1.393
90°	0.000	1.414	0.000	1.414	0.000	1.414	0.000	1.414	0.000	1.414

tersections of the new graticule along their azimuths from the center. This is accomplished by marking off on one edge of a strip of paper or plastic the radial scale of the stereographic and, on the other edge, the radial scale of the equidistant. The strip is then placed on the stereographic, and the distance of an intersection in the graticule from the center is noted. This distance is transferred to the radial scale of the equidistant and the position plotted *along the same azimuth*. Figure G.16 illustrates this procedure. Of course, the scale of the new projec-

tion may be made different, if desired, at the same time.

Positions outside the inner hemisphere of the stereographic may be located in the following manner. When the equidistant hemisphere has been completed, the stereographic is no longer needed. The position of every point on the earth is obviously 180° from its antipode; every point and its antipode lie on a great circle through the center of the projection; the diameter of the hemisphere is 180°; and the scale is uniform from the center. Thus, all that is neces-

TABLE G.10

TABLE FOR THE CONSTRUCTION OF A LAMBERT EQUAL-AREA PROJECTION CENTERED AT LATITUDE 40°. CO-ORDINATES IN UNITS OF THE EARTH'S RADIUS ($R = 1$) (Adapted from Deetz and Adams)

Lat.	Long. 0°		Long. 10°		Long. 20°		Long. 30°		Long. 40°	
	x	y	x	y	x	y	x	y	x	y
90°	0	+0.845	0.000	+0.845	0.000	+0.845	0.000	+0.845	0.000	+0.845
80°	0	+0.684	0.032	+0.686	0.063	+0.692	0.093	+0.704	0.120	+0.718
70°	0	+0.518	0.062	+0.521	0.121	+0.533	0.175	+0.553	0.231	+0.580
60°	0	+0.347	0.088	+0.352	0.174	+0.369	0.257	+0.396	0.334	+0.434
50°	0	+0.174	0.112	+0.181	0.222	+0.200	0.328	+0.234	0.427	+0.280
40°	0	0.000	0.133	+0.007	0.264	+0.029	0.391	+0.067	0.510	+0.120
30°	0	−0.174	0.151	−0.166	0.300	−0.142	0.445	−0.102	0.582	−0.045
20°	0	−0.347	0.166	−0.338	0.331	−0.314	0.489	−0.272	0.642	−0.214
10°	0	−0.518	0.177	−0.509	0.353	−0.484	0.524	−0.442	0.689	−0.383
0°	0	−0.684	0.185	−0.675	0.369	−0.651	0.548	−0.610	0.722	−0.553
−10°	0	−0.845	—	—	—	—	—	—	—	—

Lat.	Long. 50°		Long. 60°		Long. 70°		Long. 80°		Long. 90°		Long. 100°	
	x	y	x	y	x	y	x	y	x	y	x	y
90°	0.000	+0.845	0.000	+0.845	0.000	+0.845	0.000	+0.845	0.000	+0.845	0.000	+0.845
80°	0.143	+0.736	0.163	+0.758	0.178	+0.782	0.188	+0.808	0.192	+0.834	—	—
70°	0.278	+0.613	0.318	+0.646	0.349	+0.701	0.371	+0.751	0.382	+0.804	—	—
60°	0.403	+0.481	0.463	+0.538	0.511	+0.602	0.547	+0.674	0.567	+0.752	0.570	+0.833
50°	0.518	+0.338	0.583	+0.408	0.663	+0.489	0.713	+0.580	0.744	+0.679	0.755	+0.785
40°	0.620	+0.186	0.702	+0.267	0.801	+0.361	—	—	—	—	—	—
30°	0.710	+0.027	0.825	+0.115	—	—	—	—	—	—	—	—
20°	0.785	−0.138	—	—	—	—	—	—	—	—	—	—
10°	0.844	−0.307	—	—	—	—	—	—	—	—	—	—
0°	0.887	−0.478	—	—	—	—	—	—	—	—	—	—

sary is to mark on a straightedge the diameter of the hemisphere, and, keeping the edge on the center of the projection, locate all the outer intersections of the graticule from their antipodes in the inner hemisphere.

16. Shifting the Center of an Azimuthal Projection. The nomographic process for shifting the center of an azimuthal projection was devised by R. E. Harrison. The following discussion is somewhat modified from his explanation.[*]

To illustrate the procedure, the construction of an orthographic projection centered at 35° N is here described. The same procedure is followed in making any azimuthal projection, with only minor and rather obvious differences.

First, a circular nomograph consisting of an equatorial case of the chosen projection system is placed on a drawing board and covered with a

[*]Richard Edes Harrison, "The Nomograph as an Instrument in Map Making," *The Geographical Review*, 33, 655–657, (1943).

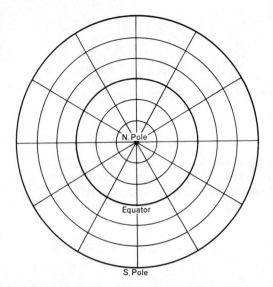

Figure G.15 The azimuthal equidistant projection centered on the pole.

by counting 90° from the new equator along the upright center line. (When making an azimuthal equidistant or equal-area projection, the South Pole is also marked.)

At this stage the North Pole and the equator (with the meridional intersections) have been located (Stage I, Fig. G.17). All the meridians may now be added to graticule. This is accomplished in the same manner as finding a great circle nomographically, namely, by rotating the circular nomograph beneath the plastic until both the North Pole and the meridional intersection of the equator are on the same nomograph meridian or occupy the same relative position between two meridians. The meridian is drawn in its entirety.

While the nomograph is in this position, the appropriate crossings of the parallels (with their proper angles of intersection) are noted on the meridian (Stage II, Fig. G.17). Since the intersections rarely coincide exactly with the parallels on the nomograph, it is generally necessary to interpolate. However, in working with a

rectangular piece of translucent plastic (or tracing paper) that overlaps the nomograph on the sides but not at the top and bottom. The plastic is fixed to the board. A needle or round-shanked thumbtack is thrust firmly through the plastic overlay and nomograph at the center of the latter.

The nomograph is positioned so that its equator is *vertical* (Stage I, Fig. G.17). The intersections of the equator and the central meridian with the bounding circle are then lightly marked on the tracing paper. Thirty-five degrees are counted off below the center along the vertical equator of the nomograph; the meridian passing through this point becomes the equator of the new graticule. This is traced, and the intersections of the parallels of the nomograph with the new equator are marked according to the desired graticule interval. This is best done so that the angles of the intersections are correctly preserved because of a remarkable feature of the nomographic method: instead of providing only horizontal and vertical coordinates to establish the points of crossing as in mathematical and other graphic procedures, it also gives the correct angle of any intersection of the projection. Before changing the orientation of the nomograph, the North Pole is marked

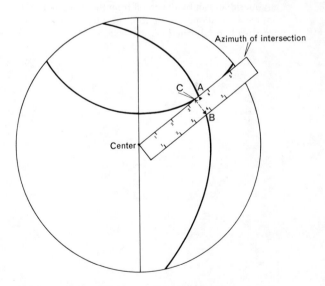

Figure G.16 The transformation of one azimuthal to another. A is the distance from the center to a point according to the radial scale of the stereographic. B is the distance of the same point according to the radial scale of the equidistant. C is the location of the point along the same azimuth on the equidistant.

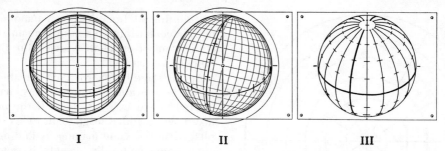

| I | II | III |

Figure G.17   Three stages in the construction with a nomograph of an orthographic graticule centered at 35°. Light lines show the nomograph, heavy lines construction on the plastic. (Courtesy of *The Geographical Review*, published by the American Geographical Society of New York.)

one-degree nomograph, the proper intersections will be so close to a parallel that they can be drawn directly with little loss of accuracy.°

°Furthermore, we often find a simple means of checking the accuracy of the interpolations. For example, on the orthographic projection, advantage can be taken of the fact that all lines connecting the intersections of parallels along any two meridians are parallel. With a parallel ruling device, the intersections along any meridian can be ticked off from the corresponding intersections along the central meridian, since these are already established.

The circular nomograph is then rotated to obtain the next meridian, and so on until all are drawn in. At this stage the graticule consists of an equator and a complete set of meridians marked with the crossings of the parallels (Stage III, Fig. G. 17); and it is a simple matter to complete the parallels, since these crossings form an almost continuous curve. The graticule can be drawn first in pencil and later in ink, or it can be inked directly.

APPENDIX

H PROOF OF TISSOT'S LAW OF DEFORMATION

The law of deformation was developed by M. A. Tissot and appears in full in his *Mémoire sur la représentation des surfaces et les projections des cartes géographiques*, Paris, 1881, which includes sixty pages of deformation tables for various projections. The late Oscar S. Adams, a noted authority on the mathematics of map projections, included an account of it in his "General Theory of Polyconic Projections," U. S. Coast and Geodetic Survey *Special Publication* 57, Washington, D.C. 1934, pp. 153-163.

The following demonstration of the proof of the law of deformation consists of slightly reworded extracts taken from the above source with the permission of the Director, United States Coast and Geodetic Survey. The modifications are those of the authors.

To represent one surface upon another it is necessary to imagine that each surface is composed of two systems of lines which divide them into infinitesimal parallelograms. To each line of the first surface is made to correspond one of the lines of the second. The intersection of two lines of the different systems upon the one surface and the intersection of the two corresponding lines upon the other, therefore, determine two corresponding points. The totality of the points of the second surface which correspond to the points of the first forms the representation or the projection of the first surface. Different methods of representation are obtained by varying the two series of lines which form the graticule upon one of the surfaces.

If two surfaces are not applicable to each other, that is, cannot be transformed without compression or expansion, it is impossible to choose a method of projection where there is similarity between every figure traced upon the first and the corresponding figure upon the second. On the other hand, whatever the two surfaces may be, there exists an infinity of systems of projection which preserve the angles; consequently, each *infinitely small* figure and its representation are similar to each other. There is also an infinity of systems which preserve the areas. These two classes of projections are exceptions, however. In a method of projection being taken by chance, the angles will be generally changed, except possibly at particular points, and the corresponding areas will not have a constant ratio to each other. The lengths on the one surface will thus be altered on the second.

Consider two curves, one on one surface and one on the other, which correspond to each other. In Fig H.1 let O and M be two points of the one, O' and M' the corresponding points of the other, and let OT be the tangent at O to the first curve. If the point M lies infinitely close to

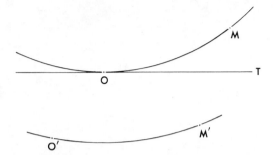

Figure H.1   A curve (above) and its projection (below).

the point O, then point M' will approach infinitely close to the point O'. The ratio of the length of the arc $O'M'$ to that of the arc OM will therefore tend toward a certain limit; this limit is what is called the ratio of lengths at the point O upon the curve OM or in the direction OT. In a system of projection preserving the angles, the ratio of lengths thus defined has the same value for all directions at a given point; but it varies with the position of this point, unless the two surfaces are applicable to each other. When the representation does not preserve the angles except at particular points, the ratio of lengths at all other points changes with the direction.

The deformation produced around each point is subjected to a law which depends neither upon the nature of the surfaces nor upon the method of projection.

Every representation of one surface upon another can be replaced by an infinity of orthogonal projections, each made upon a suitable scale.

It is noted, first, that there always exists at every point of the first surface two tangents perpendicular to each other such that the directions which correspond to them upon the second surface also intersect at right angles. In Fig. H.2 let CE and OD be two tangents perpendicular to each other at the point O on the first surface; let $C'E'$ and $O'D'$ be the corresponding tangents on the second surface. Suppose, further, that of the two angles $C'O'D'$ and $D'O'E'$ the first is acute, and imagine that a right angle having its vertex at O turns from left to right around this point in the plane CDE starting from the position COD and arriving at the position DOE. The corresponding angle in the plane tangent at O

to the second surface will first coincide with $C'O'D'$ and will be acute; in its final position it will coincide with $D'O'E'$ and will be obtuse; within the interval it will have passed through a right angle. Therefore, there exists a system of two tangents satisfying the condition stated, except at certain singular points. From this property it may be concluded that in every system of representation there is upon the first of the two surfaces a system of two series of orthogonal curves whose projections upon the second surface are also orthogonal. The two surfaces are thus divided into infinitesimal rectangles which correspond the one to the other.

With this fact established, let M be a point in Fig. H.3 infinitely near to O upon the first surface and let $OPMQ$ be that one of the infinitesimal rectangles which has just been described that has OM as a diagonal. Then let $OP'M'Q'$ be the rectangle on the second surface which corresponds to $OPMQ$ on the first. Move the second surface and place it in such a position that O on each surface coincides, and that the sides OP' and OQ' on the second surface fall upon the sides OP and OQ on the first surface. Designate as N the point of intersection of the lines OM' and PM. This point can be considered as the orthogonal projection of the point M if the plane of the rectangle $OPMQ$ were turned through a suitable angle with OP as an axis. But this angle, which depends only upon the ratio of the two lines NP and MP is the same whatever point M may be; for denoting, respectively, by a

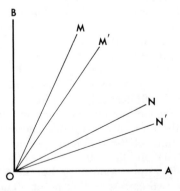

Figure H.2   Two tangents at right angles and their projections.

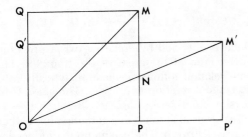

Figure H.3   Projection of infinitely near points.

and b the ratios of the lengths in the directions OP and OQ—that is, on setting

$$\frac{OP'}{OP} = a \text{ and } \frac{OQ'}{OQ} = b$$

then

$$\frac{NP}{M'P'} = \frac{OP}{OP'} = \frac{1}{a}, \text{ and } \frac{MP}{M'P'} = \frac{OQ}{OQ'} = \frac{1}{b}$$

and, consequently,

$$\frac{NP}{MP} = \frac{b}{a}$$

Thus if M moves on an infinitesimal curve traced around O, the locus described by N is obtained by turning this curve through a certain angle around OP as an axis and by then projecting orthogonally upon the plane tangent at O. On the other hand,

$$\frac{OM'}{ON} = \frac{OP'}{OP} = a$$

so that the locus of the points M' is homothetic to that of the points N; the center of similitude is O, and the ratio of similitude has the value a. The representation of the infinitesimal figure described by the point M is then in reality an orthogonal projection of this figure made on a suitable scale, or the figure formed by the points N and that formed by the points M' are formed by parallel sections of the same cone. Any map projection can therefore be considered as produced by the juxtaposition of orthogonal projections of the surface elements of the earth sphere, provided that, from one element to the other, both the scale of the reduction and the

position of the element with respect to the plane of the map are varied.

Of all the right angles which are formed by the tangents at the point O, those of the lines OP and OQ and their prolongations are the only ones one side of which remains parallel to the tangent plane after the rotation which was just described. These are, therefore, the only right-angled tangents which are projected into right angles. An addition to the proposition which has just been proved can now be stated, and the whole can be expressed in the following form: at every point of the surface which is to be represented (the earth) there are two perpendicular tangents, and, if the angles are not preserved, there are only two, such that those which correspond to them upon the other surface (the map plane) also intersect at right angles. Thus, upon each of the two surfaces, there exists a system of orthogonal trajectories, and, if the method of representation does not preserve the angles, there exists only one of them, the projections of which upon the other surface are also orthogonal.

The two perpendicular tangents, the angle between which is not altered by the projection, are designated as first and second principal tangents. The ratio of lengths in the directions of these two tangents is designated respectively by a and b. Ratio a is assumed to be greater than b.

If the infinitesimal curve drawn around the point O is a circumference of which O is the center, the representation of this curve will be an ellipse, the axes of which will fall upon the principal tangents, and these will have the values $2a$ and $2b$ when the radius of the circle is taken as unity. This ellipse constitutes at each point a sort of indicatrix of the system of projection.

In place of (1) projecting orthogonally the circumference (the locus of the points M in Fig. H.3), which gives the ellipse (the locus of the points N), and (2) then increasing this in the ratio of a to unity (which gives the locus of the points M'), one may perform these two operations in the inverse order. Then, in Fig. H.4 the point M' of the elliptic indicatrix can be obtained which corresponds to a given point M of the circle. This is done by prolonging the radius OM until it meets at R the circumference de-

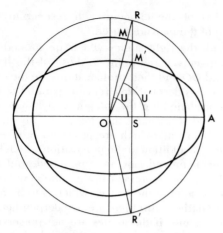

Figure H.4   Tissot's indicatrix.

$$\sin(U - U') = \frac{a - b}{a + b} \sin(U + U')$$

which is obtained by equating two expressions for the ratio of the areas of the triangles. The same relation follows at once analytically from the tangent relation first given. The angle U increasing from zero to $\pi/2$, its alteration $U - U'$ increases from zero up to a certain value ω, then decreases to zero. The maximum is produced at the moment when the sum $U + U'$ becomes equal to $\pi/2$. From the tangent formula the following are their values:

$$\tan U = \frac{\sqrt{a}}{\sqrt{b}} \text{ and } \tan U' = \frac{\sqrt{b}}{\sqrt{a}}$$

The quantity ω can be computed by any one of the formulas

$$\sin \omega = \frac{a - b}{a + b}$$

$$\cos \omega = \frac{2\sqrt{ab}}{a + b}$$

$$\tan \omega = \frac{a - b}{2\sqrt{ab}}$$

$$\tan \frac{\omega}{2} = \frac{\sqrt{a} - \sqrt{b}}{\sqrt{a} + \sqrt{b}}$$

$$\tan\left(\frac{\pi}{4} + \frac{\omega}{2}\right) = \frac{\sqrt{a}}{\sqrt{b}} \text{ and } \tan\left(\frac{\pi}{4} - \frac{\omega}{2}\right) = \frac{\sqrt{b}}{\sqrt{a}}$$

If one wishes to calculate directly the alteration which any given angle U is subject to, he may use one of the two formulas

$$\tan (U - U') = \frac{(a - b) \tan U}{a + b \tan {}^2U'}$$

$$\tan (U - U') = \frac{(a - b) \sin 2U}{a + b + (a - b) \cos 2U}$$

which follow immediately from the previous formulas.

Consider now an angle MON in Figs. H.5 and H.6, which has for sides neither one nor the other of the principal tangents OA and OB. Assume the two directions OM and ON to be to the right of OB and the one of them OM above OA. According to whether the other ON will be above OA (Fig. H.5) or below OA (Fig. H.6), the corresponding angle $M'ON'$ can be calculated by taking the difference or the sum of the angles

scribed upon the major axis as diameter, and then by dropping a perpendicular from R upon OA, the semimajor axis, and finally, by reducing this perpendicular RS, starting from its foot S in the ratio of b to a. The point M' thus determined is the required point.

If in Fig. H.4 OM' is drawn, then the angles AOM and AOM' which correspond upon the two surfaces may be designated, respectively, as U and U'. Inasmuch as the second is the smaller of the two, it may be seen that the representation diminishes all the acute angles, one side of which coincides with the first principal tangent. Between U and U' there is, moreover, the relation

$$\tan U' = \frac{b}{a} \tan U$$

since

$$\tan U = \frac{RS}{OS}$$

$$\tan U' = \frac{M'S}{OS}$$

and, consequently,

$$\tan U' = \frac{M'S}{RS} \tan U = \frac{b}{a} \tan U$$

Prolong the line RS to R' and then join O and R'. The two triangles ORM' and $OR'M'$ give

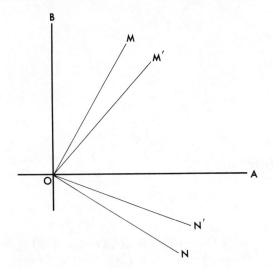

Figure H.5 Angular change in projections, first case.

AOM' and AON', which would be given by the formula stated above. The alteration MON − M'ON' would also in the first case be the difference, and in the second case would be the sum of the alterations of the angles AOM and AON. When the angle AON (Fig. H.5) is equal to the angle BOM', its alteration is the same as that of the angle AOM, so that the angle MON will then be reproduced in its true magnitude by the angle M'ON'. Thus to every given direction another can be joined and only one other, such that their angle is preserved in the projection. However, the second direction will coincide with the first when it makes with OA the angle which has been denoted by U.

The angle the most altered is that which this direction forms with the point symmetric to it with respect to OA; it is represented upon the projection by its supplement. The maximum alteration thus produced is equal to 2ω. This can never be found applicable to two directions that are perpendicular to each other.

The length OM in Fig. H.4 having been taken as unity, the ratio of lengths in the direction OM is measured by OM'. If this ratio is designated as r, it may be calculated by one of the formulas

$$r \cos U' = a \cos U$$
$$r \sin U' = b \sin U$$

or

$$r^2 = a^2 \cos{}^2U + b^2 \sin{}^2U.$$

There is also among r, U, and the alteration U − U' of the angle U the relation

$$2r \sin(U - U') = (a - b) \sin 2U$$

which expresses that, in the triangle ORM', the sines of two of the angles are to each other as the sides opposite.

The maximum and the minimum of r correspond to the principal tangents and are, respectively, a and b.

If r and r_1 are the ratios of lengths in two directions at right angles to each other and if θ is the alteration that the right angle formed by these two directions is subjected to, then from the properties of conjugate diameters in the ellipse

$$r^2 + r_1{}^2 = a^2 + b^2$$
$$rr_1 \cos \theta = ab$$

In terms of the scale along a parallel (h) and along a meridian (k) at a point on a projection, the semi-axes are given by the equations

$$a^2 + b^2 = h^2 + k^2$$
$$ab = hk \cos \theta$$

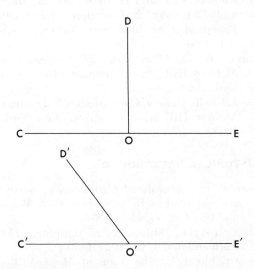

Figure H.6 Angular change in projections, second case.

B IBLIOGRAPHY

Selected List of Generally Available Materials Useful for Additional Reading

The following list includes references to only those items that are in English, and that are readily available or easily secured. They will be found useful for reference or for further reading in the topics under which they have been included.

GENERAL REFERENCES

Birch, T. W., *Maps Topographical and Statistical*, 2nd edition, The Clarendon Press, Oxford, 1964.

Greenhood, David, *Mapping*, (Phoenix Science Series), The University of Chicago Press, Chicago, 1964.

Monkhouse, F. J. and H. R. Wilkinson, *Maps and Diagrams*, 2nd edition, (University Paperbacks), Methuen and Co., Ltd., London, 1964.

Raisz, Erwin, *Principles of Cartography*, McGraw-Hill Book Company, Inc., New York, 1962.

Raisz, Erwin, *General Cartography*, 2nd edition, McGraw-Hill Book Company, New York, 1948.

HISTORICAL BACKGROUND

Bagrow, Leo, *History of Cartography*, revised and enlarged by R. A. Skelton, C. A. Watts and Co., Ltd., London, 1964.

Baldock, E. D., "Milestones of Mapping," *The Cartographer*, 3, 89–102 (1966).

Brown, Lloyd A., *The Story of Maps*, Little, Brown and Co., Boston, 1949.

Crone, G. R., *Maps and Their Makers*, Hutchin-son's University Library, London, 1953.

Fordham, H. G., *Maps, Their History, Characteristics and Uses*, Cambridge University Press, Cambridge, 1943.

Funkhouser, H. G., "Historical Development of the Graphical Representation of Statistical Data," *Osiris*, 3, 269–404 (1937).

Lynam, Edward, *The Map Maker's Art, Essays on the History of Maps*, The Batchworth Press, London, 1953.

Raisz, E., *Mapping the World*, Abelard-Schuman, New York, 1956.

Robinson, A. H. and Helen Wallis, "Humboldt's Map of Isothermal Lines," *The Cartographic Journal*, 4, 119–123 (1967).

Robinson, Arthur H., "The 1837 Maps of Henry Drury Harness," *The Geographical Journal*, 121, 440–450 (1955).

Skelton, R. A., *Decorative Printed Maps of the 15th to 18th Centuries*, Staples Press, London, 1952.

Tooley, R. V., *Maps and Map Makers*, 2nd edition, Bonanza Books, New York, 1952.

THE CARTOGRAPHIC METHOD

Balchin, W. G. V. and Alice M. Coleman, "Graphicacy Should be the Fourth Ace in the Pack," *The Cartographer*, 3, 23–27 (1966).

Garner, Clement L., "Geodesy—A Framework for Maps," *Surveying and Mapping*, 14, 154–158 (1954).

DasGupta, Sivaprasad, "Some Methodological Problems of Density Mapping," *The Geographical Review of India*, 26, 35–39 (1964).

Eckert, Max, "On the Nature of Maps and Map

Logic" (translated by W. L. G. Joerg), *Bulletin of the American Geographical Society,* **40**, 344–351 (1908).

Gould, Peter R., "On Mental Maps," *Discussion Paper No. 9,* Michigan Inter-Univ. Community of Mathematical Geographers, University of Michigan, Ann Arbor, 1966.

Heath, W. R., "Technical Problems on Thematic Mapping," *The Cartographic Journal,* **1**, 33–36 (1964).

Imhof, Eduard, "Tasks and Methods of Theoretical Cartography," *International Yearbook of Cartography,* **3**, 13–23 (1963).

Jenks, George F. and Dwight A. Brown, "Three-Dimensional Map Construction," *Science,* **154**, 857–864 (1966).

Kaminstein, A. L., "Maps, Charts and Copyright," *Special Libraries,* **51**, 241–243 (1960).

Kaminstein, A. L., "Copyright and Registration of Maps," *Surveying and Mapping,* **13**, 182–184 (1953).

Kishimoto, Haruko, *Cartometric Measurements,* (publ. by the author), Zurich, 1968.

Lee, Paul B., "Copyright—The Publisher's Viewpoint," *Special Libraries,* **51**, 244–246 (1960).

Mackay, J. Ross, "Geographic Cartography," *The Canadian Geographer,* **1**, 1–14 (1954).

Robinson, Arthur H. et al., "Geographical Cartography," Chapter 26 in *American Geography Inventory and Prospect,* Syracuse University Press, Syracuse, 1954.

Robinson, Arthur H., "The Potential Contribution of Cartography in Liberal Education," *The Cartographer,* **2**, 1–8 (1965).

Thompson, Morris M., "How Accurate Is That Map?," *Surveying and Mapping,* **16**, 164–173 (1956).

Wright, John K., "Map Makers Are Human. Comments on the Subjective in Maps," *The Geographical Review,* **32**, 527–544 (1942).

Zuber, Leo J., "What's a Map Worth?," *Surveying and Mapping,* **12**, 14–18 (1952).

COORDINATE SYSTEMS, SCALE, AND MEASUREMENT

Ballenzweig, Emanuel M., "A Practical Equal-Area Grid," *Journal of Geophysical Research,* **64**, 647–651 (1959).

Bailey, Harry P., "A Grid Formed of Meridians and Parallels for the Comparison and Measurement of Area," *The Geographical Review,* **46**, 239–245 (1956).

Colvocoresses, Alden P., "A Unified Plane Coordinate Reference System," *Surveying and Mapping,* **27**, 621–624 (1967).

Gierhart, John W., "Evaluation of Methods of Area Measurement," *Surveying and Mapping,* **14**, 460–465 (1954).

Goussinsky, B., "A Proposal for the Standardization of Scales of Maps in Cartography," *Empire Survey Review,* **14**, 298–301 (1958).

Mitchell, Hugh C., and Lansing G. Simmons, *The State Coordinate Systems,* United States Coast and Geodetic Survey, Special Publication No. 235, U.S. Government Printing Office, Washington, D.C., 1945.

O'Keefe, John A., "The Universal Transverse Mercator Grid and Projection," *The Professional Geographer,* **4**, 19–24 (1952).

Proudfoot, M., *The Measurement of Geographic Area,* Bureau of the Census, Washington, D.C., 1946.

Schmid, Erwin, "Plane Coordinate Systems and Map Projections," *Journal of the Surveying and Mapping Division,* ASCE, **90**, 17–25 (1964).

The Universal Grid Systems, TM 5-241 (Army) TO 16-1-233 (Air Force), U.S. Government Printing Office, Washington, D.C., 1951.

PHOTOINTERPRETATION AND PHOTOGRAMMETRY

Avery, Gene and Dennis Richter, "An Airphoto Index to Physical and Cultural Features in Eastern United States," *Photogrammetric Engineering,* **31**, 897–916 (1965).

Bean, Russell K., "Development of the Orthophotoscope," *Photogrammetric Engineering,* **21**, 529–535 (1955).

Bigelow, George F., "Photographic Interpretation Keys—A Reappraisal," *Photogrammetric Engineering,* **29**, 1042–1051 (1963).

Colwell, Robert N., "Aids For the Selection and Training of Photo Interpreters," *Photogrammetric Engineering,* **31**, 327–339 (1965).

Hobrough, G. L., "Automation in Photogrammetric Instruments," *Photogrammetric Engineering,* **31**, 593–603 (1965).

Kistler, Phillip S., "Continuous Strip Photography," *Photogrammetric Engineering,* 12, 219–223 (1946).

Knauf, J. William, "The Stereo Image Alternator," *Photogrammetric Engineering,* 33, 1113–1116 (1967).

Laprode, George L., "An Analytical and Experimental Study of Stereo for Radar," *Photogrammetric Engineering,* 29, 294–300 (1963).

Leonardo, Earl S., "Comparison of Imagery Geometry for Radar and Camera Photographs," *Photogrammetric Engineering,* 29, 287–293 (1963).

Lueder, Donald R., *Aerial Photographic Interpretation,* McGraw-Hill Book Company, Inc., New York, 1959.

Manual of Photographic Interpretation, American Society of Photogrammetry, Washington, D.C., 1960.

Manual of Photogrammetry, American Society of Photogrammetry, Falls Church, Virginia, 1966.

Olson, Charles E., Jr., "Photographic Interpretation in the Earth Sciences," *Photogrammetric Engineering,* 29, 968–978 (1963).

Olson, Charles E., Jr., "Accuracy of Land-Use Interpretation from Infrared Imagery in the 4.5 to 5.5 Micron Band," *Annals Association of American Geographers,* 57, 382–388 (1967).

Rosenfield, Azriel, "Automatic Imagery Interpretation," *Photogrammetric Engineering,* 31, 240–242 (1965).

Shepherd, Wesley H., "Automatic Contour Digitizer," *Photogrammetric Engineering,* 34, 75–82 (1968).

Simpson, Robert B., "Radar: Geographic Tool," *Annals Association of American Geographers,* 56, 80–96 (1966).

Smith, John T., Jr., "Color—A New Dimension in Photogrammetry," *Photogrammetric Engineering,* 29, 999–1013 (1963).

Spurr, Stephen H. *Photogrammetry and Photo-Interpretation,* 2nd edition, The Ronald Press Company, New York, 1960.

Stone, Kirk H., "A Guide to the Interpretation and Analysis of Aerial Photos," *Annals Association of American Geographers,* 54, 318–328 (1964).

Strandberg, Carl H., *Aerial Discovery Manual,* John Wiley and Sons, Inc., New York, 1967.

GENERALIZATION

Das Gupta, Sivaprasad, "Some Measures of Generalization on Thematic Maps," *Geographical Review of India,* 26, 73–78 (1964).

Jenks, George F., "Generalization in Statistical Mapping," *Annals Association of American Geographers,* 53, 15–26 (1963).

Jenks, George F. and Michael R. C. Coulson, "Class Intervals for Statistical Maps," *International Yearbook of Cartography,* 3, 119–133 (1963).

Jenks, George F., "The Data Model Concept in Statistical Mapping," *International Yearbook of Cartography,* 7, 186–188 (1967).

Lundquist, Gösta, "Generalization—A Preliminary Survey of an Important Subject," *The Canadian Surveyor,* 14, 466–470 (1959).

Miller, O. M. and Robert J. Voskuil, "Thematic-Map Generalization," *The Geographical Review,* 54, 13–19 (1964).

Pannekoek, A. J., "Generalization of Coastlines and Contours," *International Yearbook of Cartography,* 2, 55–74 (1962).

Töpfer, F. and W. Pillewizer (with notes by D. H. Maling), "The Principles of Selection, A Means of Cartographic Generalization," *The Cartographic Journal,* 3, 10–16 (1966).

SYMBOLIZATION OF DATA

Blumenstock, David I., "The Reliability Factor in the Drawing of Isarithms," *Annals of the Association of American Geographers,* 43, 289–304 (1953).

Byron, William G., "Use of the Recording Densitometer in Measuring Density from Dot Maps," *Surveying and Mapping,* 18 (1958).

Court, Arnold, "The Inter-Neighbor Interval," *Yearbook, Assoc. Pacific Coast Geographers,* 28, 180–182 (1966).

Dahlberg, Richard E., "Towards the Improvement of the Dot Map," *International Yearbook of Cartography,* 7, 157–166 (1967).

Forward, Charles N. and Charles W. Raymond, "Small-Scale Land Use Mapping from Statistical Data," *Economic Geography*, **35**, 315–321 (1959).

Mackay, J. Ross, "Dotting the Dot Map," *Surveying and Mapping*, **9**, 3–10 (1949).

Mackay, J. Ross, "Some Problems and Techniques in Isopleth Mapping," *Economic Geography*, **27**, 1–9 (1951).

Mackay, J. Ross, "An Analysis of Isopleth and Choropleth Class Intervals," *Economic Geography*, **31**, 71–81 (1955).

Mackay, J. Ross, "Isopleth Class Intervals: A Consideration in Their Selection," *The Canadian Geographer*, **7**, 42–45 (1963).

Porter, Philip W., "Putting the Isopleth in its Place," *Proceedings Minnesota Academy of Science*, **15–16**, 372–384 (1957–1958).

Schmid, Calvin F. and E. H. MacCannell, "Basic Problems, Techniques, and Theory of Isopleth Mapping," *Journal of the American Statistical Association*, **50**, 220–239 (1955).

Schultz, Gwen M., "An Experiment in Selecting Value Scales for Statistical Distribution Maps," *Surveying and Mapping*, **21**, 224–230 (1961).

Sinnhuber, K. A., "The Representation of Disputed Political Boundaries in General Atlases," *The Cartographic Journal*, **1**, 20–28 (1964).

Stewart, John Q. and William Warntz, "Macrogeography and Social Science," *The Geographical Review*, **48**, 167–184 (1958).

Thrower, Norman J. W., "Extended Uses of the Method of Orthogonal Mapping of Traces of Parallel, Inclined Planes with a Surface, Especially Terrain," *International Yearbook of Cartography*, **3**, 26–35 (1963).

Thrower, Norman J. W., "Animated Cartography in the United States," *International Yearbook of Cartography*, **1**, 20–28 (1961).

Warntz, William, "A New Map of the Surface of Population Potentials for the United States, 1960," *Geographical Review*, **54**, 170–184 (1964).

Williams, Robert L., *Statistical Symbols for Maps: Their Design and Relative Values*, Yale University Map Laboratory, New Haven, Connecticut, 1956.

Williams, Robert L., "Equal-Appearing Intervals for Printed Screens," *Annals of the Associa-tion of American Geographers*, **48**, 132–139 (1958).

Wright, John K., "The Terminology of Certain Map Symbols," *The Geographical Review*, **34**, 654–655 (1944).

Wright, John K., "A Proposed Atlas of Diseases, Appendix I, Cartographic Considerations," *The Geographical Review*, **34**, 653–654 (1944).

STATISTICS AND CARTOGRAPHY

Allcock, H. J. and J. R. Jones, *The Nomogram*, Pitman and Sons, London, 1938.

Blalock, Hubert M., *Social Statistics*, McGraw-Hill Book Company, Inc., New York, 1960.

Brooks, C. E. P. and N. Carruthers, *Handbook of Statistical Methods in Meteorology*, Her Majesty's Stationery Office, London, 1953.

Freeman, Linton C., *Elementary Applied Statistics*, John Wiley and Sons, Inc., New York, 1965.

Levens, Alexander, *Nomography*, 2nd edition, John Wiley and Sons, New York, 1959.

Moroney, M. J., *Facts from Figures*, 3rd edition, Penguin Books, Baltimore, 1956.

Robinson, Arthur H., "Mapping the Correspondence of Isarithmic Maps," *Annals Association of American Geographers*, **52**, 414–425 (1962).

Robinson, Arthur H. and Lucy Caroe, "On the Analysis and Comparison of Statistical Surfaces," in *Quantitative Geography, Part I*, Northwestern University Studies in Geography No. 13, Evanston, 1967, pp. 252–275.

Robinson, Arthur H., "The Cartographic Representation of the Statistical Surface," *International Yearbook of Cartography*, **1**, 53–61 (1961).

Robinson, Arthur H., "The Circle of Best Fit," *The Geographical Review*, **53**, 139–141 (1963).

Robinson, Arthur H. and Reid A. Bryson, "A Method for Describing Quantitatively the Correspondence of Geographical Distributions," *Annals of the Association of American Geographers*, **47**, 379–391 (1957).

Thomas, Edwin N., *Maps of Residuals from Regression: Their Characteristics and Uses in Geographic Research*, Geography Publication No. 2, State University of Iowa, Iowa City, 1960.

Thomas, Edwin N. and David L. Anderson, "Additional Comments on Weighting Values in Correlation Analysis of Areal Data," *Annals Association of American Geographers*, **55**, 492–505 (1965).

AUTOMATION

Bengtsson, B. E. and S. Nordbeck, "Construction of Isarithms and Isarithmic Maps by Computers," *B. I. T.*, **4** (1964).

Cornwell, B. with A. H. Robinson, "Possibilities for Computer-Animated Films in Cartography," *The Cartographic Journal*, **3**, 79–82 (1966).

Cude, W. C., "Automation in Mapping," *Surveying and Mapping*, **12**, 413–436 (1962).

Olliver, J. G., "Automated Cartography," *Survey Review*, **19**, 139–141 (1967).

Robertson, J. C., "The Symap Programme for Computer Mapping," *The Cartographic Journal*, **4**, 108–113 (1967).

Sherman, John C., "New Horizons in Cartography: Functions, Automation, and Presentation," *International Yearbook of Cartography*, **1**, 12–17 (1961).

Tobler, Waldo R., "Automation and Cartography," *The Geographical Review*, **49**, 526–534 (1959).

Tobler, Waldo R., "Automation in the Preparation of Thematic Maps," *The Cartographic Journal*, **2**, 32–38 (1965).

Voisin, Russell L., "Automation in Private Cartography," *Surveying and Mapping*, **28**, 77–81 (1968).

Williams, N. L. G., "The Oxford System of Automatic Cartography," *Cartography*, **6**, 17–20 (1966).

REPRESENTATION OF LAND FORM

Carmichael, L. D., "Experiments in Relief Portrayal," *The Cartographic Journal*, **1**, 11–17 (1964).

Curran, J. P., "Cartographic Relief Portrayal," *The Cartographer*, **4**, 28–37 (1967).

Directorate of Overseas Surveys, "Some Recent Developments in Hill Shading from Air Photographs in the Directorate of Overseas Surveys," *Survey Review*, **17**, 3–11 (1963).

Dornbach, John E., "An Approach to Design of Terrain Representation," *Surveying and Mapping*, **16**, 41–44 (1956).

Hammond, Edwin H., "Small Scale Continental Landform Maps," *Annals of the Association of American Geographers*, **44**, 33–42 (1954).

Hammond, Edwin H., "Analysis of Properties in Land Form Geography: An Application to Broad Scale Land Form Mapping," *Annals Association of American Geographers*, **54**, 11–19 (1964).

Keates, J. S., "Techniques of Relief Representation," *Surveying and Mapping*, **21**, 459–463 (1961).

Lobeck, Armin K., *Block Diagrams*, 2nd edition, Emerson-Trussell Book Company, Amherst, Mass., 1958.

Miller, O. M. and Charles H. Summerson, "Slope-Zone Maps," *Geographical Review*, **50**, 194–202 (1960).

Raisz, Erwin, "The Physiographic Method for Representing Scenery on Maps," *The Geographical Review*, **21**, 297–304 (1931).

Richarme, P., "The Photographic Hill Shading of Maps," *Surveying and Mapping*, **23**, 47–59 (1963).

Ridd, Merrill K., "The Proportional Relief Landform Map," *Annals Association of American Geographers*, **53**, 569–576 (1963).

Robinson, Arthur H. and Norman J. W. Thrower, "A New Method of Terrain Representation," *The Geographical Review*, **47**, 507–520 (1957).

Savigear, R. A. G., "A Technique of Morphological Mapping," *Annals Association of American Geographers*, **55**, 514–538 (1965).

Sherman, John C., "Terrain Representation and Map Function," *International Yearbook of Cartography*, **4**, 20–23 (1964).

Stacy, John R., "Terrain Diagrams in Isometric Projection-Simplified," *Annals of the Association of American Geographers*, **48**, 232–236 (1958).

Tanaka, K., "The Relief Contour Method of Representing Topography on Maps," *The Geographical Review*, **40**, 444–456 (1950).

Yoeli, P., "Relief Shading," *Surveying and Mapping*, **19**, 229–232 (1959).

Yoeli, P., "Relief and Colour," *The Cartographic Journal*, **1**, 37–38 (1964).

Yoeli, P., "Analytical Hill Shading," *Surveying and Mapping*, **25**, 573–579, (1965).

Yoeli, P., "Analytical Hill Shading and Density," *Surveying and Mapping*, **26**, 253–259 (1966).

Yoeli, P., "Mechanization in Analytical Hill-Shading," *The Cartographic Journal*, **4**, 82–88 (1967).

MAP PROJECTIONS

Balchin, W. G. V., "The Representation of True to Scale Linear Values on Map Projections," *Geography*, **36**, 120–124 (1951).

Balchin, W. G. V., "The Choice of Map Projections," *Empire Survey Review*, **12**, 263–276 (1954).

Chamberlin, Wellman, *The Round Earth on Flat Paper*, The National Geographic Society, Washington, D.C., 1947.

Dahlberg, Richard E., "Evolution of Interrupted Map Projections," *International Yearbook of Cartography*, **2**, 36–53 (1962).

Deetz, C. H. and Adams, O. S., *Elements of Map Projection,* United States Coast and Geodetic Survey Special Publication 68, Washington, D.C. (latest edition available).

Fisher, Irving and O. M. Miller, *World Maps and Globes*, Essential Books, New York, 1944.

Goussinsky, B., "On the Classification of Map Projections," *Empire Survey Review*, **11**, 75–79 (1951).

Lee, L. P., "Some Conformal Projections Based on Elliptic Functions," *The Geographical Review*, **55**, 563–580 (1965).

Leighly, John, "Extended Use of Polyconic Projection Tables," *Annals of the Association of American Geographers*, **46**, 150–173 (1956).

Kao, Richard C., "Geometric Projections of the Sphere and the Spheroid," *The Canadian Geographer,* **5**, 12–21 (1961).

Keuning, J., "The History of Geographical Map Projections until 1600," *Imago Mundi*, **12**, 1–24 (1955).

Maling, D. H., "A Review of Some Russian Map Projections," *Empire Survey Review*, **15**, 203–215, 255–266, 294–303 (1960).

Miller, O. M., "Notes on Cylindrical Map Projections," *The Geographical Review*, **32**, 424–430 (1942).

Marschner, F. J., "Structural Properties of Medium and Small Scale Maps," *Annals of the Association of American Geographers*, **34**, 1–46 (1944).

Robinson, Arthur H., "An Analytical Approach to Map Projections," *Annals of the Association of American Geographers*, **39**, 283–290 (1949).

Robinson, Arthur H., The Use of Deformational Data in Evaluating Map Projections," *Annals of the Association of American Geographers*, **41**, 58–74 (1951)

Robinson, Arthur H., "Interrupting a Map Projection: A Partial Analysis of Its Value," *Annals of the Association of American Geographers*, **43**, 216–225 (1953)

Sear, William J., "Map Projection by Transformation," *Cartography*, **2**, 23–26 (1957).

Steers, J. A., *An Introduction to the Study of Map Projections*, 14th Edition, University of London Press, London, 1965.

Stewart, J. Q., "The Use and Abuse of Map Projections," *The Geographical Review*, **33**, 589–604 (1943).

Tobler, Waldo R., "A Classification of Map Projections," *Annals Association of American Geographers*, **52**, 167–175 (1962).

Tobler, Waldo R., "Geographic Area and Map Projections," *Geographical Review*, **53**, 59–78 (1963).

LETTERING

Cornog, D. Y., F. C. Rose, and J. L. Walkowicz, *Legibility of Alphanumeric Characters and Other Symbols, 1. A Permuted Title Index and Bibliography,* National Bureau of Standards Misc, Publ. 262-1, U.S. Government Printing Office, Washington, D.C., 1964.

Gardiner, R. A., "Typographic Requirements of Cartography, A Preliminary Note," *The Cartographic Journal*, **1**, 42–44 (1964).

George, Ross F., *Speedball Text Book,* 17th edition, Hunt Pen Co., Camden, New Jersey, 1948.

Higgins Ink Company, *Lettering,* Brooklyn, New York, 1949.

Keates, John S., "The Use of Type in Cartography," *Surveying and Mapping,* **18,** 75–76 (1958).

Nesbitt, Alexander, *"The History and Technique of Lettering,* Dover Publications, New York, 1957.

Riddeford, Charles E., "On the Lettering of Maps," *The Professional Geographer,* **4,** 7–10 (1952).

Robinson, Arthur H., "The Size of Lettering for Maps and Charts," *Surveying and Mapping,* **10,** 37–44 (1950).

Updike, Daniel B., *Printing Types, Their History, Forms and Use; a Study in Survivals,* Harvard University Press, Cambridge, Mass., 1922.

U. S. Government Printing Office, *Typography and Design,* Washington, D.C., 1951.

Withcombe, J. G., "Lettering on Maps," *The Geographical Journal,* **73,** 429–446 (1929.

DESIGN

Arnheim, Rudolf, *Art and Visual Perception,* University of California Press, Berkeley and Los Angeles, 1957.

Birren, Faber, *The Story of Color,* The Crimson Press, Westport, Connecticut, 1941.

Castner, Henry W. and Arthur H. Robinson, *Dot Area Symbols in Cartography: The Influence of Pattern on Their Perception,* A.C.S.M. Monographs in Cartography No. 1, American Congress on Surveying and Mapping, Washington, D.C., 1969.

Davis, Lavoi B., "Design Criteria for Today's Aeronautical Charts," *Surveying and Mapping,* **18,** 49–57 (1958).

Harrison, Richard Edes, "Art and Common Sense in Cartography," *Surveying and Mapping,* **19,** 27–38 (1959).

International Printing Ink Corporation, *Three Monographs on Color, New York,* 1935.

Jenks, George F. and Duane S. Knos, "The Use of Shading Patterns in Graded Series,"

Annals Association of American Geographers, **51,** 316–334 (1961).

Keates, John S., "The Perception of Colour in Cartography," *Proceedings of the Cartographic Symposium,* Edinburgh, 1962.

Keates, John S., "The Small-Scale Representation of the Landscape in Colour," *International Yearbook of Cartography,* **2,** 76–82 (1962).

De Lopatecki, E., *Advertising Layout and Typography,* Ronald Press, New York, 1935.

Robinson, Arthur H., *The Look of Maps, An Examination of Cartographic Design,* University of Wisconsin Press, Madison, Wisconsin, 1952.

Robinson, Arthur H., "Psychological Aspects of Color in Cartography," *International Yearbook of Cartography,* **7,** 50–59 (1967).

MAP REPRODUCTION AND CONSTRUCTION

Clare, W. G., "Map Reproduction," *The Cartographic Journal,* **1,** 42–48 (1964).

Kantrowitz, M. S., "Printing Papers," *Surveying and Mapping,* **10,** 119–126 (1950).

Merriam, M., "The Conversion of Aerial Photography to Symbolized Maps," *The Cartographic Journal,* **2,** 9–14 (1965).

Mertle, J. S. and Gordon L. Monsen, *Photomechanics and Printing,* Mertle Publishing Company, Chicago, Illinois, 1957.

Moore, Lionel C., *Cartographic Scribing Materials, Instruments, and Techniques,* 2nd edition, Technical Publication No. 3, American Congress on Surveying and Mapping, Cartography Division, Washington, D. C., 1968.

Ovington, J. J., "An Outline of Map Reproduction," *Cartography,* **4,** 150–155 (1962).

Raymond, J. P., "Reproduction Materials," *The Cartographer,* **2,** 92–96 (1965).

Line Photography for the Lithographic Process (revised by Karl Davis Robinson), Lithographic Technical Foundation, New York, 1956.

Sale, Randall D., "A Technique for Producing Colored Wall Maps," *The Professional Geographer,* **12,** 19–21 (1961).

Schlemmer, Richard M., *Handbook of Advertising Art Production,* Prentice Hall, Inc., Englewood Cliffs, New Jersey, 1966.

Striefler, Josef J., "Plastic Scribing," *Photogrammetric Engineering*, **23**, 330–335 (1957).

GOVERNMENT PUBLICATIONS

Branches of the U.S. Government issue a large number of useful publications concerning many aspects of cartography. Publications in series, such as the *Technical or Training Manuals* of the Departments of the Army and Air Force, the *Special Publications* of the Coast and Geodetic Survey and the Geological Survey, etc., are published by the U.S. Government Printing Office, Washington, D.C. 20402, from which listings are available. In addition, the major Defense mapping agencies, such as the Army Map Service, the Naval Oceanographic Office, and the Aeronautical Chart and Information Service, themselves publish *Technical Bulletins,* etc., which are ordinarily obtainable directly from the issuing agency.

INDEX